Recovering Energy from Waste
Various Aspects

Editors

Velma I. Grover
Vaneeta Kaur Grover
Natural Resource Consultants,
Hamilton, ON
Canada

William Hogland .
Department of Technology,
University of Kalmar,
Kalmar
Sweden

Science Publishers, Inc.
Enfield (NH), USA   Plymouth, UK

SCIENCE PUBLISHERS, Inc.
Post Office Box 699
Enfield, New Hampshire 03748
United States of America

Internet site: *http://www.scipub.net*

sales@scipub.net (marketing department)
editor@scipub.net (editorial department)
info@scipub.net (for all other enquiries)

Cover Photograph: Reproduced with permission from Spittelau, Spittelauer Lände 45, A 1090 Wien, Austria

Library of Congress Cataloging-in-Publication Data

Recovering energy from waste: various aspects/editors, Velma I. Grover, Vaneeta Kaur Grover, William Hogland.
    p. cm.
  Includes bibliographical references and index.
  ISBN 1-57808-200-5
  1. Waste products as fuel. I. Grover, Velma I. II Grover, Vaneeta Kaur. III. Hogland, William.

TP360.R393 2002
662'.87--dc21

2002072667

Published by Science Publishers, Inc., Enfield, NH, USA
Printed in India.

Foreword

Some decades ago Marshall McLuhan, a provocative and visionary Canadian, observed that "The new clothing of the planet is garbage." McLuhan became famous for his articulation of the impact of the emerging electronic information media in creating a global village. Today his insight about a new interrelationship among people of the earth has been realized. We live in a world where ideas cross borders unimpeded, where cyberspace is beyond national control and where the speed and magnitude of capital flows is incredible. We are ever more interconnected, and environment has surely become the quintessential global issue.

The state of the world is showing signs of stress. Two-thirds of humankind fall far short of having a decent quality of life. The world's population has exceeded six billion and the numbers living in poverty increase daily. Meanwhile the share of the planet's resources being used by the affluent is also growing. The United Nations Environment Programme concludes that these two issues - "the poverty of the majority and the excessive consumption of the minority" - are the driving forces of environmental degradation.

The statistics are mind-numbing. Suffice it to say that on almost any indicator about our air, water, land or species, there is no room for complacency. We are surrounded by symptoms of a world in wobbly disequilibrium. Clearly we cannot continue on this trajectory of increasing demand for energy and phenomenal increases in waste.

A sustainable planet is not an unreachable goal. Thankfully some of the best minds have turned their attention to solutions, perhaps motivated by the advice "Where there are hurricanes, build windmills." This book chronicles the experiences of those who believe that new technologies and changes in mind-set can transform waste to energy. For countries of the North and South, for the private and public sectors, evidence is mounting that it can be profitable to reduce the amount of waste produced. If we can gradually shift the thinking of industry and municipal leaders away from the traditional waste management responses of landfill sites, incineration, composting and recycling, towards eliminating the production of waste material, we can significantly reduce man's ecological footprint. If we can shift the behaviour of consumers, it is not unrealistic to envision preservation of the integrity of the earth.

Cleaning up the unfinished business of the 20th century demands the best efforts of all—scientists, industrialists, citizens and policy-makers. If we are to avoid a collision between the powerful forces of growing ecological pressures and economic expansion our policy responses must be strengthened. Acting on the lessons learned from technological innovation in addressing waste can perhaps render McLuhan's observation obsolete.

July 2002   Elizabeth Dowdeswell

Preface

This book examines various aspects involved in recovering energy from waste. It presents a mosaic of technologies and country experiences. Instead of giving an exhaustive coverage of all the technologies, it focusses on some success and some failure stories from developed as well as developing nations. A policy maker or a project manager can, by reading this book, glean valuable information about the experiences of persons in other countries while choosing a given technology or, for that matter, in choosing the appropriate technology from the several available.

The continuing growth in demand for fossil fuel has necessitated the search for alternative sources of renewable energy. One source of renewable energy, which has not been given adequate attention, is waste, particularly solid waste. Different ways of recovering energy are discussed in this book and, in addition to incineration, the book describes a number of technologies such as landfill gas recovery, biogas as an energy source, bio-reactors for wastes fermentation and fuels production, electricity from biomass and biological wastes and electrokinetic soil processing for energy from waste. For example, in Holland, excess heat from urban industry or high rise buildings is used for warming green houses; in Vancouver, B.C., from the city waste landfill, methane gas, in addition to generating electricity for sale, also provides heat for greenhouses. Case studies from developing as well as developed countries are discussed: from landfill gas utilization in Canada, landfill biogas management in Chile, a case study from Brazil, current situation of waste incineration and energy recovery in Germany, thermal treatment of waste in Vienna, enhanced acidification and methanation process for biomethanation of organic municipal solid waste in India, industrial biomethanation practices for decentralized energy from waste from Sri Lanka. These case studies cover both successes and failures.

One of the most used technologies is incineration—for reduction of volume or for producing heat or generation of electricity. Unfortunately the drawback with incineration is that it replaces one evil with another—from waste pollution to air pollution. Vienna has come up with a solution to deal with emissions which meets European emission standards. The Austrians were so proud of this incineration plant that they requested Austrian artist and architect, Friedensreich Hundertwasser, to design the façade of the plant. After studying the facts, he was impressed and agreed to design the façade free of charge. The artist successfully demonstrated that industrial construction is just as important as residential buildings in determining the quality of life. Visual pollution by 'aggressive' and 'brutal' constructions, he says, is the most dangerous form of pollution because it destroys man's dignity and essential nature. He managed to turn this absolutely repugnant piece of architecture into a site of such playful delight, that it has become a focal point for tourists from around the world. It also won public acceptability for the incinerator. This is a model which can be emulated by others.

Choosing the right technology is very important. Selection of the right technology is dependent on a number of factors such as quantity and quality of waste, climatic conditions, ability to adopt certain technology and technical skills available etc. It is not always advisable to use developed-country technology in developing countries. This book gives examples of technologies developed in developing nations such as that developed by Tata Energy Research Institute, New Delhi, which can be easily adapted in other developing nations.

The technical design of MSW incinerators has improved in recent years and regulatory controls have reduced emissions but, the degree of health impacts depends, at least in part, on the effectiveness of the system of regulation imposed on these installations. With new emission standards and equipment, pollution from incinerators can be controlled but not eliminated, and thus the concern about emissions on human health remains. One of the chapters contributed by Greenpeace International discusses this issue at length.

Besides incineration, waste has other repercussions on climate change on account of methane gas emissions from landfill site. Some of the authors have discussed the impact of different waste management options on climate change.

Another issue addressed in the book is the social factor. The technology adopted has not only to be technically viable but has to take into consideration social cost-benefit factors. In third world countries, there are many people whose livelihood is made by handling solid waste, i.e., scavengers or junk dealers. Care should be taken that such persons do not lose their livelihood.

Usually incineration plants are centrally located and huge. In order to meet European emission standards, the U.K. has developed small incinerators to be used by large companies to get rid of the packaging material locally. One of the British companies carried out a market survey about the acceptability of such incinerators by large companies, municipalities and landfill site operators. We are fortunate to have an author who has reported the results of such a survey. Such an experiment can be emulated by others to design a better incinerator, or, to carry out such a survey in their jurisdiction.

The book presents a mix of technologies, country reports and issues dealing with socio-economic and health issues, making it unique publication.

February 2002   **Velma I. Grover**

Contents

Foreword by *Elizabeth Dowdeswell*   iii

Preface   v

List of Contributors   ix

SECTION I: INTRODUCTION

1. Introduction   3
   Velma I. Grover

2. Waste-to-Energy: Comprehensive Recycling's Best Chance?   15
   John A. Merritt

3. Energy-from-waste: A Perspective from Developing Countries   21
   A.A. Oladimeji, S. Kumar, C.T. Parekh and M.J. Bhala

SECTION II: TECHNOLOGIES

4. Biomass as a Source of Energy   33
   J. Mata-Alvarez and S. Macé

5. Small Diameter Forest Residues for Soil Rehabilitation   49
   Tatjana Stevanovic Janezic

6. Bioreactors for Wastes Fermentation and Fuels Production—   75
   Relevant Thermodynamic and Process Parameters, Knowledge
   and Engineering Data
   Marija S. Todorovic, Franc Kosi and Ljiljana Simic

7. Electricity from Biomass and Biological Wastes: European Scenario   95
   Emmanuel G. Koukios and Kyriakos D. Panopoulos

8. Electrokinetic Soil Processing for Energy from Waste   107
   Kyoung-Woong Kim and Soon-Oh Kim

SECTION III: CASE STUDIES

9. The TEAM (TERI's Enhanced Acidification and Methanation)   143
   Process for Biomethanation of Organic Municipal Solid Waste
   Bindiya Goel, Dinesh Pant, Kusum Jain and V.V.N. Kishore

10. Industrial Biomethanation Practices for Decentralized Energy from   151
    Waste: Options for Sri Lanka
    Ajith de Alwis

11. Energy from Landfill Gas Utilization in Canada   167
     Edward A. McBean, Rick Mosher and Alain David

12. Landfill Biogas Management. Case of   177
     Chilean Sanitary Landfills
     Jose Arellano

13. Energy from Waste: Case Study from Brazil   183
     T. Cássia De Brito Galvão and Willer M. Pos

14. Current Situation of Waste Incineration and Energy Recovery   195
     in Germany
     Bernt Johnke

15. Thermal Treatment of Waste in Vienna: An Ecological Solution?   201
     Senatsrat Dipl.-Ing. Helmut Löffler

16. Co-firing: Primary Energy Savings and Carbon Dioxide Avoidance   215
     Ivo Bouwmans

17. Baled MSW and Associated Problem, in the Context of Fire Hazard   223
     William Hogland, Diauddin R. Nammari, Sven Nimmermark,
     Marcia Marques and Viatcheslav Moutavtchi

     SECTION IV: WASTE TO ENERGY AND CLIMATE CHANGE ISSUES

18. Waste, Energy and Climate Change   249
     Joyeeta Gupta and Frédéric Gagnon-Lebrun

19. Waste, Recycling and Climate Change: US Perspective   261
     Frank Ackerman

     SECTION V: HEALTH IMPACTS

20. Incineration and Human Health: Characterization and Monitoring of 273
     Incinerator Releases and their Impact
     Michelle Allsopp, Pat Costner, Paul Johnston and David Santillo

     SECTION VI: SOCIAL IMPACTS

21. Socially-responsive Energy From Urban Solid Wastes in   307
     Developing Countries
     Christine Furedy and Alison Doig

     SECTION VII: MARKETING ISSUES OF WASTE TO
     ENERGY TECHNOLOGIES

22. The Market for Small-Scale Waste Gasification/Pyrolysis Projects— 317
     Preliminary Scoping Study
     Sarah Knapp

Index   331

List of Contributors

Ackerman, Frank, Global Development and Environment Institute, Cobot Intercultural Center, Tufts University, Medford MA, USA

Allsopp, Michelle, Greenpeace Research Laboratories, University of Exeter, Prince of Wales Road, Exeter, Ex4 4PS, UK.

Alwis, Ajith de, Department of Chemical and Process Engineering, University of Moratuwa, Moratuwa, Sri Lanka

Arellano, Jose, Av. Larrain 9975, La Reina, Santiago, Chile

Bhala, M.J., Faculty of Applied Sciences, National University of Science and Technology, P O Box AC939, Ascot, Bulawayo, Zimbabwe.

Bouwmans, Ivo, TBM/E&I, Jaffalaans 5, 2628 BX Delft, The Netherlands

Costner, Pat, Greenpeace Research Laboratories, University of Exeter, Prince of Wales Road, Exeter, Ex4 4PS, UK.

David, Alain, Environment Canada, 351st. Joseph, Hull, Ovebec, KIAOH3, Canada

Doig, Alison, ITDG: The Schumacher Centre for Technology and Development Boutron Hall, Bourton-on-Dunsmore, Rugby, Warwickshire, CV23 902, UK

Furedy, Christine, ITDG: The Schumacher Centre for Technology and Development Boutron Hall, Bourton-on-Dunsmore, Rugby, Warwickshire, CV23 902, UK

Gagnon-Lebrun, Frédéric, Fruitlaan 6, 2292 BB Wateringen, The Netherlands

Galvão, T. Cássia de Brito, School of Engineering—Federal University of Minas Gerais, Av. Contorno 842/104, 30110-060, Belo Horizonte, Brazil

Goel, Bindiya, Research Associate, Tata Energy Research Institute, Darbari Seth Block, Habitat Place, Lodhi Road, New Delhi-110003, India

Grover, Velma I., 411 – 981 Main St West, Hamilton, On, L8S 1A8, Canada.

Gupta, Joyeeta, Institute for Environmental Studies, Vrije Universiteit Amsterdam, The Netherlands

Hogland, William, Department of Technology, University of Kalmar, Kalmar P.O. Box 905 SE- 39129, Sweden

Janezic, Tatjana Stevanovic, Université Laval, Département des sciences du bois et de la forêt, Ste Foy, Quebec, Canada

Johnke, Bernt, Umweltbundesamt Berlin, Seecktstr. 6-10 D-13581, Berlin, Germany

Johnston, Paul Greenpeace Research Laboratories, University of Exeter, Prince of Wales Road, Exeter, Ex4 4PS, UK.

Kim, Kyoung-Woong, Department of Environmental Science and Engineering, Kwangju Institute of Science and Technology (K-JIST), 1 Oryong-dong, Puk-gu, Kwangju 500-712, Republic of Korea.

Kim, Soon-Oh, Department of Environmental Science and Engineering, Kwangju Institute of Science and Technology (K-JIST), 1 Oryong-dong, Puk-gu, Kwangju 500-712, Republic of Korea.

Knapp, Sarah, M.E.L. Research Ltd., Birmingham, UK.

Kosi, Franc, Laboratory for Thermodynamics and Thermotechnics of The Division for Energy Efficiency and Renewable Energy Sources, Agricultural Faculty, University of Belgrade, Yugoslavia.

Koukios, Emmanuel G., Bioresource Technology Unit, Department of Chemical Engineering, National Technical University of Athens, Zografou Campus Heroon Polytexniou 9 GR-15700 Athens, Greece.

Kumar Santosh, Department of Mathematics and Statistics, University of Melbourne, Parkille, Victoria 3052, Australia.

Löffler, Senatsrat Dipl.-Ing. Helmut, Head of the Environmental Protection Department of the Municipality of Vienna, Florianigasse 47, A 1080 Wien Austria.

Macé, S., University of Barcelona, Dept. of Chemical Engineering, Martí i Franquès, 1, pta. 608028 Barcelona, Spain.

Marques, Marcia, Department of Technology, University of Kalmar, Kalmar P.O. Box 905 SE- 39129, Sweden.

Mata-Alvarez, J., University of Barcelona, Dept. of Chemical Engineering, Martí i Franquès, 1, pta. 608028 Barcelona, Spain.

McBean, Edward A., Conestoga-Rovers & Associates, 651 Colby Drive, Waterloo, On NZVICZ Canada.

Merritt, John A., President, Merritt Communications, Inc., Environmental Solutions Division, 340 Woodland st Holliston, Massachusetts, 01746, USA.

Mosher, Rick, Conestoga-Rovers & Associates, 651 Colby Drive, Waterloo, ON, NZVICZ, Canada.

Moutavtchi, Viatcheslav, Department of Technology, University of Kalmar, Kalmar P.O. Box 905 SE- 39129, Sweden

Nammari, Diauddin R., Department of Technology, University of Kalmar, Kalmar P.O. Box 905 SE- 39129, Sweden

Nimmermark, Sven, Department of JBT, University of Agricultural Sciences, Alnarp, Sweden

Oladimeji, A.A., Dept. of Biological Sciences, Federal University of Technology, P.O. Box 65, Minna, Nigeria.

Panopoulos, Kyriakos D., Bioresource Technology Unit, Department of Chemical Engineering, National Technical University of Athens, GR-15700 Athens, Greece.

Parekh, C.T., Faculty of Applied Sciences, National University of Science and Technology, P O Box AC939, Ascot, Bulawayo, Zimbabwe.

Pos, Willer H., Supervisor of Environmental Politics—State of Minas Gerais Government, Environmental and Sustainable Development Secretary, Instituto Mineiro de Gestão das Aguas-Head Rua Santa Catarina 1354, 4° andar, 30170-081 Belo Horizonte, Brazil.

Satillo David, Greenpeace Research Laboratories, University of Exeter, Prince of Wales Rd, Exeter, Ex44PS, UK.

Simic, Ljiljana, Laboratory for Thermodynamics and Thermotechnics of The Division for Energy Efficiency and Renewable Energy Sources, Agricultural Faculty, University of Belgrade, Yugoslavia.

Tatjana Stevanovic Janezic, Université Laval, Département des sciences du bois et de la forêt, stefoy, Quebec, Canada.

Todorovic, Marija S., Laboratory for Thermodynamics and Thermotechnics of The Division for Energy Efficiency and Renewable Energy Sources, Agricultural Faculty, University of Belgrade, Yugoslavia.

Section I

Introduction

1

Introduction

VELMA I. GROVER

411 – 981 Main St West, Hamilton, On, L8S 1A8, Canada.

The continuous rise and competitive demand for fossil fuels is depleting the non-renewable energy sources. This is leading to search for energy from renewable sources and one of the renewable sources is "waste". Recovering energy from waste has a long history. Earlier processes of recovering energy were technically unsophisticated and not well researched. But now with increasing generation of solid waste and decreasing space for landfill sites, waste-to-energy is emerging as a promising option for the future. A lot of research is being conducted to recover energy from waste and biomass. Attempting to treat such a subject adequately in one volume, or even many volumes, is not an easy task. Volumes can be written on different technologies available, different case studies (failure and success), experiences from developing/ developed countries, examples from industry or research undertaken by academicians. The purpose of this book is to give a brief introduction to some of the technologies available with focus on case studies. Some of the chapters also correlate solid waste management options and energy recovery options to climate change – an important issue at this turn in the history of mankind.

The obvious question is why recover energy from waste. Apart from providing an alternative source of energy the other environmental advantages to be had from using waste are:

(1) Reducing volume of MSW by 90% to an inactive ash residue.
(2) Avoiding leachate formation from landfills.
(3) Conservation of fossil fuels, thus displacing the pollution caused by conventional generation from these sources.
(4) Reduction of greenhouse gas emissions, and the resultant global warming. Conversion of waste-to-energy, prevents the release ot greenhouse gases such as methane (approximately 1 million tons per annum), carbon dioxide (approximately 10 million tons annually), nitrogen oxides, and volatile organic compounds. Waste-to-energy power as an alternative to coal prevents the release of nearly 25,000 tons of nitrogen oxides and 5,000 tons of volatile organic compounds.

It might appear that waste-to-energy plants would interfere with much better options like recycling and waste minimization and would cause pollution. But with stricter emission limits the pollution issue is being resolved slowly and the fact that conversion of waste-to-energy and recycling are compatible and are mutually supportive is also getting established.

This book is about recovering energy and looks at various aspects involved with recovering energy from waste. Different ways of recovering energy are discussed and in addition to incineration, the book describes a number of technologies like landfill gas recovery, biogas recovery. Besides technologies, it describes a number of case studies—both from developing countries as well as developed countries. What works for developed countries may not work for developing countries—simply because the quantity and quality (composition) might be altogether different. The case studies highlight these facts and draw examples of both success and failure from real life. In addition to technologies and case studies, one of the chapters discusses impact on human health and one of the sections of the book is also on the marketing research experiment on one of the incinerators—which meets emission criteria required under European Commission's regulation. One of the important issues today is climate change and impact of waste management options on climate change. A number of authors have expressed their views on the impact of different waste management options on climate change. One of the chapters looks at the sociological aspect. Thus this book is a sort of mosaic of technologies, case studies, impact of incineration on health, social issues and marketing experiments related to energy from waste – that is what makes this book unique.

After a **Foreword by Elizabeth Dowdeswell,** Section I gives a general introduction to the book. The lead article, **"Waste to Energy: Comprehensive Recycling's Best Chance?"** by John A. Merritt, discusses key elements required to make "waste to energy" a complete recycling option—that is it should not only be an option in dealing with the waste but there should also be a method to re-use or recycle the waste produced by such an option. The author looks at the compatibility of waste-to-energy facilities with pre-combustion and post-combustion material separation, combustion practices, air pollution control techology. This is followed by a more detailed focus on ash recycling—the missing link keeping Waste-to-Energy technology from achieving its full recycling potential. He feels that although bioconversion may be a good solution for small-scale waste management projects, Waste-to-Energy might be the best way to achieve comprehensive recycling for large-scale waste management projects.

The continuous rise and competitive demands for fossil fuel necessitates the need for a search for alternative sources of renewable energy. One source of renewable energy, which has not been given adequate attention, is waste, particularly solid waste. The next chapter by A. A. Oladimeji, S. Kumar, C. T. Parekh and M. J. Bhala, **"Energy from waste: a perspective from developing countries"** reviews the status of 'waste' as a renewable source of energy in developing countries. The authors conclude that, in view of the level of technology in most developing countries, waste at present has no singnificant value, and at times it costs money to remove unwanted waste from the site of production. But in future, with advancement in technology, more use could be found for the various categories of solid waste. In that situation, energy from waste will be an issue, which will be subject to optimization and biomass based fuels could be less harmful to the environment than fossil fuels. These technologies must however be used on site-specific bases in order to reduce the negative impacts on the environment and make the use of biomass energy sustainable.

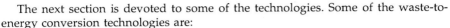

The next section is devoted to some of the technologies. Some of the waste-to-energy conversion technologies are:

(1) *Mass burn facilities:* In this sort of facility mixed municipal waste is fed into large furnaces dedicated solely to burning trash. The resulting energy produces steam and/or electricity. Many mass burn facilities have nearby material recovery facilities, or material recovery facilities that separate and recycle trash prior to processing.

(2) *Refuse-derived fuel or RDF plants:* In this type of facility recyclable or unburnable materials are removed and the rest of the trash is shredded or processed into a uniform fuel. Sometimes, RDF powers a generating plant on site, and sometimes the fuel is burned off-site for generating energy. The experiments on pellitization have been carried out by a number of countries including India. Department of Science and Technology, India, set up a pilot plant in Mumbai (Deonar) for processing garbage into fuel pellets using indigenous technology. A similar project was also established by M/s Shivashankar Engineering Company Pvt. Ltd. in Bangalore. This company has been producing pellets since 1989—compacting 50 tons of garbage into 5 tons of pellets each day. Besides, there are reports from African countries, though use of briquettes seem to be more popular in Africa.

(3) *Modular facilities:* These are similar to mass burn plants, but these smaller plants are prefabricated and can be quickly assembled where needed.

(4) *Biogas plants* are normally used to recover biogas from animal manure or waste composted anaerobically (in absence of air or oxygen). This has been discussed in the case studies in the book.

(5) *Recovering energy from biomass:* Biomass is, naturally occurring energy-containing carbon resource and constitutes all non-fossil organic materials. This would include all water-based and land-based vegetation and trees, and all waste biomass such as municipal solid waste, municipal bio-solids (sewage) and animal wastes, forestry and agricultural residues, and certain types of industrial wastes. Unlike fossil fuels, biomass is renewable in the sense that only a short period of time is needed to replace what is used as an energy resource. In fact, this category of recovering energy from biomass is a vast field, covering many technologies like biogas recovery from the treatment of bio-solids in municipal wastewater treatment plants by anaerobic digestion, recovery of landfill gas from municipal solid waste landfills, and the conversion of municipal solid waste to refuse-derived fuel. In addition to municipal waste, waste from farm, forestry, and certain industries (such as black liquor generated by the paper industry), is also used to recover energy.

(6) *Capturing gas from landfill sites:* This is another way energy can be recovered from waste and this gas (depending on the quantity) can be used for different purposes like generating steam and/or electricity. This book discusses case studies related to recovering/capturing gas from landfill site from developing as well as developed nations.

(7) *Electrolysis:* Another technology discussed in this book is electrolysis. Improper waste management like dumping of waste instead of engineered landfill sites cause leachate formation and contaminate the soil and ground water. Besides waste from other sources like mining, industries (solid as well as liquid) also contaminate the soil and water. This is becoming a threat to the very survival

of human beings as there would be no land for agriculture and no pure drinking water. This technology as described by the authors in the chapter later in the book mainly deals with how soil can be decontaminated with the process known as electrolysis. Electrokinetic soil processing is an effective technology for removal of contaminants in low–permeability soils ranging from clay to clayey sand. The advantages of this technology are its low cost of operation and its potential applicability to a wide range of contaminant types. Electrokinetic soil processing is envisioned for the removal/separation of organic and inorganic contaminants and radionuclides. Electrokinetic remediation technology has recently made significant strides and been tested for commercial application in the United States and Netherlands.

(8) *Bailing:* It is in fact not exactly a energy recovery technology but it is a way in which waste can be stored safely to recover energy whenever and wherever needed.

It is a known fact that the combustion process carries risk of releasing air pollutants. Emissions from incinerators can include toxic metals and toxic organics. The primary goals of modern waste-to-energy incineration are to maximize combustion and minimize pollution. Two other goals are high plant availability and low facility maintenance costs. If these goals are achieved by stricter emission limits then waste to energy plants do offer a good option to waste management planners.

The first chapter under this section is **"Biomass As An Energy Resource"** by J. Mata-Alvarez and S. Macé. This chapter discusses that anaerobic digestion plants compared to other technologies to treat solid bio-wastes is better from an ecological point of view. They have also discussed the environmental aspects of using biomass as an energy source and point out that the use of biomass as an energy resource has lesser contribution to the increase in green house gases rather than just sending it to landfill site. They further discuss the potential biomass as a source of energy in Europe, its present use in Europe and use of municipal solid waste as an energy source in Europe followed by relevant methods (two methods) for obtaining bio-fuels from biomass. One such methods is bio-ethanol, which has been exploited mainly from crops especially rich in sugar, and the second method is biogas. They conclude that recovery of energy from biomass can be profitable in future.

Another un-conventional way of recovering energy is use of forest waste to retain fertility of soil. The fertility of the soil depends on the organic matter content— and is derived mainly from polyphenols. The chapter **"Small diameter forest residues for soil rehabilitation"** by Tatjana Stevanoric Janexic discusses the experiment conducted in which they have applied forest residue (in form of chipped small branches and twigs) to soil. Experiments have indicated that this increases soil productivity and better water economy.

"Bio-Reactors for Wastes Fermentation and Fuels Production - Relevant Thermodynamic and Process Parameters, Knowledge and Engineering Data" by Marija S. Todorovic, Franc Kosi and Ljiljana Simic gives basics for improving understanding of biophysics of synergetic microbiological and physicochemical processes. Relevant phenomena occurrence and interaction have been analytically described and parametrically investigated. Reactor phenomenon is used at both micro- (i.e., pore- and particle-size) and macro- (i.e., reactor) levels, to describe synergetics between bio-dynamics and physicochemical dynamics as well as their intrinsic micro-macro relations. Different formulations of fluid flow, mass and heat transfer in two or multiphase reactor media - content, which exist in the literature as a result of the importance of this topic to bio-engineering and bio-processing - including aerobic and anaerobic systems, have been reviewed. To describe the occurrence and correspondence of micro and macro phenomena and synergetism of micro-bio-dynamic and dynamical changes in physicochemical fields, a mathematical formulation based on a hierarchical volume averaging method, has been applied in defining system of governing equations. A path of analysis that originates with the continuum axioms for the mass and momentum of multi-component systems, is discussed. Corresponding mathematical methods of closure have also been reviewed. Regarding the interwoven nature of wastes as one of the most critical environmental problems, as well as the number of possible processing routes and technologies, development of new more efficient multi-parametric system approach is also discussed. This chapter also describes project and results of interactively connected, integrated research aimed at advancing fundamental knowledge and technologically intrinsic system approach.

According to the EU "White Paper for Renewable Energy" biomass is among the most promising renewable energy sources for power production, and should contribute 230 TWhe per annum by the year 2010. This corresponds to 10 % of the 1995 total electricity production in Europe or 8 % for the target year 2010. The chapter, **"Electricity from biomass and biological wastes: European scenario"** by Emmanuel G. Koukios, and Kyriakos D. Panopoulos attempts to allocate the total amount of electricity energy from biomass to the 15 EU member countries. The authors have investigated four different scenarios according to typical power plants sizes that might suit better each country's infrastructure. They have based the allocation of bioelectricity among the EU countries on a simple algorithm which takes into account each EU-member state present progress of biomass-to-energy utilization, their overall biomass potential and the extent of their CO_2 emissions deriving from electricity production. This factor represents the different greenhouse gas emissions abatement efforts allocated within the EU member states to achieve the overall Kyoto protocol CO_2 reduction obligation of the EU (8% in year 2010 compared to 1990).

Soils can be contaminated with heavy metals derived from various sources including abandoned mining wastes, improper treatment of industrial wastes, incomplete collection of used batteries, leakage of landfill leachate, accidental spills and military activities. Kyoung-Woong Kim and Soon-Oh Kim in **"Electrokinetic Soil Processing for Energy from Waste"** – describe various electrokinetic phenomena in soils and principles of electrokinetic soil processing followed by the application of this technology. The authors also present the principles and applications of various metal recovery technologies.

Every time one talks about the introduction of any technology, the first question which comes to mind (especially to a policymaker) is—does this technology actually work? Are there any projects where the successful use of the technology has been demonstrated? In the third section of the book some of these technologies with actual case studies have been discussed.

The following series of chapters give case studies of some technologies. In fact if the lessons learnt from these projects are kept in mind before implementing the technologies in some other place, it can improve the success rate of a technology. The first chapter in this section is by Bindya Gupta **"The TEAM (TERI's Enhanced**

Acidification and Methanation) process for biomethanation of organic municipal solid waste". It deals with waste-to-resource technology options for municipal solid waste processing, followed by discussion on TERI's TEAM process for biomethanation of municipal solid waste. It also describes various other European technologies existing in India and problems associated with them. The chapter also emphasizes on policy and research needs in the field of solid waste management.

"Industrial Biomethanation Practices for Decentralised Energy from Waste: Options for Sri Lanka" by Ajith de Alwis looks at waste from different industries and gives a rough idea about how much waste is generated from those industries and their potential for biomethanation for energy recovery. He feels that industrial biomethanation offers twin benefits of energy and environmental management. He also discusses that there is a commonly held belief that anaerobic technology is not appropriate for treatment of most industrial wastewaters due to 'toxic' compounds frequently present in such wastes, and that this assumption has greatly hindered widespread application of these processes to industrial wastewater. Today however, with introduction of the concepts of bioaugmentation, and with more understanding of the methanogenic bacteria and their metabolic needs, anaerobic digestion is increasingly applied to industrial wastes. The author also discusses the various methods of anaerobic degradation being investigated and applied in reducing or eliminating the many toxic compounds found in chemical and industrial effluents.

"Energy from Landfill Gas Utilization in Canada" by Edward A. McBean, Rick Mosher, and Alain David give an account of energy from landfill gas utilization in Canada. In the beginning historical basis for these projects are described, followed by potential for a number of additional projects to be developed in the near future, due in part to new initiatives including greenhouse gas credits and the possible designation of it as green power. Some of the constraints and possible resolutions of these issues are also described. The authors describe the major constraint influencing the development of additional sites as economics, availability of greenhouse gas credits, greater access to green power etc. They conclude that it is very likely that the next few years will see significant increase in the numbers of landfill gas-to-energy projects coming on-line in Canada.

Jose Arellano in "The Landfill Biogas Management. The Case of the Chilean Sanitary Landfills" gives a brief account of different projects undertaken in Chile since 1980s where biogas has been utilized as an energy source for domestic (cooking and heating), as well as industrial use. The chapter gives both success stories (at Santiago and Valparaiso) and failures (at Rancagua and Concepción) followed by typical extraction systems used in Chile. In the end, case study of extraction of biogas at the Lo Errazuriz sanitary landfill is discussed.

The next chapter by T. Cássia De Brito Galvão on "Energy From Waste: Case Study From Brazil" describes a state-of-the-art method in developing clean technologies. Clean technologies in Brazil were developed mainly due to economic needs, especially during the Petroleum crisis. For example the Alcohol Programme – in which, biomass (for example sugarcane residue) is transformed to alcohol, to be used as combustible. This programme was developed in 1973 due to Petroleum crisis, which increased the price of a barrel four-fold, and there was need for an alternative fuel. The author also gives a discussion on the potential to develop clean technologies for energy production (recovering energy from waste), which emphasizes the environmental aspects too rather than just the economical needs.

The main focus is on the Brazilian experiences related to biomass and landfilling process and the use of landfill gas. It concludes that, although all these technologies benefit the environment, and represent a necessary condition towards safer living conditions on this planet, yet they hardly will shake (compete with) the traditional electric power generating industry—one of the world's largest, in the coming years.

After giving a brief introduction on the status of incineration in different EU countries, Bernt Johnke in **"Current Situation of Waste Incineration and Energy Recovery in Germany"** discusses municipal solid waste incineration followed by hazardous waste incineration in Germany. He discusses the different ways employed in incineration for municipal and hazardous waste, and other types of waste. He further gives energy utilization efficiency of (energy supplied by) different municipal solid waste incineration plants. He has compared the electricity produced in municipal solid waste plants and in the case of the co-generation of electricity and heat. The author also illustrates that the energy transformation efficiency as well as the proportion of energy actually supplied by waste incinerators to substitute for fossil energy sources, as estimated on the basis of it, and the resultant emissions, are of major importance for the calculation of climate-relevant emissions. Citing the example of Germany, he has shown calculations for, the credit for energy from waste incineration.

Vienna has been environmentally conscious since the 1970s when steps to improve air quality were taken. Keeping in mind similar trend of environmental awareness and for improving the environment, the city has also looked for ecologically and environmentally friendly ways to deal with wastes. Helmut Löffler in **"Thermal Treatment of Waste in Vienna: An Ecological Solution?"** discusses the fact that waste is produced as a part of our culture and it is clear that the waste which can neither be avoided nor recycled nor reused has to be subjected to a special treatment before its ultimate disposal to prevent any future environmental burden; i.e. it must neither end up in abandoned polluted areas nor must a long-term supervision of the disposal site be required. The author feels that the most flexible and commonly used process to meet with these ecological requirements is the thermal treatment of unavoidable and unrecoverable residues in modern waste incineration plants equipped with special flue gas cleaning processes. The chapter also gives waste management laws in Vienna, Viennese waste management policy and new standards for waste incineration plants in addition to this technology. He discusses how the incinerator in Vienna meets tough environmental standards and lives up to high emission standards. Famous Austrian artist Friedensreich Hundertwasser was so impressed by the high standards achieved by the plant that he agreed to undertake the new architectural design of the plant's facade without accepting remuneration. The author elaborates on the emissions from waste incineration plants and compares it with the emissions from household stove heating and gaseous emissions from household waste in landfill. In the end he has given survey results, about the acceptability of the plant, which show that opposition against municipal waste incineration in Vienna is relatively low and has remained between 3% and 4%, whereas acceptance among the population has risen from 73% to 81%.

Following the Kyoto Agreement, where the stress is on reducing greenhouse gas emissions, waste disposal options will need to be addressed more carefully. Keeping this in mind, waste co-firing (or co-incineration) may play an important role. If waste is burnt in a power plant or an industrial plant rather than in a waste incinerator, its calorific value avoids the use of fossil fuels and therefore results in an overall reduction of carbon dioxide emission. In **"Co-firing - Primary Energy Savings and Carbon Dioxide Avoidance"** - Ivo Bouwman, compares three installations that can process waste: a dedicated waste incinerator, a power plant, and a cement kiln, where both the calorific and the material content of the waste are used. The effects of the incineration of three typical examples of waste (mixed plastic waste, rubber, and sludge from a waste water treatment plant) are evaluated. The calorific value of the waste and the carbon dioxide emissions are compared with those of the primary fuel that would otherwise have been used in the processes. He concludes that this integrated approach makes clear that for the system as a whole the net gain may be smaller than often claimed. However, provided that other environmental criteria are met, the substitution value can be substantial and application of waste in a power plant or a cement kiln can have considerable advantages.

According to investigations carried out both in laboratories and outdoors on a large scale, baling technique has been shown to be the most promising storage method for waste fuels. It is easier to store baled waste to be used whenever and wherever needed. Normally studies have shown that storage of such bales affords no risk of self-ignition. The last chapter in this section is **"Baled MSW and associated problem, in contexts of fire hazard"** by William Hogland, Marcia Marques, Viatcheslav Moutavtchi, Diauddin R Nammari, Sven Nimmermark describes the results of a study in which storage tests were carried out under high temperature conditions and the physical and chemical processes operating during storage of round and square bales were studied. In addition, results of incineration tests carried out to assess the flammability and the concentration of a variety of pollutants in the solid and gaseous parts of the smoke are also described.

The next section deals with solid waste management options and climate change issues. The first chapter by Joyeeta Gupta and Frédéric Gagnon-Lebrun **"Waste, Energy and Climate Change, General Introduction"** examines a number of issues related to both solid waste management and climate change. It starts with a discussion on the fact that environmental problems manifest themselves at different geographical levels and scales, and decisions to deal with these problems have to be taken at different levels. This is followed by arguments that there are many material linkages between the two problems and that in many ways addressing some aspects of one problem also incidentally addresses some aspects of the other problem. The authors further examine the similarities in the root cause of both the climate change and the waste problems, which lies in our system of production, consumption and trade. The authors feel that, if the issues of production and consumption patterns are addressed, it will become much easier to deal with both problems although we may inadvertently create a new problem. Finally, the paper argues that if a political link can be made effectively between the two problems, this may give a boost to addressing both global and local problems. The authors conclude that there appears to be increasing reason to try to connect these two fields in order to give the global problem of climate change a direct local dimension and to infuse financial and political resources to the local priority of waste management.

The other chapter in this section **"Waste, Recycling and Climate Change: A US Perspective"** by Frank Ackerman presents a framework for analysis of the greenhouse impacts of climate change, then offers some estimates of the size of the impacts and finally gives approximate calculations of the importance of these impacts for the U.S. The author describes five types of impacts waste management has on climate change. He concludes that in spite of some uncertainties the effect of waste management choices on climate change is large enough and it is well worth studying it in greater detail. Paper recycling and the analysis of the paper life cycle appears to be of particular importance. It is remarkable to think that we have the potential to achieve one-tenth of the Kyoto targets through an activity that already has widespread grassroot support.

The fifth section of this book deals with health impacts of energy to waste options specifically incineration. The technical design of MSW incinerators have improved in recent years and regulatory controls have reduced emissions, but the impact on health depends, at least in part, on the effectiveness of the system of regulation imposed on these installations. The only contribution in this section is a chapter by Greenpeace **"Incineration and Human Health - State of Knowledge of the Impacts of Waste Incineration and Human Health"** authored by Michelle Allsopp, Pat Costner and Paul Johnston gives the findings of research on human exposure to pollutants from incinerators and health impact studies on workers and local populations. A broad range of health impacts have been documented in these two groups, including adverse effects on children living in local populations near incinerators. In addition to discussion of impact of incineration on human health and environment, authors also discuss about Adoption of the Precautionary Principle and Zero Emissions Strategy. They also point out that the way forward for waste management in line with a zero emission strategy and hence towards sustainability, lies in waste prevention, re-use and recycling. In other words, the adoption of the well known principle of "REDUCE, RE-USE AND RECYCLE".

The sixth section deals with the social aspects related to energy recovery projects. Although passing of legislations and selection of optimal technology is necessary for better solid waste management but actual legislation means little until social sentiments change. The mere passing of a law cannot alter practices of centuries. Therefore it is necessary to incorporate public involvement and social issues in all aspects of planning for Solid Waste Management, which includes disposal and "waste to energy" options. In this section, Christine Furedy, and Alison Doig in **"Socially–Responsive Energy from Urban Solid Wastes in Developing Countries"** illustrate an approach that gives initial emphasis to social and economic cosiderations in designing and implementing small or intermediate technologies in cities of developing countries. According to the authors, because of increasing waste and also increase in demand for energy, extracting the maximum value/energy from solid waste before going for disposal is gaining importance. But the breakthrough in WtE in urban areas will depend more on planning based waste–generator co–operation and on social insight, than on advances in technology. Also the energy recovery options would be more sustainable if they are sort of linked to poverty alleviation (because livelihood of many people depend on garbage) and environmental management. The main questions addressed in the chapter are: Are the 'wastes' designed to be exploited truly going to waste? What wastes in the area are accumulating and appear to be of no perceived value to local people? If wastes are being used, are there ways to increase the efficiency and income-earning potential of this use, with affordable technology, or by training in accessing raw materials or in marking? Are there alternative ways of using wasted or under-utilized resources that would serve poverty alleviation better? Project work done by ITDG, in Nairobi (part of the large interntional project done for Small Scale Energy Production plants in which study is done to investigate the opportunities for applying waste to energy technology for livelihood of urban poor) on briquettes from charcoal/saw dust is used to illustrate the social methodology. Under this study after considering the other options of waste disposal and livelihood and stakeholder analysis if a WtE technique appears applicable, there is a detailed assessment of it in the light of the data from the components, including technical/physical requirements, health aspects, financial feasibility, and markets for the energy produced. This chapter concludes that, for the developing countries, bioconversion of wastes (mainly through composting) is more feasible for small–scale projects than Waste to Energy. However, there are specific options that are worth exploring. The most promising small–scale projects are those that capture single source or separate, relatively uncontaminated wastes such as charcoal, cinders, lumber–yard sawdust, organic market wastes, and food wastes from factories, hotels and restaurants.

The seventh and the final section concentrates on the marketing issues of waste to energy technology—waste gasification plant in U.K. Sarah Knapp Dent in **"The Market for Small Scale Waste Gasification/Pyrolysis Projects - Preliminary Scoping Study"** presents the findings of a market analysis for small waste gasification plants in the UK. This chapter is based on a scoping study, which involved carrying out a broad but comprehensive market research exercise, with the aim of identifying the likely demand for purchasing of small gasification/pyrolysis plants. The research was broken down into appropriate market segments and sub-segments e.g. local authority and inter-authority or waste management company partnerships. The chapter also gives the approach and methodology of the survey followed by views of technology suppliers, details of organizations contacted and typical buyer profile among other things.

It can be concluded that even with the most optimistic assumptions about waste minimization, recycling and composting, there would be residual waste, which requires disposal. Incineration, or energy from waste, plants are likely to appear a necessary element of local waste management strategies and some authors also suggest that with little more research it can be seen as a good replacement for fossil fuels! Although, nothing would be able to replace landfill sites, but probably waste-to-energy plants (with stricter emission controls) would appear in the list of waste management options for municipal waste planners. While making the decision all the technical, economic and social factors should be integrated for it to be more sustainable and successful.

It needs to be emphasized that the problems of developed and developing nations are different and so are the solutions, therefore, applying the solutions blindly will not help the developing nations in achieving a sustainable or even workable solution. As the quantity and composition of the waste differs in developing and developed countries (within a country even within rural and urban areas) waste management practices need to be site specific. To start with, in developing countries organic waste being the main component should be the target for energy recovery (it has been observed that recovering energy from rural agricultural waste, manure and sludge is more successful than from pure urban solid waste. A few experiments with mixed urban waste have also been performed). (However with globalization, as the composition of waste in metropolitan cities in developing nations are changing, at some point in future the case studies from developed nations can be tailored to waste management solutions in developing nations). Another word of caution needs to be added is that while planning for waste management options and strategies, all the above mentioned factors are to be kept in mind otherwise the solutions like incinerator in Delhi, India are doomed for failure. In this particular case the calorific value of waster of Delhi is not fit for the incinerator obtained from DANIDA, may be because calorific value from the analysis of waste in the laboratory has never been observed in reality. But the basic fact is that the actual quality, quantity and composition of waster reflect that this technology (an incinerator with waste recovery option) is/was not appropriate for the local conditions.

Also, though waste management seems to be a local problem but combining it with global problems like climate change would make planners think more critically about waste management options including waste to energy plants in view of green house gas emissions from landfill site.

Above all, if the real issues surrounding modern integrated waste management are to be widely understood by local communities so that long-term, sustainable solutions are supported and adopted, a genuine partnership must be created between industry, central and local government and other interested parties. Last but not the least, keeping in mind the philosophy of "small is beautiful" the solutions (especially for developing nations) should be simple and straightforward keeping health, socio-economical and other factors like appropriate technology, in mind.

2

Waste-to-Energy: Comprehensive Recycling's Best Chance?

JOHN A. MERRITT

President, Merritt Communications, Inc., Environmental Solutions Division
340 Woodland St, Holliston, MA 01746, USA

Thermal conversion of solid waste, or "waste-to-energy" (WTE), as the process is known in the USA, is implemented in many areas of the world. Through controlled incineration of discarded materials, useful steam is generated which is either used directly for industrial purposes such as heating or channeled to turbines, which generate electricity, which is delivered to power grids. During that process, bacteria and pathogens are destroyed. In Europe, Japan and certain regions of the USA, WTE represents a prominent alternative to the landfilling of waste materials. In all of these areas, open land, available for waste disposal is very limited, making WTE an attractive alternative. Currently, there are 102 such facilities in the USA.[1] These facilities burn about 14% of the waste, generated by about 37 million people, in 37 of the 50 States in the USA, generating more than 2,800 megawatts, enough to power nearly 2.5 million homes. The value of that energy is in excess of $850 million and is produced in facilities that represent a total capital investment of over $10 billion.[2]

Additional products of these facilities are ash and air emissions, two elements that have brought this technology under intense scrutiny by some in the environmental advocacy population. It is the case that both ash and air emissions have negative environmental impacts if not managed appropriately. For this reason, in all jurisdictions where WTE is used, there is stringent regulatory oversight of these facilities and their operation regarding these potential pollutants. As a further result, these elements have been very carefully studied and analyzed around the world for over twenty years.

Based on my experience working with WTE projects directly, my review of the literature and participation in the public debate internationally, I believe that WTE may very well represent the very best chance, today, to achieve comprehensive recycling of waste materials possible, for large-scale waste management projects. Bioconversion may represent a smaller scale solution for some locations, as the in-vessel technology advances in future. However, today and for projects with a very large capacity requirement, WTE can offer a nearly complete recycling option if implemented properly. In this chapter I will discuss the key elements required to achieve this end. Topics will include the compatibility of WTE facilities with pre-

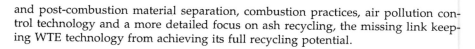

and post-combustion material separation, combustion practices, air pollution control technology and a more detailed focus on ash recycling, the missing link keeping WTE technology from achieving its full recycling potential.

Pre-combustion Material Separation

In the USA, it is quite common for certain materials to be separated from the waste material stream before either incinerating or landfilling. In fact, in many jurisdictions, recycling is made mandatory by regulation or ordinance. The most common materials separated and recycled into other products include fiber, such as cardboard, newspaper and office papers; metals including ferrous and aluminum; glass of all colors and certain plastics, including generally, at least the single polymer plastics polyethylene terephthalate (PET) and High Density Polyethylene (HDPE). The markets for most of these materials vary over time, but some, like aluminum, always bring a high enough price to ensure a very high degree of separation.

Mercury has become an important focus of pre-combustion separation programs. Button and other household batteries containing mercury are collected and kept out of the combustion stream in order to reduce the strain on air pollution control systems to remove this toxic heavy metal.

Also bulky appliances, car batteries, furniture and carpeting, as well as construction and demolition debris (C&D) are frequently separated from the waste materials to be burned or buried. Sometimes these separation and collection activities take place near the homes of waste generators and other times it is incorporated into an integrated solid waste management (ISWM) facility which incorporates a landfill or WTE facility as the next disposition for post-separation materials. It is at this point, however, where WTE begins to differ dramatically from landfilling with respect to further recycling.

Combustion Practices

At WTE facilities, the recycling continues after traditional separation of recyclable materials is over, by converting the remaining waste materials into energy. The conversion of this fuel into steam or electricity represents a tremendous recycling value. Some WTE facilities are "mass burn" (unprocessed waste as fuel) while others are fed "refuse-derived" fuel (RDF). At an RDF facility, very substantial pre-processing step is employed. All delivered waste material typically is first passed through equipment that batters it into a fluff, out of which magnets remove ferrous metal. This provides a fairly consistent and uniform fuel stock that can lead to more controlled and less variable combustion parameters. Irrespective of fuel type, all WTE facilities require careful monitoring and maintenance of combustion that maximizes burnout of fuel and maximum conversion of heat to energy, while minimizing potentially toxic air emissions and ash.

Most modern WTE facilities have elaborate digital monitoring and control systems to help keep track of all aspects of the combustion process, allowing a very high level of control on an ongoing basis. In addition, regulatory agencies typically require constant air emission monitoring, which provides additional information about the quality of combustion or burnout.

Air Pollution Control (APC) Technology

By the end of the year 2000, WTE facilities in the USA are required by the US Environmental Protection Agency (EPA) to meet very stringent air emission standards that are based upon the "maximum available control technology", the so-called MACT standard. (Fig. 1). These standards are derived from measuring the maximum toxin reduction that is achieved on average by the most modern plants with the most advanced technology available. While this will cost the majority of older WTE facilities in the USA a great deal of money to retrofit earlier APC equipment with technology to meet the newest standards, when accomplished the quality of WTE emissions will reduce toxic discharges below levels that can be demonstrated to have negative health impacts. I believe this is an important step in bringing the WTE technology closer to being a fully safe, as well as comprehensive, waste recycling technology.

The typical elements of APC technology required by the MACT standards are as follows:[3]

- Bag House—works like a giant vacuum cleaner with hundreds of fabric filter bags, which clean the air of soot, smoke and metals.
- Scrubber—sprays a slurry of lime into the hot exhaust, neutralizing acid gases and reducing some of the mercury in the exhaust.
- Selective Non-Catalytic Reduction (SNCR)—sprays ammonia or urea into the hot furnace to convert nitrogen oxides, a cause of urban smog, into harmless nitrogen.
- Carbon Injection Systems—blows charcoal into the exhaust gas to absorb mercury, as well as controlling organic emissions such as dioxins.

Post-combustion Separation

The ash residue from WTE facilities represents 25% of the incoming fuel by weight, but only about 10-15% by volume. It consists of two major components: bottom or grate ash and fly or APC residue. In some facilities, these streams are combined, while at others they are kept separate, as they are produced. The bottom ash or combined ash is typically passed under strong magnets to remove more ferrous metal and over grates to remove bulky remainder pieces or "clinkers." This becomes the third stage of recycling in the WTE process.

Ash Recycling

Because of the high concentrations of toxic heavy metals in the fly ash fraction of WTE residue, many European facilities and some American facilities keep it separate from bottom ash and specifically treat it to permanently capture or neutralize those toxins. That "conditioned" fly ash, representing only about 1% of the original fuel input, is typically landfilled as there are presently no commercially viable re-use or conversion technologies available. However, there are new ideas presented regularly about potential uses, which are still in research stages.

At most American facilities, the fly and bottom ash are typically combined, resulting in a dilution of the toxins in the fly ash sufficient to allow the combined

MACT RULES FOR EXISTING LARGE UNITS AT FACILITIES

Emission	Limit	Emission	Limit
Dioxin/Furans Large	(/dscm)	SO$_2$	
ESP–equipped units	60 ng	Large	29 ppm or 75% removal
All others	30 ng		
Cd Large	(/dscm) 0.04 mg	HCl Large	29 ppm or 95% removal
Pb Large	(/dscm) 0.44 mg	Opacity	10%
		Hg (/decm)	0.080 mg or 85% removal
PM Large	27 mg	Fugitives	Visible less 5% of the time

NOx: Operator of large plants may select one of the two options
    Option A: Units must meet the following standards:

Mass Bum/Water Wal'	205 ppmv	Fluidize Bed	180 ppmv
RDF	250 ppmv		
Mass Burn Rotary	250 ppmv		

    Option B: Plants may "bubble" units wihtin the plant to meet the following standards:

Mass Burn/Water Wall	180 ppmv	Fluidize	160 ppmv
RDF	230 ppmv		
Mass Burn Rotary	220 ppmv		

CO Modular	50 ppmv	Spreader Stoker Coal/ RDF Mixed, RDF Stocker	200 ppmv
Mass Burn Rotary Refractories, Waterwall, Rafractory, Fluidized Bed	100 ppmv	Mass Burn Rotary Waterwall	250 ppmv
Pulverized Coal,RDF Mixed	150 ppmv		

NEW SOURCE RULES FOR NEW PLANTS

Emission	Limit	Emission	Limit
Dioxin/Furans	13 ng/dscm	SO$_2$	30 ppm or 80% removal
Cd	0.020 mg/dscm	HCI	25 ppm or 95% removal
Pb	0.20 mg/dscm	Opacity	10%
PM	24 mg	Hg (/dscm)	0.080 mg or 85% removal
Fugitives Visible less than 5% of the time		NOx First year After lst year	180 ppm 150 ppm
CO Modular/Mass Burn, Fluidized Bed	100 ppmv	Complete: ■ Siting Analysis	
RDF	150 ppmv	■ Materials Senaration Plan	

Fig. 2.1 MACT Standards

ash to regularly test as non-hazardous waste, suitable for disposal in a landfill the same as household solid waste. Federal and state regulators generally require ash disposed in this fashion to first pass the Toxicity Characteristic Leaching Procedure (TCLP) or similar test designed to exaggerate the effects of acid rain, potentially leaching heavy metals into the environment. This ash routinely passes these tests.[4] In fact, when test results are taken in actual field conditions, leachate metal levels are found to be far below TCLP standards and actually approach drinking water standards.[5] In practice today, the vast majority of ash in the USA is landfilled in this fashion, with no recycling at all. There are a number of reasons for this, not the least of which is the general public's lack of understanding about the nature and potential value of WTE bottom ash residue.

There are, however, a number of approaches to recycling bottom ash which have been investigated, some of which I believe show some promise in future, particularly as the reduction in land available for ash landfills diminishes and hence, it will become more expensive to landfill. Currently, there are few, if any, ash recycling technologies that would be profitable enough in the current environment to compete with landfill disposal. However, there are clear applications for ash that have been demonstrated to work well.

The key categories of recycling possibilities for WTE ash, from my perspective, include natural aggregate substitutes, cementaceous applications and vitrified products. There are some well-established trial projects in New Hampshire and New York in the USA and elsewhere internationally, using some percentage of WTE bottom ash as a substitute for natural aggregates in road base construction.[6] After substantial and continuous monitoring with respect to any negative environmental impacts, none have been found. There are also demonstrated applications in flowable fill with ash as a substitute for natural aggregate, though WTE ash has yet to be used in this application, though it is physically suitable.

Cementaceous applications depend upon the pozzolanic or binding properties, particularly of the fly ash, in combined WTE ash. Concrete can be made substituting WTE ash for expensive cement in the mix. An underwater reef constructed of this material on Long Island has been monitored for years with no adverse environmental impacts recorded.[7] There are many applications for such concrete that would be a better use of this valuable material, than burying with no use, except to remove the land in which it's buried from productive future use.

Vitrification, or melting ash at extremely high temperatures, is a relatively expensive process for recycling ash, given current sources of energy and cost. However, the molten ash can be spun into mineral wool, useful as industrial insulation; cooled and fractured into a grit that can be used in "sandblasting" applications among others or cast into forms to create tiles or curbing material.

As a practical matter, the small fraction of WTE ash that sees any useful purpose in the USA today, is found in landfills as well, employed as gas vent layers or alternative cover material for household waste landfills, a very slight change from simply being landfilled directly. However, if economics, regulators and/or the public evolve and change, higher recycling options for WTE ash may become substantially more common. If this happens, I believe WTE technology will have lived up to my belief that it represents a nearly complete recycling option, with a tiny percentage of what comes in the front door not being put to good use and subject to landfilling. I believe this is a very worthwhile effort with the potential of

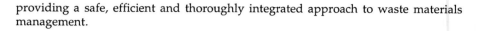

providing a safe, efficient and thoroughly integrated approach to waste materials management.

Conclusions

At a minimum, WTE management of waste materials already is completely compatible with all current pre- and post-combustion recycling separation processes and represents tremendous re-use by converting the remaining materials to clean energy. Air emissions are already undergoing careful scrutiny and with modern retrofits, the quality of emissions will be very good indeed. With respect to recycling programs and waste-to-energy facilities, local communities using this technology actually recycle at a rate of 33% of the total waste stream, or a full 5% higher than the national average.[8] When ash recycling matures and is also accepted widely, WTE will, indeed, become a comprehensive recycling process for household waste materials.

END NOTES

1. Jonathan V. L. Kiser and Maria Zannes, Integrated Waste Services Association, The ISWA Directory of Waste–To–Energy Plants, May 2000.

2. Maria Zannes, Integrated Waster Services Association, *Waste–To–Energy Industry Fact Sheet*, August 2000.

3. Maria Zannes. Integrated Waste Services Association, *New Clean Air Rules For Waste–to Energy Facilities*, August 2000.

4. U. S. Environmental Protection Agency and the Coalition on Resource Recovery and the Environment, "Characterization of Municipal Combustion Ash, Ash Extracts and Leachates," March 1990.

5. Ibid., U.S. EPA, C.A. Andrews, "Analysis of Laboratory and Field Leachate Test Data for Ash From Twelve Municipal Solid Waste Combustors," In: *Municipal Waste Combustion: Conference Papers and Abstracts for the Second Annual International Specialty Conference*, April 1991.

6. W.C. Ormsby, Federal Highway Administration, "Paving with Municipal Incinerator Residue," In: *Proceedings of the First International Conference on Municipal Solid Waste Combustor Ash Utilization*, at p. 49 May 1988; C.N. Musselman et. al., "Utilizing Waste–to–Energy Bottom Ash as an aggregate Substitute in Asphalt Paving," in *Proceedings of the Eighth Interational Conferece on Municipal Solid Waste Combustor Ash Utilization*, at p. 59 November 1995.

7. F.J. Roethal and V.T. Breslin, "Stony Brook's MSW Combustor Ash Demonstration programs," In: *Proceedings of the Third International Conference on Municipal Solid Waste Combustor Ash Utilization*, p. 237, November 2000.

8. Katie Cullen, Integrated Waste Services Association, *Recycling and Waste–to–Energy: Compatible Options for Solid Waste Management*, August 2000.

3

Energy-from-Waste: A Perspective from Developing Countries[1]

A.A. Oladimeji[2], S. Kumar[3], C.T. Parekh and M.J. Bhala

Faculty of Applied Sciences, National University of Science and Technology
P O Box AC939, Ascot, Bulawayo, Zimbabwe

Introduction

Developing countries depend mostly on wood, agricultural wastes, straws and cow dung for energy. In these countries, use of imported fossil fuels, coal, oil and natural gases has been confined to urban areas due to the prohibitive cost. Further, all these sources of energy require some kind of infrastructure which in developing countries may not be as developed as in the advanced countries. Traditional fuel, firewood, is also becoming scarce due to over exploitation of forests. When available, it is collected from places that are too far.

The continuous rise and competitive demand for fossil fuel necessitates the need for a search for alternative sources of renewable energy. Reference is made here of the escalating cost of crude oil observed in the USA, which rose from $10 (US) in 1974 to $30 (US) per barrel in the year 2000. The solution is to develop energy sources based on local materials and labour; and should be massive enough to take care of heating, lighting and power generation on a small scale for light industries. One source of renewable energy, which has not been given adequate attention, is waste, particularly organic solid waste comprising crop residues and animal wastes which are abundant in rural agrarian communities. Methane produced from dung and other agricultural or industrial wastes could play a very important role as a form of energy.

Tebo (1997) says: "Waste is really a mindset. All we really have are ingredients that haven't yet found a home". For Tebo and many likeminded people, waste is but a transitional stage of material yet to be converted into some useful application. This ultimately suggests that if a high level of technological development is attained, waste can be put into some use. The World Health Organisation (WHO)

---

[1] Correspondence to S. Kumar
[2] Present address: Department of Biological Sciences, Federal University of Technology, P.O. Box 65, Minna, Nigeria
[3] Present address: 1. School of Communications and Informatics, Victoria University, P.O. Box 14428, Melbourne City, MC 8001, Australia. (skumar@sci.vu.edu.au) and 2. Department of Mathematics and Statistics, University of Melbourne, Parkville, Victoria 3052, Australia.

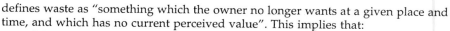

defines waste as "something which the owner no longer wants at a given place and time, and which has no current perceived value". This implies that:

- Waste is a burden to the one who generates it,
- Waste occupies useful space which would otherwise be used as storage for value-giving materials, and
- Waste has no value i.e. cannot, in its existing form, bring the much needed monetary gain to the owner (Suess and Huismans, 1983).

Industrial solid waste, depending on the nature of industry, can be hazardous or non-hazardous, toxic or non-toxic, complex in structure or single component material. Non-hazardous solid waste comprises such material as paper and plastic packaging material, glass, food scraps, domestic, animal waste and other residues, and garden materials. It can also contains small quantities of hazardous substances, such as paints, medicines solvents, cleaning materials and batteries, making waste management more difficult and increasing the environmental risks that waste presents.

In general, developed and industrialized countries produce a higher proportion of toxic, non-organic and non-biodegradable waste, particularly packaging materials and paper. Conversely, in less developed and low-income countries, a large proportion of municipal waste consists of organic matter and ashes or grit.

The health risks of uncollected wastes have been most obvious in less developed countries. Notably, pre-school children are at risk of injury, intoxication or infection since they are likely to be exposed to uncollected waste in streets. Organic domestic wastes in particular pose serious health risks since they ferment, creating conditions favourable for the survival and growth of microbial pathogens, and especially if they are mixed with human excreta due to poor sanitation. Organic wastes also provide feed stock and a natural environment for insects, rodents and other animals, which are potential carriers of enteric pathogens. Uncollected solid wastes can also obstruct storm water run-offs, resulting in flooding or creation of stagnant water bodies, which become habitats and breeding places for waterborne vectors of tropical diseases. However, the risk to health posed by waste, particularly organic waste, can be averted or minimized if technology is harnessed to produce energy from the so-called waste material.

Ethanol from Industrial Wastes

Alcohol production from biomass is not an entirely new concept; it can be traced back at least 200 years to the Egyptians, who produced ethanol as a beverage (Karekezi and Ranja, 1997) early in the 19th century. Henry Ford, inventor of Ford vehicles, invented an engine that could use both gasoline and ethanol. However, the technology was not put into use for sometime due to the discovery of petroleum, which was much cheaper at that time. Renewed interest in ethanol developed in the 1970s as the cost of petroleum fuels increased.

Ethanol (grain alcohol) and methanol (wood alcohol), are produced by anaerobic digestion of plant materials with high sugar content, mainly grain and sugarcane. Ethanol can be burnt directly in automobile engines adapted to its use or it can be mixed with gasoline of up to 10% and be used in any normal automobile engine. A mixture of gasoline and ethanol is called gasohol. When used in gasohol, alcohol increases the octane number of gasoline blend. This allows the substitution of the environmentally harmful lead compounds in gasoline, which are added to boost octane numbers. (Karezeki and Ranja, 1997). Ethanol production can also be used as a solution to grain crop surpluses and earn a higher price for grain crops than the food market offers. It offers reduced dependence on gasoline and conserves foreign exchange.

The Zimbabwe ethanol programme is one of the most successful programmes for the production of transportation fuel from renewable sources such as sugar, molasses, and starch. Like most petroleum-importing countries, Zimbabwe has had to cope with incessant economic and social dislocation brought about by spiralling oil prices and a drain on vital foreign exchange. Ethanol production was therefore a conscientious effort to establish an essential and indigenous source of liquid fuel supply and a greater measure of self-sufficiency in meeting national energy demand (Scurlock *et. al.*, 1991).

The ethanol plant at Triangle in southeast Zimbabwe is an example of production of energy from biomass. The ethanol was produced from molasses, a sugar-refining by-product comprising uncrystallizable sugars and solubles. It was otherwise an industrial waste but thanks to advances in technology, it can now be put to use. The triangle ethanol plant was designed to operate on a variety of feedstocks: different grades of molasses or even raw sugar could be fermented. The by-products of the sugar industry were disposed of with due economic consideration. The carbon dioxide from the fermentation process was sold to soft drink bottling industries while bagasse is burnt to produce steam in boilers for generating electricity, some of which is sold to the National Electrical Supply Authority. During the cane-growing season, the plant used stored and imported molasses for ethanol production powered by coal.

Approximately 120,000 litres of anhydrous ethanol were produced daily, with an estimated maximum annual capacity of 40 million litres. This turned out to be a success story because Zimbabwe has a good agricultural base, which produces surplus food. Therefore the question of producing fuel at the expense of food for the masses did not arise. Moreover the technology was assembled using local labour; 60% of the plant was fabricated and constructed locally (Scurlock *et. al.*, 1991). The final cost of (US$6.4 million) constructing the Zimbabwe ethanol plant made the plant one of the cheapest of its capacity in the world. Within one year of its operation the plant had met its goal of showing a net savings in foreign exchange (Scurlock *et al.*, 1991). In Zimbabwe, the ethanol gasoline blend (target blend, 13%) was the only fuel available for spark-ignition engines for a period (Scurlock *et. al.*, 1991). In the wake of the recent fuel shortages, plans are underway to resuscitate the ethanol plant for a new range of fuel products.

Brazil embarked on a national programme to substitute crop-based ethanol for a good proportion of imported petroleum. In 1985, the Brazilian sugar harvest produced 2.5 billion gallons of ethanol and plans were to produce 4 billion gallons in a few years (Cunningham and Saigo, 1992). Brazil is currently the world's leading producer of fermentation alcohol, which provides approximately 50% of gasoline requirement in the country (Goldemberg *et al.*, 1994).

Like Zimbabwe, Malawi is also landlocked with no indigenous oil resources or refineries. As a result a good proportion of its foreign exchange earning is spent on importing (Karekezi and Ranja, 1997) of refined petroleum. In 1982, Malawi commissioned a molasses-based ethanol plant. Like in Zimbabwe, molasses from the sugar mill was used, with production averaging 13 million litres per annum (Karekezi and Ranja, 1997). The intended ratio of ethanol with petroleum was 20:80. In 1983, the ratio was 15:85 (Karekezi and Ranja, 1997). However, the blend was not maintained for long due to disagreements between the Ethanol Company Limited (ETHCO) and the oil industry as to the acceptable market shares and the pricing of ethanol in relation to imported gasoline. Critics of the ethanol project in Malawi argue that it was no longer economical to produce ethanol as the molasses used were imported and that after the Gulf war, it was cheaper to import fuel. There was however evident that the blending of ethanol with fuel reduced foreign exchange spending substantially for Malawian economy (Karekezi and Ranja, 1997).

Kenya's interest in the production of ethanol was prompted by the increase in oil price in the early 1970s. The Kenyan Agro-Chemical and Food Corporation (ACFC) was established in 1978 with the main objective of utilizing the surplus molasses from the sugar mills. According to Karekezi and Ranja (1997), 145, 000 tonnes of molasses were produced by the ACFC in 1988, of which 30,000 tonnes were exported to Europe, while 20,000 tonnes were used as animal feed. The surplus, which was dumped in pits, posed environmental problems.

The ACFC plant, sited at Muhoroni near three sugar factories (Omondi, 1991), used the surplus molasses. The average daily production of ethanol was 45,000 litres (Karekezi and Ranja, 1997). The use of the surplus molasses in the manufacture of ethanol did not only reduce the environmental threat, but also benefited both the economy and the community as it employed about 200 people. In addition, it helped conserve foreign exchange in the country.

Lack of commitment by the government and the absence of a clear-cut production, blending and marketing policies led to the collapse of the ethanol/gasoline programme (Okwatch, 1994). In addition, low sugarcane yields in 1992, resulting from drought, affected the operation of the plants.

Biogas Energy

Another renewable energy source is biogas. Biogas can be generated from agricultural residues, which are readily available in rural areas. Such residues include coffee pulp, sisal slurry, maize husk, bean pods, and even animal dung. Biogas is a combustible gas produced by the fermentation of organic residues in the absence of oxygen and comprises about 60% methane (CH_4) and 40% carbon dioxide (CO_2). The gas finds application in gas lamps, in running refrigerators, in generating electricity and to power engines.

Most conventional biogas plants, which operate on cow dung face shortages in dung availability especially in developing nations like India, and many African countries where cows are allowed to graze over a free range, letting the dung scatter. There is therefore need to restrict movement of cows in stables so as to be able to collect enough dung to feed the digester or seek alternative source of organic matter. Aquatic weeds might be a good renewable and probably inexhaustible alternative.

Biogas Energy from Aquatic Weeds

It has been estimated (Hall, 1978) that the total world's petroleum energy is equivalent to the total plant photosynthesis for only 10 years. Further estimations show that the biological solar energy conversion via photosynthesis produces every year, an amount of stored energy in the form of biomass 10 times the world's annual use of energy. One of the methods of biological transformation of this solar energy into renewable, storable and ecological fuels is bacterial digestion of organic materials into flammable biogas under anoxic conditions. Biogas so obtained can be used for heating, lighting, cooking and powering generators.

Problems associated with nuisance weeds have worsened because water bodies are becoming more enriched by industrial and domestic effluents as well as fertilizer runoff from farmlands. Water hyacinth *(Eichhornia crassipes)* is a nuisance weed in many tropical and semitropical water bodies. High growth potentials and yields have been reported by Pillai *et al*. (1984). Water hyacinth productivity of up to 85 tons dry weight per acre per annum was reported by Young and Crossman, 1970. In recent years many important water bodies in Africa have been infested with nuisance weeds as the water hyacinth, *Salvinia molesta* and the water lettuce (*Pistia stratiotes*). On the Kenyan side of lake Victoria, water hyacinth had become such a nuisance that its clearance had to be contracted out to an American company in 1999. Weeds harvested during such an exercise were normally used for composts. Such harvested weeds could also be used in the production of biogas.

In Bangladesh and India, water hyacinth infests large areas of stagnant water bodies (Sarker *et al*. *1984)*. Research by (Sarker *et al*. 1984) has shown that in-view of the cheap labour available for collecting the weed and in manning the digester in developing countries, the production of biogas might be cheaper and economical than other sources of energy. Other researchers (Guha *et al*. 1984) have demonstrated the production of biogas at three different stages, namely laboratory scale, enlarged scale and working plant scale and arrived at a workable design of 3 cubic metre gas plant. One kg of water hyacinth (dry weight) evolved 370 litres of gas with average of about 5 litres per day per kg of feed. Murty (1984) reported the development of a low cost biogas plant in Kodumunja village, in Karimnagar District, India. Although the production was based on dung from 300 cattle, a similar low cost plant with biomass of water hyacinth: dung of 89.5: 10.5 demonstrated by Dhahiyat and Siregar (1984) could be feasible and economical.

Odeyemi (1980) produced biogas from *Eupatorium odoratum*, water lettuce and water hyacinth seeded with microbial population by addition of fresh cow-dung in digesters incubated at 29° C for 36 days. The peak production of biogas per kg of plant per day was about 256 litres for *E odoratum* and 108 litres for water lettuce. 98 litres of biogas were produced per kg of water hyacinth. In many parts of Nigeria, freshly cleared lands are overgrown with *Eupatorium odoratum* and it is therefore readily available. In view of its digestibility and high biogas generating capacity, it has a good potential as a source of fuel gas for household cooking and lighting, especially in rural areas.

Although the production of biogas from water hyacinth under varying conditions has been studied in various countries, the cost of collecting the plant and lack of constant supply of the weed has hindered production on a large scale in most of these countries.

Dung and Biogas as Fuels

In many developing countries, approximately 70-80% of the population live in rural areas where the bulk of their fuel demand is met by dry cattle dung, firewood, leaves and other plant residues. Although burning of cow dung maybe a logical use of waste biomass where wood is in short supply, it can intensify food shortages that could have been averted if the waste had been put back on land as fertilizers to improve crop production.

Cows in India produce more than 800 million tons of dung per year, more than half of which is dried and burnt in cooking fires (Cunningham and Saigo, 1992). When cow dung is burnt in open fires, more than 90% of the potential heat energy is lost. Studies have shown that using the dung to produce methane is much more efficient. In the 1950s, simple cheap methane digesters were designed for villages and homes in India, though they were not widely used. There are currently over 50,000 Gobar (cow dung) anaerobic gas plants in India and more than 6 million plants in China (Ikhu-Omoregbe and Idahosa, 1999).

In Italy, biogas plants were developed for controlling odours as well as for energy production. Piccinini *et al.* (1988) reported that more than a hundred farm biogas plants were built and were in operation over a ten-year period in the eighties.

Similarly, a number of institutions in Zimbabwe obtain energy from biogas plants which operate on cow dung or even human faeces. The simple biogas digester located at Minda (Roman Catholic) Secondary School in Maphisa is a success story. Maphisa is at a distance of approximately 100 km from Bulawayo, the centre of development in southern Zimbabwe. The biogas plant, installed by Rainer K R Wesenberg, Germany, became operational in 1995. The plant comprises a floating drum, water jacket and a mixing chamber. Biomass, comprising cowdung obtained from 15 cows, is fed into the mixing chamber (digester) and the gas produced is led directly to the students' kitchen or stored in a tarpaulin balloon for the Convent kitchen. According to the Headmaster, Mr. B. Moyo, the biogas supply provides energy for cooking food in a canteen serving 389 students and staff of the school.

The design of the digester is simple, and the technology required for the day to day running is such that, local semi-skilled operators could handle. According to Wenman (1985), 'In a developing country, it was necessary to design and build a plant appropriate to the abilities of the people who were to run it' and this set-up certainly meets that requirement.

After the biogas is obtained, the slurry (digested material) is led to an orchard of pawpaw, orange, onion and garlic where it is used for manure. According to Moyo, a similar type of digester which uses human faeces as biomass is operational at the 'Brunaberg Hospital' in Kezi Rural Council, SW Zimbabwe and Empandeni Secondary School in the same area. This had made it possible for the hospital and the school to overcome the problem of sewage blockage.

Landfill

Sanitary landfill can provide a source of energy generated by the natural anaerobic digestion of solid wastes—placed within the landfill. Landfill siting requires careful planning if energy benefits are to be exploited from it. Landfill should be located more than 100 m from unusable area and should not be near residential areas. It should also be, as far away from schools, hospitals, clinics, food-processing facilities, churches, hotels, public parks etc, as the emissions of odourous gases can become a nuisance. A landfill area should be fenced and also monitored by a supervisor to prevent entry by unauthorized persons. Designing of new landfill must also take into consideration (i) ground and surface water to avoid contamination, (ii) leachate generation and its impact on the environment and (iii) gas venting and/or recovery.

Recovery of Landfill Gas

Recovery of landfill gas methane is very inefficient because of lack of control of important environmental conditions such as moisture, pH, temperature, nutrients and air intrusion. Hence in the conventional landfill (or open landfill) the amount of methane gas produced is very low due to lack of optimum anaerobic conditions. Recovery of landfill gas requires the construction of extraction wells. This operation is usually difficult as well as expensive because the wells must be driven through the solid waste and must be gas tight to avoid methane escaping and exploding. The amount of methane gas produced in a landfill depends on the nature of the solid waste imbedded. It has been estimated that about 270 million tons of municipal solid waste annually generated in the U.S. contain 75% biomass with potential gas production of 1500 cubic feet of landfill gas from one ton of refuse. (Onu et.al.,1991).

The Municipal Solid Waste in many African countries is very low in comparison with United States or any other first world country. This will make it more difficult to recover methane gas as a viable source of energy. Nevertheless, methane gas can be recovered for energy consumption from landfill dumps.

Emission of methane gas has been observed from the abandoned Pumula landfill, in Bulawayo, Zimbabwe. Some of the pipes laid to release methane and other gases are therefore still in place. If the technology is available, the methane can be extracted through wells sunk into the landfill. This will reduce the emission of methane gas into the atmosphere and cut down on the much-needed foreign exchange spent on import of fuel.

Environmental Impacts

Most environmental scientists believe that biomass fuels used in a sustainable manner result in no net increase in atmospheric CO_2. This is based on the assumption that all the CO_2 given off by the use of biomass fuels is absorbed from the atmosphere by photosynthesis. Substitution of more fossil fuels with energy from solid waste would therefore help to reduce the threat of global warming due to increased atmospheric concentration of carbon dioxide. However, things are not as simple, as biomass is currently being used in an unsustainable manner and the long-term effect of basing energy on biomass on the plantations is yet to be determined.

In less developed countries, many people are faced with starvation and hunger. Therefore the question arises as to the justification of producing biomass energy at the expense of food for the masses. It may therefore be argued that use of land to grow "plantations for fuel" instead of food will aggravate this problem. Moreover, using organic material which otherwise would have served as fertilizer for the soil and increase food production as a means of generating energy, would no doubt increase the food problem in developing countries. However, in developing countries where there is surplus of certain grains, the surplus could be used for ethanol production and earn a higher price for grain crops than the food market offers. It should however, be noted that the transformation of natural ecosystems into monoculture plantations, like sugarcane, for energy production may lead to a reduction in biodiversity. Such ecosystems are normally more susceptible to damage by pests and diseases.

The replacement of fossil fuels with energy from biomass reduces the potential for acid rain, which results from the sulphur dioxide and nitrous oxides emitted from the fuels. It is also a well-established fact that methane gas is a contributor to ozone depletion and its uncontrolled use may lead to an escalation of the problem. The recovery of methane gas therefore has two fold advantages i.e. recovery for energy use and reducing pollution. There is therefore an eminent danger associated with the use of inflammable gases by communities that comprise mostly illiterates and thus concerted effort should be made to ensure the users are adequately instructed in the use of such energy.

Conclusion

This chapter has reviewed some usage of solid waste as a potential source of energy for developing countries. In view of the level of technology in most developing countries, waste at present has no significant value, and at times it costs money to remove unwanted waste from the site of production. The potential is therefore not fully tapped. In future, with advancement in technology, more use could be found for the various categories of solid waste. In that situation, energy from waste will be an issue, which will be subject to optimization. If the available technologies for deriving energy from waste are properly harnessed, biomass based fuels can be less harmful to the environment than fossil fuels. These technologies must however be used on site-specific bases in order to reduce the negative impacts on the environment and make the use of biomass energy sustainable.

ACKNOWLEDGEMENTS

We wish to thank Mr. B. Moyo, Principal Minda Secondary School for access to their biogas plant and to Mr. G. Manager for linking us with the school.

REFERENCES

1. Cunningham, W.P. and Saigo B. (1992). Environmental Science, Wm. G. Brown Publishers, pp 405-406.
2. Dhahiyat, Y and Siregar, H. (1984). Studies on the usage of water hyacinth as biogas energy resource in the dam of Curug (West Java), In: *Water Hyacinth*. Ed. Thyagarajan G. UNEP Reports and Proceedings Series 7, pp 604-614.
3. Goldemberg, J., Monaco, L.C. and Macedo, I.C. (1994). The Brazilian alcohol programme, Renewable Energy: Sources for fuels and electricity, Washington, D C, Eds. Johnson, T.B., Kelly, H., Reddy, A.K.N., Williams, R.H. and Burnham, L., Island Press, pp 842-62.
4. Guha, B.R., Bandyopadhyay, T.K. and Basu S.K. (1984). Fuel gas from water hyacinth, . In: *Water Hyacinth*. Ed. Thyagarajan G., UNEP Reports and Proceedings Series 7, pp 615-623.

5. Hall, D.O. (1978). The present status of solar energy for fuel,—UNESCO 2nd International Forum on Fundamental World Energy Problems, Solar Energy Report, October 1978.

6. Ikhu-Omoregbe, D.O. and Idahosa, P.I. (1999). Energy from cassava wastes, *Botswana Journal of Technology*, October 1999, pp 47-53.

7. Karekezi, S and Ranja, T. (1997). Renewable Energy Technologies in Africa, Zed Books Ltd. in Association with African Energy Policy Research Network (Afrepen) and the Stockholm Environment Institute (SEI).

8. Murty, Y.S. (1984). Development of low cost biogas plants under Karimnagar project. In: *Water Hyacinth*. Ed. Thyagarajan G., UNEP Reports and Proceedings Series 7, pp 593-614.

9. Odeyemi, O. (1980). Biogas from *Eupatorium odaratum.*, Global Impacts of Applied Microbiology (GIAM VI), *Proceedings of the Sixth International conference*, 30th August- 7th September 1980, Lagos, Nigeria pp 245-252.

10. Okwatch, D. (1994). Molasses Put to good use. In: *Standard Newspaper*, Nairobi, Karekezi, S and Ranja, T. (1997).

11. Omondi, R. (1991). Molasses – In: Major Raw material, Kenyan Times, Nairobi, Karekezi, S. and Ranja, T. 1997.

12. Onu, C, Gidley J. and Sack, W. (1991). Enhanced energy recovery from municipal solid wastes in sanitary landfills by two phase digestion of biomass. *Proceedings of 7th international conference on solid waste management and secondary materials*. Philadelphia, 1991, PA USA.

13. Piccinini, S., Fabbri, C. and Verzelles, F. (1998). Integrated Bio-systems for Biogas recovery from pig slurry: Two examples of simplified plants in Italy, *Internet Conference on Integrated Bio-systems*, April-December. 1998.

14. Pillai, K., Unni, B.G., Borthakur, A, Singh, H. D. and Baruah, J.N. (1984). Production of Biogas from water hyacinth, In: *Water Hyacinth*. Ed. Thyagarajan, G., UNEP Reports and Proceedings Series 7, pp 507-525.

15. Sarker, H., Haque K.A., and Alam A.K.M.K. (1984). Biogas from water hyacinth and cow dung. In: *Water Hyacinth*. Ed. Thyagarajan G., UNEP Reports and Proceedings Series 7pp 500-505.

16. Scurlock, A., Rosenchein, A, and Hall, D.O. (1991). *Fuelling the future, Power Alcohol in Zimbabwe*, ACTS Press, Nairobi.

17. Suess, M. and Huismans, J.W. (1983). *Management of hazardous waste*, WHO Regional Publication, European Series No. 14.

18. Tebo, P.V. (1997). Going green. *Chemical Engineering News*. August 4, p35.

19. Wenmann, C. (1985). The Production and use of fuel ethanol in Zimbabwe, *Energy from Biomass*, Eds. Coombs, J., Hall, D.O. Elsevier Applied Science Publishers, London.

Section II

Technologies

4

Biomass as a Source of Energy

J. MATA-ALVAREZ AND S. MACÉ

University of Barcelona, Dept. of Chemical Engineering, Martí i Franquès,
1, pta. 6 08028 Barcelona, Spain

The use of biomass as a resource for energy is not new. In fact, the first fuels used by humanity came directly from biomass. With the growth of population and the development of industry, needs of energy grew and other sources of energy were sought. Today, society looks at the sustainability of the present model or resource consumption and the use of renewable sources appears as a feasible alternative which should be developed adequately (Grassi, 1996). Present consumption of petroleum-derived product is approximately equivalent to ca 10^{18} tep per year, which represents the geological accumulation of biomass of 2.10^6 years. Following this pattern, reserves will finish within the present century. The key lies in developing for technologies obtaining renewable energies in the following years. Figure 4.1 shows the past trend with respect to the biomass energy exploitation and the expected one in the future.

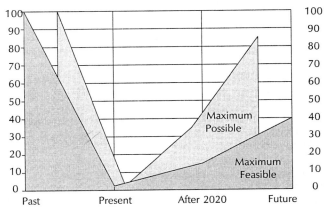

Fig. 4.1 Past trend and expected one in the future of biomass energy exploitation.

It should also be taken into account that biomass is one of the renewable sources, in fact, the oldest, which has been continuously used, although in some cases only domestically. Recently there have been developments that offer a large range of opportunities of energetic use of biomass. In this way, biomass can be considered a strategic resource within the European Union. Present utilization of biomass is slightly over 2.5% and the use is basically centred on wood and coal (Kumazaki, 2000, Geletukha, 1999).

The concept of biomass is linked to organic material, either of vegetal or animal origin (biological in general). Biomass can be residual, as for instance, organic matter in pig manure or the organic fraction of municipal solid waste (OFMSW), or coming directly from crops. The use of biomass for energetic purposes will usually require a previous conditioning, the complexity of which will be a function of the type of industrial or domestic use and the material itself. This condition will normally include physico and biochemical steps.

Before taking any decision on the possible use of a residual biomass as an energetic resource, it is advisable to follow the normally accepted hierarchy of waste management, that is considering, from maximum to minimum priority, the following possibilities:

(a) reduction of wastes at source
(b) recycling or reuse
(c) energetic use
(d) final disposal

Thus, it is clear that in order to achieve the above-mentioned sustainability, it is important to consider other ways of waste biomass use, that eventually, can result in a more profitable method of managing waste, not only from the environmental point of view, but also from the economic one.

Despite the fact that reduction of wastes at source is the first step towards managing waste, the fact is that biomass waste both from domestic (basically municipal solid waste) and industrial origin has notably increased during recent years. (European regulations do not encourage using landfills for dumping organic matter, preferring energetic use of waste biomass). Because the environmental implications of this problem are significant due to its large volume, appropriate management of these wastes can contribute to the development of this energetic source. However, it must be taken into account that there are many (multidisciplinary and complex) socio-economic problems. Here, it is good to remember that exploitation of biomass as an energetic resource can benefit rural areas in many socio-economic aspects.

Biomass Generation

Biomass has its origin in the photosynthetic function carried out by solar energy as the sole source. This function, which can be carried out by autotrophic plants or autotrophic bacteria, fixes carbon dioxide from atmosphere and transforms solar energy in chemical-bond energy of organic molecules of the plant (celluloses, hemicelluloses, etc.). The efficiency of molecules responsible for photosynthesis (chlorophylls, red carotene, yellow xanothophylls, which are located inside the cells of the plants) depends on the characteristics of solar radiation, that means, on the particular location, the weather, conditions, etc. (Fujita, 1998).

Figure 4.2 shows a scheme of this process, which can be represented by the following reaction:

$$CO_2 + H_2O + \text{Solar energy} \rightarrow \text{(Molecules composed of}$$
$$\text{C, O, H in plants)} + O_2$$

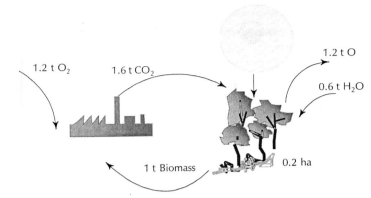

Fig. 4.2 Scheme of the photosynthesis process.

The chemical energy within these bonds is transferred through the vegetarian food chain to the animals. These animals generate wastes which have organic matter with chemical energy in their bonds. These wastes are characterized by a high contents of volatile material (chains of C-H-O) which concentrates this chemical energy. This energy can be released, for instance as heat by combustion, which essentially would be the reverse reaction of the previous one:

(Molecules composed of C, O, H in plants) $+ O_2 \rightarrow CO_2 + H_2O +$ Energy

It is important to note that the value of the ratio of solar energy/energy transformed into biomass is, in average, equal to 1%. Figure 3 shows the energy distribution during the photosynthetic function: through photosynthesis, solar energy is transformed into biomass, with an average yield of 2%. As mentioned before, two main types of biomass are of interest from an energetic point of view: energetic crops and agricultural crops. Agricultural crops represent food for herbivores and carnivores, who, with a conversion energy ratio of 10% for herbivores and 20% for carnivores convert it into work power. Energetic crops, together with industrial and urban wastes, are used for their conversion into energy for boilers, motors, etc.

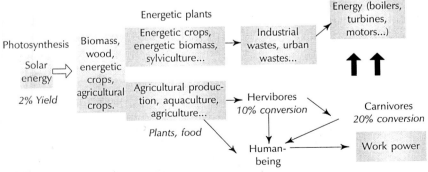

Fig. 4.3 Energetic distribution during photosynthetic function.

Environmental Aspects

The gravest environmental problem faced by our planet today is global warming, to which the greenhouse effect contributes in a definite way. Emissions of different gases – CO_2 among them – due to the anthropogenic activities are, of major concern. At this point it is important to point out that the use of biomass as an energetic resource does not contribute to the growth of gases with such polluting effects. As can be seen in the two reactions and in Fig. 2, the CO_2 released to the atmosphere comes from the atmosphere itself, in a cyclic process in which CO_2 is continuously renewed. On the contrary, with the use of fossil fuels, the released CO_2 comes from organic matter that has been fixed over thousands of years (Bauen et al., 1999).

More benefits for environment derived from the use of biomass as a resource for energy are the reduction in nitrogen oxide emissions, particles, sulphur dioxides (Forsberg, 2000; Yamamoto et al., 2000) (in Europe, around 20 millions tons coming from fossil combustibles are released into the atmosphere) and the preservation of the rural landscape (abandonment, forest fires: it has to be taken into account that in Europe, around 500.000 hectares burn annually).

On the other hand, biomass exploitation as a renewable resource can bring social benefits like creation of employment (Lal and Singh, 2000). The necessary man-power for energy exploitation of biomass can represent around 50 to 60 % of the total costs. If this is done in a systematic way, it could lead to the creation of 6,00,000 jobs, at medium term in the European Union. For instance, in Brazil, the annual production of around 12 millions of m^3 of ethanol coming from sugarcane (Dos Santos, 1999), has created in a direct way 4,50,000 fixed and 1,00,000 temporary jobs. It is estimated that biomass exploitation can create 1.8 million jobs.

Main Biomass Resources

Biomass sources can be classified into two broad groups: vegetal and animal. With respect to vegetal biomass, it is important to consider agricultural and forest wastes, which come from agricultural industry (straw, leaves, etc.) and from forest exploitations (bushes, branches, etc). Animal wastes constitute another important source of residual biomass, specially animal (dejections). Energetic use of these wastes is related to their water content. Thus, bovine or (beard-rearing wastes) can be the objective of a direct energetic exploitation. However, the most common procedure is through a biological procedure, with the extraction of biogas (which is a gas produced by the biological activity with a methane percentage around 65 %) within the anaerobic digestion process, also called biomethanization.

Domestic and Industrial Wastes

These kinds of wastes include organic fractions of municipal solid wastes (OFMSW) and biomass from the biological treatment of wastewaters (sewage sludges, SS). MSW, including fractions that cannot properly be considered biomass (for instance, non-biodegradable plastics, artificial textile fibers), can be sent for energetic exploitation through combustion processes in incinerating plants designed for this purpose. Nevertheless, the OFMSW has very low calorific power (sometimes negative) and, therefore it is not susceptible for this direct energetic exploitation, specially when the OFMSW comes from separated collection. In this case, the energetic exploitation goes necessarily through a biomethanization process, carried on in anaerobic digesters (later in the chapter Section 4).

Energetic Crops

The main objective of producing these kind of crops is to use them for its energetic exploitation (Perez, 1999). For instance, in the European Union, one of the most interesting plantations is sweet sorghum. This species grows under different climatic conditions and can bring a large variety of products like starch, ethanol, sugar, carbon, mill pulp, compost, etc. Sweet sorghum is an annual plant that requires few fertilizers and pesticides; its water requirements are minimum. Besides, it has a high photosynthetic efficiency (2-3%) and its productivity is high. These characteristics make sweet sorghum a very interesting energetic crop. Other plantations that can be found in other regions are sugarcane (Florida, Porto Rico), pine, sequoia, etc.

Table 4.1 summarizes the characteristics of some bio-energetic products (agricultural productions), like humidity, the ratio Kcal/kg, the place of cultivation and the yield of biomass production. Sugarcane and eucalyptus are two crops which have the higher yield of annual biomass production. Table 4.2 shows the average energetic productivity of some bio-energetic products, particularly of animal origin.

Table 4.1. Characteristics of some bio-energetic products
(adapted from *http://www.univ-pou.fr/~scholle/ecosystemes*)

Plant		Place of cultivation	Kcal/kg	Humidity (%)	Yield of biomass production (t/ha-year)
Eucalyptus		Ethiopia			48
		India			39
Sugarcane (bagasse)		Florida, Puerto Rico	3860	12	36
Water hyacinth		Florida, Puerto Rico			36
Sorghum	Grain	In irrigated, Kansas			27
	Straw	France			13
Coconut (husk)			4010	13	
Pine	Wood		4230	12	
	Bark		4790	0	
Bamboo (stick)		Thailand	3925	11	11
Kenaf		Florida			45
Wheat	Straw				3.5
	Seed				2.2

Table 4.2. Average energy productivity of some bio-energetic products, particularly of animal origin.

Bio-energy product	Productivity
Domestic refuse	0.5 tep/t refuse
Muds from sewers	0.2 tep/t-Dry Material
Dairy waste	0.45 Tep / t-Dry Material
Broom	0.22 Tep / t-Dry Material
Waste from slaughter houses	0.23 Tep / t-Dry Material
Bovine excrements	6500 kcal/day-animal
Bovine fertilizer	6800 kcal/day-animal
Fade of lentils	0.14 Tep /t-Dry Material

On the other hand, sewage sludge with a high water content is susceptible of being incinerated if previously it undergoes a process of thermal drying. However, the final energetic balance is very poor. For this reason, the most viable option from an energetic point of view is its anaerobic digestion (see later in the chapter 4).

Industrial wastes come from agrofood industries, production of dairy products, preserve makers, meat industries, etc. When their recycling is not possible (for instance, to obtain fodder specially after the recent problems related to animal illness), biomethanization has to be considered again as a profitable alternative for its disposal. Figure 4.4 shows a scheme of the biomass sources for its energetic exploitation.

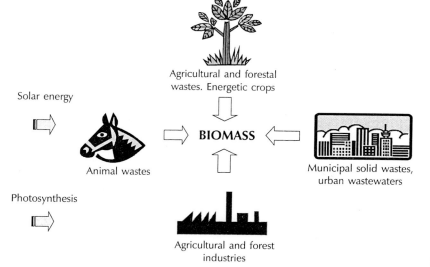

Fig. 4.4 Scheme of the biomass sources for its energetic exploitation.

Possible Biomass uses as Energetic Resources

After transportation and the necessary pretreatment, which can vary depending on the concrete utilization, biomass will be processed following one of the following four types of treatments:

(a) *Thermochemical processes*: In this area are included combustion, gasification and pyrolisis processes. The later two are characterized by processing with a very limited amount of air. Its main advantages are the production of a fuel which does not need to be used "in situ" (Storm *et al.*, 2000; Andries and Buhre, 2000; Yin *et al.*, 2000; Morris and Waldheim, 1998; Lede, 1999). Furthermore, its environmental impact is more reduced. However, its industrial development is still incipient.

(b) *Chemical processes*: Processes related to chemical transformations of oils and other products extracted from plants in order to convert them into fuels (called biofuels).

(c) *Physical processes*: Basically, they include pressing processes and extraction of vegetable oils, which can be used directly or indirectly as biofuels.

(d) *Biological processes*: Among biological processes, there are two relevant methods for obtaining biofuels (Sawayama, 1999):

    (d1) Alcoholic fermentation, from sugars obtained by rich crops in these compounds (the so named bioethanol) and (d2) biomethanization, from the biodegradable organic matter for obtaining biogas. Both possibilities are detailed in Section 4.

Figure 4.5 shows the main ways of energetic exploitation of biomass.

Biomass Potential as an Energetic Source in Europe

Figure 4.6 shows the potential of different biomass sources comparing present and future within the European Union.

For instance, for forest plantations of rapid cycle like eucalyptus, in 3-5 year cycles, they can bring raw ligno-cellulosic material at a cost lower than 50 euros/dry ton. Sweet sorghum plantations have double potential than eucalyptus.

The biomass potential for energetic purposes has been calculated around the following figures:

    720 millions of tons of dry material/year

    320 millions of agricultural residues and wood/year

    250 millions of crops in agricultural soil/year

    150 millions of crops in marginal soils/year

It is also estimated that potential energetic crops increase when they are multipurpose. In fact, this must be the attitude in order to accelerate the incorporation of biomass products in the energetic and industrial sectors. This is because the commercialization of two or more products reduces the cost of the individual production and because the energetic balance of the crop is substantially positive. For instance, if the crop is exclusively dedicated to ethanol production, the value of energy-consumed-in-the-crop/recovered-energy is 1/2. If the waste pulp of Sorghum (2/3 of the energetic content) and sugar (1/3) are used for energetic purposes, this value becomes equal to 1/5. More examples of obtaining different products with the same consumption are showed in Fig. 4.7.

If we look at the possible annual production t/h, it appears that combusting wood produces more energy (2.637 tep/ha-year) than other plants or other technologies, but this does not mean that wood combustion is the best option. If, for instance, we compare its combustion with the one of wheat, heat is the only resulting production in the case of wood, whereas wheat produces heat from straw combustion and bio-ethanol from the distillation of grains. It is also important to take into account the possible annual production, which is the highest in the case of wood (6.7 t). See also Table 4.3, where the energetic productivity of different methods of energy recovery is presented.

Table 4.3. Energetic productivity of different methods of energy recovering.

Procedure	Tep/ha-year
Combustion (straw)	1.87
Combustion (wood)	2.637
Distillation (wheat grain)	0.353
Distillation (wood)	0.474
Fermentation (wheat straw)	0.413
Chimiurgy (wheat straw)	0.137

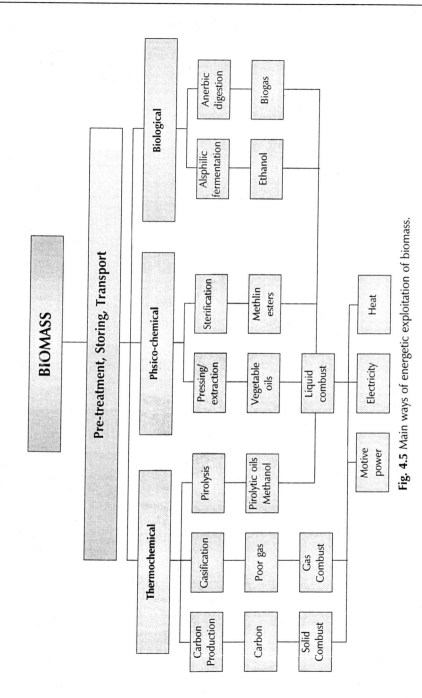

Fig. 4.5 Main ways of energetic exploitation of biomass.

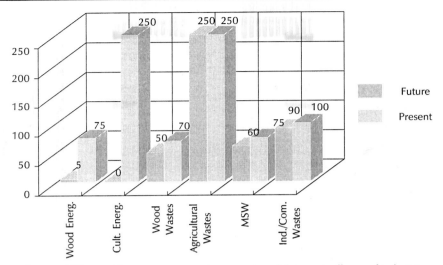

Fig. 4.6 Different biomass resources in present and future. (millions of to/year)

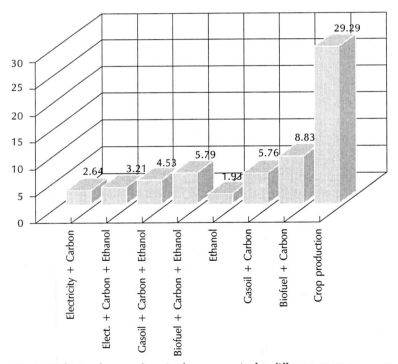

Fig. 4.7 Obtained energy/required energy ratio for different energy sources.

Present uses of Biomass. An Example from Europe

In the European region Catalonia/Midi-Pyrénées/Languedoc-Roussillon, the consumption of renewable energies in the year 1992 was 21 millions of tep; in Catalonia, the industrial and transport sectors shared the first place, whereas in France this place was occupied by the tertiary and residential sectors. These 21 millions of tep represent the 4.5% of the total energy consumption. Exploitation of biomass has a remarkable relevance in the regions cited above due to the importance of the furniture industry and forest activity, which at present contribute of the industrial wastes near zero but will probably increase significantly due to the new regulations about the disposal of organic wastes.

Clearly, there is potential in forest resources as being sources of renewable energy which can be used in cogeneration projects or in small firewood boilers has to be taken into account. (Viesturs et al., 1999).

About 50% of the consumption is represented by firewood and bushes (domestic consumption). The other 50% is constituted by triturated forest and industrial wastes. Table 4.4 presents a classification of boilers adapted to heating needs and to different types of fuel.

Table 4.4. Classification of boilers adapted at heating needs and at the different types of fuel.

Power	Type of biomass use as fuel
Lower than 200 kW	Fuels with a big range of humidity with a little granulometry
200 to 800 kW	Dry or semi-dry fuels
Higher than 800 kW	Dry or wet fuels

Use of MSW as Energy Source in Europe

The annual production of MSW in the European region is estimated around 4.5 annual tons, which is equivalent to a potential of 150 annual ktep. At present, MSW destination is different depending on the location. Landfills with biogas exploitation are a desirable option. In France, near Toulouse, the landfill of Montech represents the most important project of biomass exploitation, with a gas compression station for its utilization as a fuel. However, the most typical use of produced biogas is cogeneration. At this point, it can be mentioned that the use of the biogas produced in wastewater treatment plants from sewage sludge anaerobic digestion is another option less important than the use of biogas from MSW, because the installation of the necessary equipment for biomass digestion is presently only justified from around 50.000 population equivalent. Policy regarding taxes should favour these energetic uses.

Biological Process to Obtain Biofuels

As mentioned earlier, there are two relevant methods for obtaining biofuels from biomass. The first one is bioethanol, which has been exploited mainly from energetic crops (Sheehan, 2000), and the second is biogas, which comes from biomass wastes and has a big potential for solving environmental problems.

Bioethanol as a Biofuel

Bioethanol is another biofuel that has been included in many research programs. Four basic steps produce bioethanol:

— Production of biomass from solar energy (biomass comes from different sources: energetic crops, waste from food crops, biomass waste in general)
— Conversion of biomass to fermentable products (there are basically four processes, which differentiate technologies for bioethanol production, which will be commented below)
— Fermentation of biomass intermediates to ethanol
— Recovery of ethanol and byproducts.

Fermentation – conversion - of sugars to bioethanol and separation of fermentation products is a known and established process. Major improvements of these processes have already been made. As mentioned basic differences among bioethanol technologies lie in the hydrolysis processes for obtaining sugars. Basically there are four: (a) Concentrated acid hydrolysis (b) Diluted acid hydrolysis (c) Enzymatic hydrolysis (d) Gasification and fermentation.

Concentrated Acid Hydrolysis

This is an old process, with patents in 1937 (Germany), 1948 (Japan) and 1980 (USA). They rely on the breaking of cellulose hydrogen bonds carried out by concentrated sulphuric acid, which brings cellulose to an amorphous state, which is easily hydrolysed by water at mild temperatures. This process is penalized by the high amounts of acid required. The bottleneck of the economy is the acid recovery and reconcentration, where developments are still in course. Figure 4.8 presents a basic scheme of the process.

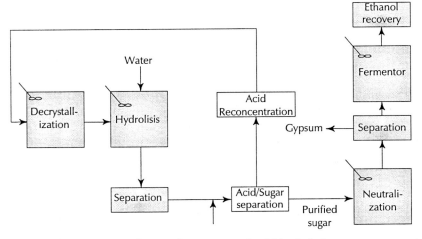

Fig. 4.8 Basic scheme of concentrated acid hydrolysis process.

Dilute Acid Hydrolysis

In this process hydrolysis is carried out in two steps: the first is a mild attack for hemicelluloses hydrolysis; The second is designed to attack the more resistant cellulose. Hydrolyzates are recovered, neutralized and converted to ethanol. This is the oldest technology and Germany, Japan and Russia have been operating percolating systems of this type since the last 50 years. However, today these processes are not competitive as there is too much energetic consumption because of the high-diluted operating conditions. To increase the yield, new equipment are under study such as plug flow and batch reactors, but they are facing problems with solid handling. Figure 4.9 presents a basic scheme of the process.

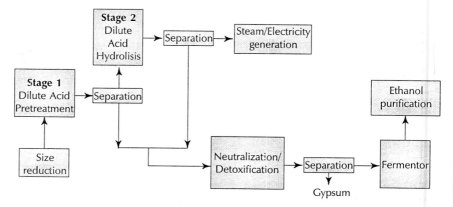

Fig. 4.9 Basic scheme of dilute acid hydrolysis process.

Enzymatic Hydrolysis

Basically the first enzymatic technology consisted in substituting the acid hydrolysis step by an enzymatic one. More modern methods perform at the same reactor the hydrolysis and the fermentation, avoiding in this way the problems of product inhibition, as sugars are converted to ethanol, not being accumulated in the system. Basically there are three major classes of enzymes: (i) endoglucanases, which act on soluble and insoluble glucose chains; (ii) exoglucanases, which liberate glucose monomers and cellobiose from the end of the cellulose chains, and (iii) β-glucosidases which liberate D-glucose from cellobiose dimmers and soluble cellodextrins. However the understanding of the mechanisms of the enzymatic synthesis is far from complete, and more development is expected in this area. Development of this technology is not very large because cellulases, which are commercially available, are applied to systems that do not require complete hydrolysis as in this case. Figure 4.10 presents a basic scheme of the process.

Biomass Gasification and Fermentation

This process consists in the conversion of biomass to synthesis gas using gasification process. This gas formed by CO, CO_2 and H_2, can be converted to ethanol by anaerobic bacteria such as *Clostridium ljugdahlii*. This technology is still in development although there is already an American patent by BRI.

Biogas as Biofuel. Biomethanization of the OFMSW

Biogas production from anaerobic digestion of wastes is an attractive option for

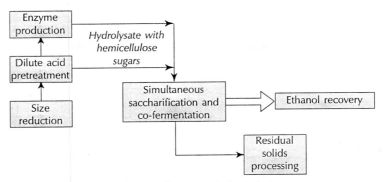

Fig. 4.10 Basic scheme of enzymatic hydrolysis process.

treating biomass wastes. Wastes such as the organic fraction of municipal solid waste (OFMSW) with a large concentration of organics are particularly suitable for this treatment.

Anaerobic digestion (AD) is a biological process in which the organic matter is metabolized to biogas (a mixture basically of methane and CO_2) in the absence of oxygen. Among the obvious advantages of a positive energy balance, the residual sludge production is low and its contribution to global warming is zero. AD can be applied both to liquid and semi-solid wastes. Thus, in dry systems, TS content has reached 40 % or even higher values.

Anaerobic digestion stabilizes the organic material to a high degree. However, some of the slowly degradable material remains (as celluloses) due to the practical hydraulic retention time used. This material needs further stabilization for certain applications to the soil. AD combined with composting of the digester effluent constitutes a good solution, as this treatment completes biodegradation and also fixes ammonium nitrogen, which assures the procurement of a high quality product.

AD releases NH_4^+ coming from the organic matter decomposition and is an operation free of odours as it is carried out in closed vessels. Anaerobic digesters can be operated at three ranges of temperature (psycrophilic, at temperatures lower than 30°C, mesophilic, between 35-40°C and thermophilic between 50 and 65°C). Only this last range assures complete hygienization of the digester effluent. AD technology is not a simple biological treatment: It requires careful monitoring of the process in order to avoid digester failures. This is specially true because start-up is rather slow. It can take from 1 to 3 months. In relation to the other technologies, investment costs of AD are higher than composting for solids treatment. On the contrary, operating costs are much lower due to the positive energy balance.

Anaerobic digestion has been used for more than 120 years for sludge digestion. Later and because of the energy crisis of 1973, it was applied to livestock wastes. More recently, in the mid-80's it was applied to the stabilization of the organic fraction of municipal solid wastes. Finally, Figure 4.11 shows a basic balance of the solids in a sewage sludge AD process.

Aspects related to additional pre-treatments, which can alter the final product value, have to be included, as they can be of importance in future. With effluents, the methanogenic reaction is usually considered as the rate-limiting step of the

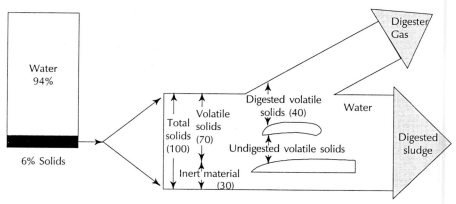

Fig. 4.11 Solids breakdown path during anaerobic digestion.

overall process. When considering particulate substrate like solid wastes, both accessibility of hydrolytic microorganisms to the solid matter and hydrolysis of complex polymeric components constitute the rate-limiting step. As a consequence, it is possible to improve the performance of digesters treating solid wastes by reducing the size of the particles. Therefore, pretreatment of the substrate by mechanical disintegration should have positive effects on the anaerobic biodegradability of the substrate, through an increase of the available specific surface to the medium. The other way of improving performance is to promote hydrolysis of organic matter by a pre-treatment of the substrate. Such pre-treatments can be mainly biological or physico-chemical and they break the polymer chains into soluble components. The main objective is to obtain an extension and an acceleration of the anaerobic process, an increased amount of biogas as well as a reduction of the amount of anaerobic sludge and of the digestion time (Delgenes, 2001). All these pre-treatment possibilities are mainly at the research level, but can modify substantially the yields of AD and composting processes.

For OFMSW, which is characterized for high volatile solids contents, AD seems to be the best selection, whether as a final treatment or as a previous step of a combined composting system to reach a high quality product standard. This is true because composting, the other possibility, consumes energy which can be important when the amounts of waste are high. Small communities should consider the idea of treating a series of wastes (co treating) such as OFMSW, garden wastes, some agricultural residues and the sewage sludge of the wastewater treatment plant of the community to reach a capacity such as to be practical for AD. In addition and although this suggestion lies out of the scope of this chapter, other sources of energy should also be sought, such as solar, eolic, etc, which, together with passive measures, can guarantee the attainment of an energetic autonomy for the community. In this respect, balances of consumptions and productions should be carefully performed.

It has also to be considered that there are some organic wastes that are suitable for composting only (for instance, lignocellulosic wastes) whereas other highly wet and biodegradable wastes are more appropriate for AD. However, both types of wastes have the same biomass potentials. It should also be said that AD and composting are complementary in most of the situations, and not competitive with the other. Combination of both technologies has a number of advantages (see Table 5) on an energy and quality basis and this approach is therefore highly recommended. Figure 4 shows a simple flow diagram of such a combined treatment.

Regarding sewage sludge (SS), the results obtained by Folch *et al.* (1999) show that the anaerobic digestion provides lower contents of nitrogen and organic matter, as well as a reduction of the amounts of sludge obtained, related with the lower synthesis rate of anaerobic microorganisms, when compared with aerobic digestion.

Conclusions

There are many possibilities regarding the energetic exploitation of biomass. Today this use is still marginal, but it is clear that the use of biomass as an energy source has to increase to achieve an environmental equilibrium with consumption and resources. Energetic autonomy should be a general objective for a sustainable society. When comparing different technologies, energy plays a predominant role. Not only energy from biomass should be sought, but many other ecological options should be put in this scenario.

Management of wastes is another problem faced by society. This gives an opportunity to obtain biofuels, through adequate processing of this residual biomass. In this sense, and as discussed in this chapter, anaerobic digestion plants compared to other technologies for treating solid biowastes are better from an ecological point of view, because they do not need external fossil and electrical energy. If only one quarter of biogenic waste is digested, a plant can be self-sufficient in energy (Edelmann *et al.*, 2000). The production of renewable energy has positive consequences on nearly all impact categories, because of saving or compensation for nuclear and fossil energy. This reduces the impacts of parameters such as radioactivity, dust, SO_2, CO, NO_x, greenhouse gases, ozone depletion, acidification or carcinogenic substances. Thus, the higher costs for anaerobic operation are compensated with energetic use diminution and energy return (Genon, 1999).

In this sense and on a long-term basis, it is important to seriously consider the possibility of adopting AD as a single treatment or, if quality requirements are imperative, followed by a final composting treatment. This can be applied for most of the wastes considered in this paper, as AD is very flexible in treating all types of organic residues commented here. Furthermore, pure composting technologies appear to be less ecological than digestion: the higher the percentage of digestion, the better the ecological score (Edelmann *et al.*, 2000).

Bioethanol is also another possibility specially when the biomass source is biowaste. This use can be profitable in future, if environmental regulations are more stringent and favours this type of solution.

REFERENCES

1. Andries, J.; Buhre, B.J.P., (2000). Small-scale, distributed generation of electricity and heat using integrated biomass gasification-gas turbine-fuel cell systems. *Tagungsber.*, **2000**-1L 115-125.
2. Bauen, Ausilio; Kaltschmitt, Martin. (1998). Contribution of biomass toward CO2 reduction in Europe (EU). *Biomass, Proc. Biomass Conf. Am.*, 4th, Volume 1, 371-376.

3. Delgenès J.P., (2001). *Anaerobic Digestion of Municipal Solid Wastes*. Edited by J. Mata-Alvarez. Ed. IWA Publishing Company. London 2001. (In press).

4. Dos Santos, Marco Aurelio, (1999). A brief history of energy biomass in Brazil. Biomass, *Proc. Biomass Conf. Am.*, 4th, Volume 2, 1673-1678. Overend, Ralph P.; Chornet, Esteban. Elsevier Science: Oxford, UK. 1999.

5. Edelmann W., Schleiss K., Joss A. (2000). Ecological, energetic and economic comparison of anaerobic digestion with different competing technologies to treat biogenic wastes. *Water Science and Technology*, **41**(3): 263-274.

6. Forsberg, Goran, (2000). Biomass energy transport analysis of bioenergy transport chains using life cycle inventory method. *Biomass Bioenergy*, **19**(1): 17-30.

7. Folch M., García M., Salgot M., Pigem J., Caus J.M., López D., Herrero N. (1999). Comparative study of the anaerobically and aerobically stabilised sludge quality. In: *Proceedings of the II International Symposium on Anaerobic Digestion of Solid Wastes*, Vol II, pp. 290-293.

8. Fujita, Masanori, (1998). Generation of electricity by biomass. Nippon Kikai Gakkaishi, **101** (959): 59-60.

9. Geletukha, G. G.; Zheleznaya, T. A., (1999). Review of current wood combustion technologies for production of heat and electric power. *Ekotekhnol. Resursosberezhenie*, 5: 3-12.

10. Genon G. (1999). Economic assessment of MSW Anaerobic Digestion in comparison with composting plants. In: *Proceedings of the II International Symposium on Anaerobic Digestion of Solid Wastes*, Vol I, pp.282-289. Barcelona, June 1999.

11. Grassi, G., (1996). Future perspectives of bioenergy activities in the European Union. Bio-Oil Prod. Util., *Proc. EU-Can. Workshop Therm. Biomass Process.*, 2nd, Meeting Date 1995, 17-21.

12. Kumazaki, Minoru, (2000). Role of bioenergy in a society based on sustainable resource use. Kami Parupu Kenkyu Happyokai Koen Yoshishu, 67th, 68-71 (Japanese).

13. Lal, Murari; Singh, Roma, (2000). Sustainable forestry as a source of bio-energy for fossil fuel substitution. Adv. Global Change Res., 3(Biomass Burning and Its Inter-Relationships with the Climate System), 281-298.

14. Lede, Jacques, (1999). Solar thermochemical conversion of biomass. *Sol. Energy*, **65**(1): 3-13.

15. Morris, Michael and Waldheim, Lars, (1998). Energy recovery from solid waste fuels using advanced gasification technology. *Proc. Int. Conf. Incineration Therm. Treat. Technol.*, 141-147.

16. Perez, E. Menendez, (1999). Possibilities of the energy use of biomass. *Ing. Quim.* (Madrid), **31**(361): 252-256.

17. Perez, J. A., (1999), Biomass, an energy source that is difficult to manage but necessary. Quim. Ind., 46(1), 28-32.

18. Sawayama, Shigeki, (1999). Biological energy conversion technology from biomass. *Nippon Enerugi Gakkaishi*, **78**(4): 259-264.

19. Sheehan, John, (2000). The road to bioethanol: a strategic perspective of the U.S. Department of Energy's national ethanol program. Ser., 769(Glycosyl Hydrolases for Biomass Conversion), 2-25.

20. Storm, C.; Spliethoff, H.; Hein, K. R. G., (2000). Gasification and pyrolysis of biomass for generation of a reburn gas for combustion in coal-fired boilers. *Tagungsber.*, **2000-1**: 87-97.

21. Viesturs, Uldis; Telysheva, Galina; Dizhbite, Tatiana, (1999). Chemical processing of photosynthesized biomass for industry and energy. *Bulg. Chem. Commun.*, **31**(3/4): 596-614.

22. Yamamoto, H.; Yamaji, K.; Fujino, J., (2000). Scenario analysis of bioenergy resources and CO2 emissions with a global land use and energy model. *Appl. Energy*, **66**(4): 325-337.

23. Yin, Xiuli; Wu, Chuangzhi; Xu, Bingyan; Chen, Yong, (2000). Effect of biomass gasification on reducing CO2 emission. Taiyangneng Xuebao, **21**(1): 40-44.

5

Small Diameter Forest Residues for Soil Rehabilitation

Tatjana Stevanovic Janezic

Université Laval, Département des sciences du bois et de la forêt,
Ste Foy, Quebec, Canada

Introduction

The soils beneath our feet enclose the treasures of organic compounds within their humus layers and an enormous number of living organisms (it is estimated that a pinch of soil contains close to one billion living organisms!) which are essential for their fertility. The soil micro-organisms alone are represented by 10,000 distinct species, many of which are not yet named, catalogued or understood. Current estimates suggest that the total biomass contained in the soils of the earth is greater than that above the ground! The similar seems to be true for soil organic matter.

Humus or soil organic matter (SOM) as it is alternatively designated, represents one of the most important natural resources, which is of fundamental importance for the fertility of soils. It contains about the same quantity of organically bound carbon (22 • 10^{14} kg) as do together the total of living and fossil biomass on the earth's surface (about 21 • 10^{14} kg) (Senesi, Loffredo, 1998). By the process of mineralization, part of organic carbon from the plant (and animal) residues is converted into carbon dioxide, part of it is metabolized by soil (micro-) organisms and thereafter converted into their biomass and part is transformed through the processes of biodegradation, re-polymerization and reactions with soil inorganic constituents to humus, or soil organic matter. Resistant organic compounds with aromatic structure, such as plant polyphenols, belong to the last mentioned category. More knowledge about these compounds and their relation to the humic substances formed is required as there seem to be many details about soils and their functioning yet to be clarified and explained. The words of Leonardo da Vinci, pronounced more than five centuries ago, seem still to correctly describe the actual status "We know more about the celestial bodies than about the soil underfoot". This research therefore was aimed to utilize some wood chemistry concepts to contribute to clarification and better understanding of the structures of forest soil humus biomolecules.

The forest operations leave behind huge amounts of small branches, twigs and foliage which cannot be utilized for the manufacture of any of traditional forest products. These forest residues comprising branches with diameter smaller than 7 cm and twigs, when fragmented by conventional knife-chipper into pulpwood size chips are named rameal chipped wood (RCW). A technology has been developed during the past two decades at the University of Laval in Quebec, Canada which briefly consists of reconstituting the humus layer by treatments of fallow land with RCW (Lemieux, 1986, 1990; 1998). Several researches have been performed on RCW application in agriculture (N'Dayegamiye, Dubé, 1986; Beauchemin *et al.*, 1990; Larochelle, 1994) and in forestry (StAmand, Lauzon, 1995; Pettigrew, 1998; Tremblay, Beauchamp, 1998; Tissaux, J.C., 2000). The synergistic effects of application of mixtures of forest trees residues in form of RCW to forest soils have been noted in a recent study (Tissaux, J.C., 2000).

There are several theories of humus formation (Stevenson, 1982), to cite lignin-protein theory or its modification in which lignin fragments released during microbial attack may or may not combine with amino compounds as just one of them. Yet another polyphenol theory implies participation of polyphenols which are the results of non-lignin carbon sources transformation by micro-organisms which undergo afterwards the same oxidation changes as was the case with lignin fragments, leading to quinone functions, the source of dark coloration of humic substances. Numerous precursors of humic substances have been discussed in an effort to explain and clarify the variety of structures identified in humic substances of different origin. Bacteria and fungi are known to synthesize polysaccharides, proteins, nucleic acids, carotenoids, and lipids. Some of these are easily degraded upon their death and autolysis. They are also known to synthesize one type of black pigments—melanins, which being the recalcitrant substances were regarded as likely contributors to humic substances formation, which has been however questioned recently (Saiz-Jimenez, 1996). The phototrophic micro-organisms, the cyanobacteria and green algae have also been considered as contributors to the building blocks for humic substances, especially on stones and wet soil surfaces. (Saiz-Jimenez, 1996). The mosses (bryophyta) are usually associated with cyanobacteria and green algae. Their chemical composition resembles that of the higher plants, their cuticular components being particularly similar to those of vascular plants. The mosses are indicated as important contributors to humus formation in cold, wet climate (Stevenson, 1982). The lichens are yet another class of non-lignified plants in which different types of carbohydrates (sugar alcohols, cyclytols, monosaccharides and polysaccharides) are the major intracellular constituents.

Contribution of lichens to soils is considered to be restricted to sparsely distributed soil-inhabiting lichen communities, while the contribution of saxicolous lichens (those inhabiting the bare rocks) is regarded as less important. However, one should keep in mind that extracellular products of lichens, such as oxalic acid, may lead to complexation reactions with cations from the rocks, such as calcium, which can then lead to rock weathering.

The contribution of vascular plants to soil humus formation has been mainly related to the recalcitrant plant components, the lignins because of their abundance, being the most important, especially for forest soils.

When discussing the biodegradation of plant constituents in soils, numerous factors are to be considered, such as climate conditions, type of plant, soil pH, parent rock, etc., which can also influence microbial activity necessary for the biodegradation of plant constituents. The importance of aliphatic structures identified in some humic acids has inspired some authors to consider lignin contribution to humus formation overestimated (Saiz-Jimenez, 1996). It is true that the persistence of some structural blocks of lignin should not be exclusively attributed to their recalcitrance but also to the adverse environmental conditions which under given circumstances may impede microbial activity. However, in the context of the research interest of this study which is dealing with lignin (and other polyphenols) rich forest residues, the polyphenol theory seems to be of particular pertinence, which does not deny the importance of other structures from different plant recourses in humus formation.

The small diameter wood material of RCW of interest for this study is characterized by high proportion of bark, which is known to contain large quantities of different types of polyphenols. There is an important contribution of structural wood polyphenols -lignins, and those extraneous (to cell walls, situated in porous structure of wood) in nature, which are biosynthetically related to lignins, to humus formation. Therefore an overview of the biochemistry of lignins and related extractive polyphenols will be presented, with an emphasis on common features linking lignins as structural polyphenols with those extraneous in nature, sharing common biosynthetic routes with them.

Lignins-Structural Polyphenols of Higher Plants

In botanical sense, lignification is accompanying the specialization of tissues for transport of fluids, which means the development of vascular system in plants. Lignin is found in terrestrial plants such as lycopods, ferns, softwoods and hardwoods and is absent from algae, mosses, fungi and lichens.

Development of a more perfect excretory system has certainly influenced the development of lignification of plant cell walls. Higher plants can perform the excretion through their root system, through the foliage or they can also dispose of such metabolites into the vacuoles. This explains a variety of secondary metabolites that are found in higher plants. Soluble polyphenols, such as anthocyanidins, are accumulated in vacuoles or in the pores of cell walls, while the insoluble polyphenols, such as lignins, are deposited directly into cell walls and are therefore designated as structural polyphenols, as they construct the cell walls together with cellulose and hemicelluloses.

Lignins are designed to provide mechanical support to land plants which are characterized by unparalleled growth in height. By imparting hydrophobic properties to cell walls of tracheary elements, lignins are enabling them for long-distance water conduction. Being tridimensional amorphous macromolecules, lignins are providing the cell walls the impenetrable barriers against the attacks of encroaching micro-organisms and herbivores and the compression strength necessary for plants growth in height.

Woody tissues are in general characterized by high lignin contents since they are composed to a large extent of vascular and sclerenchyma cells, such as tracheids, vessels and fibres. In herbaceous plants, for instance, these elements are very much "diluted" by other cell types, such as parenchyma cells, which do not have strongly lignified walls and consequently the grasses contain less lignin. Woods contain 18-30% of lignin. Typical lignin contents for hardwood is 18-25%, while for softwood it is 25-33%. Forest residues of interest for this study are extremely lignin-rich materials.

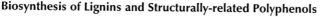

Biosynthesis of Lignins and Structurally-related Polyphenols

Lignins originate from a metabolic pathway that exists only in land plants, which evolved, following the lignification of their cell walls as tall, upright forms, hundreds of millions of years ago.

Lignin macromolecules are formed by dehydrogenative polymerization of three cinnamic alcohols, which are in lignin chemistry nomenclature designated as monolignols: trans or E-p-coumaryl, E-coniferyl and E-sinapyl alcohol. These monolignols are formed by the same general phenylpropanoid metabolic route in all vascular plants: CO_2 → carbohydrates → shikimic acid → phenylpropanoid amino acids: (L- tyrosine- important for grasses) and L-phenylanine (Fig. 5.1). L-phenylanine's and L- tyrosine's (important for grasses) conversion to cinnamic acid derivatives and further to cinnamic alcohol derivatives (monolignols) is specific to lignin biosynthesis.

The precursors to the aromatic amino acids with a phenylpropane skeleton are the intermediate from glycolysis, phospoenol pyruvate (PEP) and erythrose-5-phosphate, which is an intermediate from pentose phosphate pathway. The C-7 compound, the 2-keto-3-deoxy-arabinoheptulosonate- 7 phosphate is formed upon their condensation reaction, which after the cyclisation is ultimately transformed into chorismate, the branching point between the tryptophan and prephenate synthesis, the latter being the intermediate leading to the amino acids with phenylpropane skeleton of interest for our study. The synthesis of aromatic amino acids occurs only in plants.

Concepts of Direct Reduction of Cinnamic Acids to Monolignols and Their Random Polymerization in Lignin Biosynthesis are Currently being Modified

L- phenylalanine is an essential amino-acid in primary metabolism, and it is converted by phenylanine ammonia-lyase (PAL) to trans-cinnamic acid. Phenylalanine ammonia-lyase (PAL) is a key enzyme in the synthesis of various phenolic secondary metabolites, lignin being the most important of them, because of its natural abundance (second most abundant natural polymer after cellulose). The deamination step catalysed by PAL is a principal step that marks the transition from primary to secondary metabolism in higher plants.

Trans-cinnamic acid formed by the mediation of (PAL) from phenylalanine is further converted to p-coumaric, ferulic and synapic acids by successive hydroxylation and methylation reactions, as presented in Fig. 5.2. These hydroxycinnamic acids are first converted to their CoA esters which are then reduced, according to a classical theory, to monolignols, that is to p-coumaryl alcohol, coniferyl and sinapyl alcohols, via the respective hydroxycinnamic aldehydes. Numerous enzymes catalysing the reduction of acyl-CoA derivatives of cinnamic acids have been isolated from bacteria, algae and higher plants, which has been taken as an evidence of gradual reduction of cinnamic acids to corresponding cinnamic alcohols (monolignols) (Gross, 1985).

This classical view of direct reduction of each of the cinnamic acids p-coumaryl, ferulic and sinapyl, through corresponding aldehydes, to the cinnamic alcohols (p-coumaryl, coniferyl and sinapyl, the true lignin precursors (monolignols according to Freudenberg's nomenclature) has been seriously challenged recently (Li *et al.*, 2000).

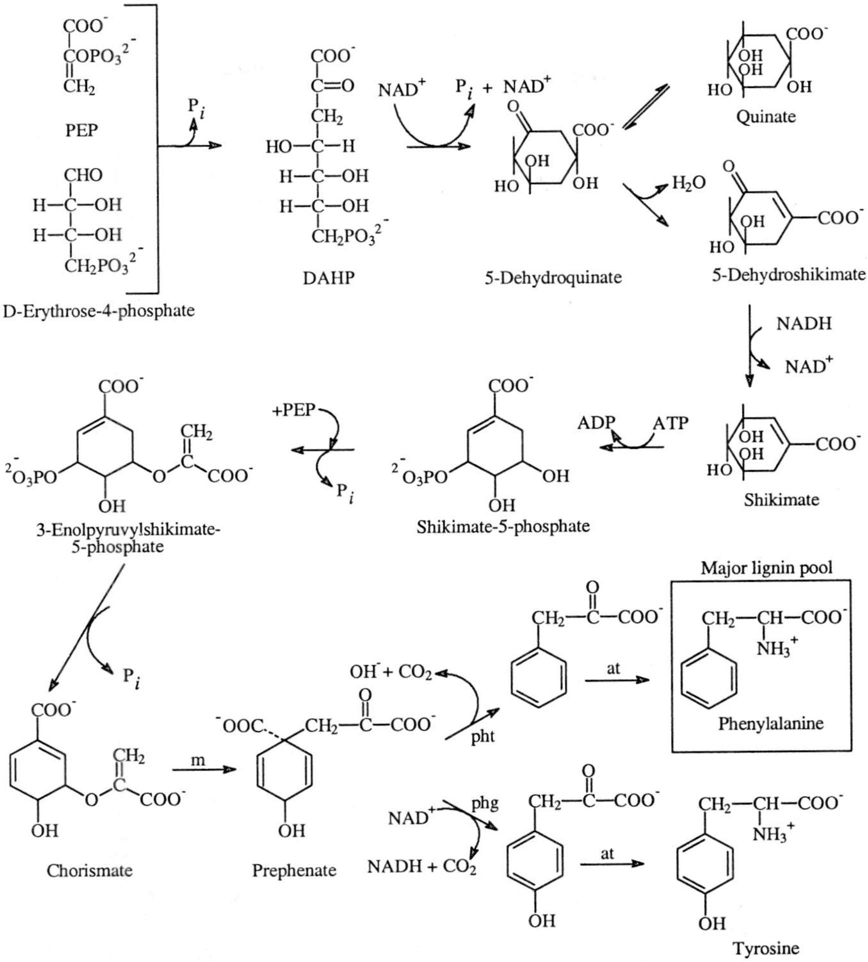

Fig. 5.1 General phenylpropanoid metabolism in plants.

PEP- phospoenol pyruvate; DAHP- 2-keto-3-deoxy-arabinoheptulosonate- 7 phosphate; Pi - phosphate at amino transferase; m - chorismate mutose; phg - prephenate dehydrogenase; pht - prepbenate dehydratase.

The evidence was provided that the syringyl monolignol biosynthesis is independent of caffeate and 5-hydroxyferulate, the methylation of which is furthermore inhibited by the presence of coniferyl aldehyde. The coniferyl aldehyde ensures a coniferyl aldehyde 5-hydroxylase mediated biosynthesis of 5-hydroxyconiferyl aldehyde. The ferulate hydroxylation/methylation is sequestered from coniferyl aldehyde hydroxylation/ methylation and only the latter has been shown to be specific to lignification. This view challenged the traditional concept of ferulate 5-hydroxylation/methylation in syringyl lignin generation.

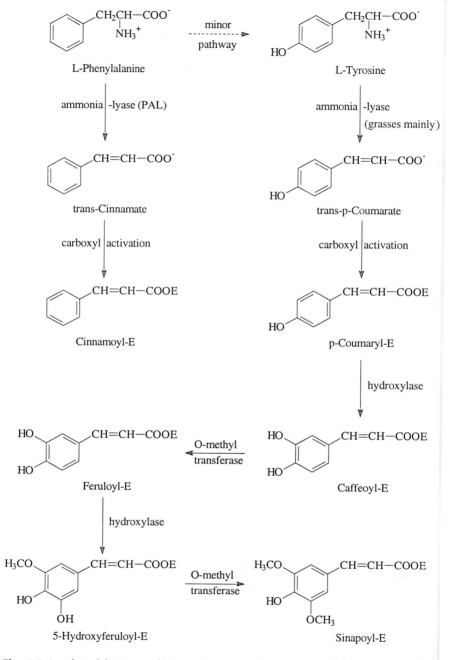

Fig. 5.2 L- phenylalanine and L-tyrosine transformation to cinnamic acids which are further converted into cinnamic alcohols (monolignols).

The 5-hydroxyconiferyl aldehyde and the coniferyl aldehyde modulation appear to be widely distributed in angiosperms. In the presence of coniferyl aldehyde and 5-hydroxyconiferyl aldehyde, therefore, the 4-coumarate and caffeate seem to be the only two cinnamates that are likely to be metabolised to monolignols in biosynthesis of syringyl - guaiacyl lignins of angiosperms for which the sinapyl acid does not seem to be important, as previously postulated (Li et al.,2000).

Lignins are formed in vascular plants by dehydrogenative polymerisation of cinnamic alcohols derivatives, which is in the classical theory of lignin formation designated as random polymerisation. The heterogeneity of lignins encountered in nature reflects the varying ratios of the monolignols themselves, but reveals also an evidence of a variety of cinnamic acids derivatives which are participating in lignin biosynthesis.

It is interesting to note that homogeneous wood species such as softwoods are characterized by "homogeneous" lignins which are commonly designated as guaiacyl G lignins, referring to guaiacyl rings introduced by polymerisation of almost exclusively coniferyl alcohols during lignin formation in softwoods. Meanwhile, the heterogeneous hardwoods are characterized by heterogeneous G-S (guaiacyl- syringyl) lignins, formed through participation of both sinapyl and coniferyl alcohols in dehydrogenative polymerisation.

Several enzymes have been proposed to be involved in oxidative (dehydrogenative) polymerization of monolignols: hydrogen peroxide dependent peroxidases and oxygen dependent laccases (Higuchi, 1957; Freudenberg, 1968) and coniferyl alcohol oxidases (Savidge, Udegama-Randeniya, 1992, 1994). The capability of these enzymes to oxidise monolignols in vitro and/or their association with vascular tissues has usually been taken as the evidence of their implication in lignification.

The quoted enzymes are certainly necessary to produce radicals (through one electron oxidation) from monolignols which then, according to the classical view in lignin chemistry, undergo random coupling which yields lignin macromolecules. The resonance hybrids of radicals produced by oxidation of p-coumaryl, coniferyl and sinapyl alcohols are presented in Fig. 5.3. Their number for coumaryl, coniferyl and sinapyl alcohol and their structural particularities are already an indication of a variety of different structures identified in lignins, which are resulting from their coupling reactions.

The conclusions on random coupling were based primarily on comparisons of the structures of artificial lignins (dehydrogenative polymer, DHP, according to nomenclature of Freudenberg) prepared by action of, for example, molecular oxygen and laccase on mixture of monolignols with milled wood lignin (MWL) isolated from the studied wood type (Freudenberg, 1968). The MWLs are still regarded as the best lignin preparations available for structural studies.

Random coupling concept cannot explain, however, various particularities of lignification in vivo. For instance, the frequency of the intermolecular linkage of the type β-O-4 is very low in DHP artificial lignins, while it is invariably determined as the major linkage connecting the phenylpropane residues in natural lignins (50-70 %) macromolecules. The heterogeneity of lignins encountered in different plants and in different tissues within the same plant (fibres and vessels in hardwoods) is yet another issue difficult to explain simply by random polymerization.

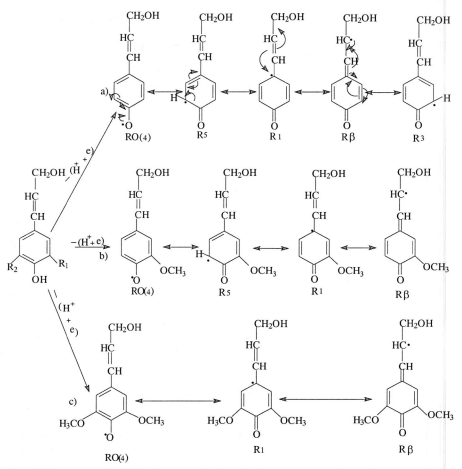

Fig. 5.3 Resonance hybrids for monolignol radicals: the coumaryl, coniferyl and sinapyl alcohol radicals (from the top to the bottom).

General Structural Characteristics of Lignins from Hardwoods and Softwoods

Lignin contents, their monomeric compositions as well as types of linkages between phenylpropane units vary between different species of land plants and between different tissues and different cell wall layers of the same wood species.

Lignins in gymnosperms are mainly derived from coniferyl alcohol and are designated as guaiacyl (G) lignins. Lignins in angiosperms are derived from coniferyl and sinapyl alcohol monolignols in roughly equal proportions and are designated as guaiacyl-syringyl (G-S) lignins.

Lignins in angiosperms are further diversified between different tissues for hardwoods. It is now well documented that lignins in vessels and fibre cell walls of hardwoods are different. The former are guaiacyl type (G) lignins while the latter are guaiacyl-syringyl- (G-S) type of lignins (Saka, Thomas, 1982; Saka, Goring, 1985).

For the biosynthesis of guaiacyl lignin in gymnosperms and guaiacyl- syringyl lignin in angiosperms except for grasses, only L-phenylalanine is used as substrate. For grass lignins, which beside guaiacyl, syringyl and p-hydroxyphenylpropane (H) units contain also p-coumaric acid units esterified to terminal hydroxyl units of propane side chains (γ– carbons), both L-phenylalanine and L-tyrosine are used as substrates in lignin formation. [14]C – labelled L-tyrosine is incorporated in H, G and S units of the lignin polymer and also into the esterified p-coumaric acid via mediation of L-tyrosine ammonia- lyase, which is only found in grasses (Gross, 1985).

Particularities of Lignins from Some Plants and Tissues

There are many known lignins which cannot be described simply as polymers derived from three cinnamic alcohols but their formation implicates the participation of some derivatives other than monolignols.

Grasses, for example, utilize p-coumarates, hardwoods and some dicotyledons such as kenaf utilize acetates, while poplars and willows utilize p-hydroxybenzoates as "monomers" for lignification. Ferulates and diferulates are found in grasses (cereals) and some dicots, in which they represent equal partner for radical polymerization and may even present the nucleation sites for lignification (Ralph *et al.*, 2000).

Esterified *p-coumaric acid* can comprise 5 to 10% of the total weight of isolated lignin from bamboo and *grasses*. The participation of p-hydroxyphenyl-β-aryl ether structures is of minor importance in grass lignins, contrary to what was previously believed (Gross, 1985). *Poplar* (aspen) lignins *contain p-hydroxy benzoic* acid linked through γ- carbon by an ester bond (Higuchi, 1985).

Ferulic acid is mainly associated with hemicelluloses in cell walls of *cereals fibers*. Ferulates play important role in cross-linking between polysaccharides and lignin. Polysaccharide-polysaccharide cross-linking is achieved by ferulate dimerisation either by photochemical, or more importantly, by radical coupling reactions of ferulate-polysaccharide esters. Radical cross-coupling of (polysaccharide linked) ferulates results in lignin-polysaccharide cross-linking in cell walls of cereals. Nine diferulate products were identified recently upon saponification of insoluble cereal fibre (Ralph *et al.*, 2000). Not all of the ferulic acid, however, serves for linking lignin with carbohydrates. Most of the acid may be present as terminal ester linked residues on arabinoglucuronoxylan in cereals.

Compression wood of conifers is characterized by an abnormally high lignin content, the biosynthesis of which implicates participation of high amounts of *p-coumaryl* alcohol, which is otherwise negligible in lignin from normal wood in conifers. Lignin with higher participation of "condensed" substructures (containing more frequently C-C bonds linking the phenylpropane units of the macromolecule), due to participation of p-coumaryl alcohol, is deposited in the tissues of compression wood of conifers (Timell, 1986).

All mentioned structural particularities depending on lignin origin are indicating to potential biochemical control of lignin formation in plants. If this proves to be true, the formation of lignin in land plants would be in harmony with the way of formation of all other known biopolymers in nature. One would expect that the same rules govern the formation of the nature's second most abundant biopolymer, as is the case with other natural polymers.

The recent discovery of dirigent-protein involvement in biosynthesis of lignans (optically active phenylpropanoids derived mainly from two molecules of monolignols through $\beta-\beta'$ C-C linkage) and their potential implication in the process of lignification, favoured a new concept of biochemical control of lignification which questioned the traditional view about the random polymerization (Kwon *et al.*, 1999). This new concept is a matter of huge dispute between the lignin scientists, some of whom still favour the classical concept of random polymerization.

Several oxidising enzymes have been associated with lignin formation, just on the grounds of their capacity to oxidise the monolignols. Oxidising capacity alone cannot, however, explain the regiospecificity of lignins discussed above, nor the optical activity of lignans. With an oxidising enzyme such as laccase, any of the monolignols produces a whole range of racemic coupling products, since the radicals formed generate many coupling sites (see resonance hybrids of radicals in Fig. 5.3). The occurrence of enantiomerically pure (+)- pinoresinol in *Forsythia intermedia*, a dimeric lignan derived from E-coniferyl alcohol, has inspired the isolation of the "dirigent" protein for the first time. This protein had no oxidising capacity, but it contained sites which bind either the monomers or the monomer radicals in specific orientations which lead to selective coupling (Davin *et al.*, 1997).

It can be suspected that other dirigent proteins exist in other plants which engender other specific coupling modes, such as the one producing only (-) -pinoresinol in flaxseed. Davin and co-workers have indeed found dirigent proteins, as well as their homologs, in all major land plant groups that they have examined so far, each of which yielded just one lignan enantiomer. The experimental results accumulated so far indicate that a large class of dirigent proteins is present in nature, which can both engender distinct coupling modes and use specific substrates in the course of lignification in higher plants (Davin, Lewis, 2000).

So far, the proteins comparable to dirigent proteins have not been found outside of land plants. This land plant exclusivity, much in a comparable way to the exclusivity of lignification in land plants, could, if confirmed, have evolutionary implications.

The monolignols evidently differentially make their way to different parts of cell walls. Furthermore, which monomers go where obviously also varies with the cell type. The fact that the specific linkage type, the β-O-4, which dominates the natural lignins, is less important for synthetic lignins (true random dehydrogenative polymers), could already be taken as a proof that regiospecific preference in lignin formation exists naturally. Lewis and his co-workers have therefore proposed a hypothesis that lignin biosynthesis is controlled by an array of dirigent (monomer specific binding) proteins similar to those controlling individual lignan (dimeric phenylpropanoid polyphenols) biosynthesis. Their identification is, however, still to come.

These proteins should be, in the case of lignins, located at the lignin initiation sites, and they would determine which monomers will be incorporated in lignin macromolecule and through which type of linkage. The linkages that are formed would be defined by the specificity and orientation of the binding sites of dirigent proteins. These assemble the progenitor macromolecules from the monolignols through a series of oxidations (radical formation) and radical couplings. Once the progenitor is formed, it is replaced by a direct template polymerisation mechanism. This suggests that macromolecular lignin could be formed *in vivo* by such a direct template polymerisation mechanism, which in turn could provide an explanation as to how the β-O-4 linkages predominate in native lignins. Therefore the dirigent proteins are proposed to have complementary functions: to control the lignification and the biosynthesis of lignans in land plants (Kwon *et al.*, 1999).

Importance of Cinnamyl Alcohol Glycosides for Lignification is Questioned

Involvement of glycosides of cinnamyl alcohols in lignification has been early postulated by Freudenberg, who proposed the glycosides as "storage reserves" of monolignols from which they are regularly supplied for the lignification process (Freudenberg, 1968). An alternative role of monolignol glycosides has been discussed recently. It has been demonstrated that despite of a wide distribution of the cinnamyl alcohol glycosides in the plant kingdom, their occurrence in vascular plants is limited. They are found in conifers and are mainly restricted in *Magnoliaceae* and *Oleaeceae* of angiosperms. If these compounds were merely contributing to lignification, they would have had much wider occurrence in woody plants. The role of these glycosides originating from the lignin – branch pathway has therefore recently been postulated rather in a defence mechanism of plants, denying their importance as storage reserve (Förster *et al.*, 1999).

Covalent Bond between Lignin and Carbohydrates in Wood Cell Walls

Dehydrogenative polymerisation is enzymatically initiated by laccase and molecular oxygen or peroxidase and peroxide, which leads to formation of resonance hybrids of monolignol radicals (Fig. 5.3). These then combine to produce first the dilignol intermediates among which the quinone methide intermediate containing the major interunit linkage of lignin macromolecules, the β-O-4, is of the central importance for lignin chemistry (see top left corner of Figure 5.4). It can be attacked by different nucleophiles present in lignification system, contributing at a time, to growth of macromolecule and to covalent bonding with polysaccharides (and/or other hydroxyl groups bearing species).

Various hydroxyl containing nucleophiles such as molecules of water, monolignols, but also the hemicelluloses are the candidates for the reaction with this quinone methide. The addition of hydroxyl groups from uronic acid residues or from primary hydroxyl groups of monosaccharide residues in hemicelluloses to methide carbons in this quinone methide yields benzyl ester and benzyl ether bonds respectively, the most commonly proposed covalent bonds between lignin and carbohydrates in lignin- carbohydrate complexes (LCC), as illustrated in Figure 5.4.

The evidence of other types of bonds between lignin and carbohydrates confirms further the already discussed concept of specificities of lignin formation in land plants.

Hydroxycinnamates (mainly ferulates and p-coumarates) are accumulated in cell walls of monocots and some caryophyllaceous dicots in form of esters of polysaccharides, as mentioned previously. Grasses contain substantial amounts of hydroxycinnamic acids intimately associated with the cell wall.

Ferulic acid esters form cross links between lignin and hemicelluloses by simultaneous esterification of their carboxyl group to arabinose substituents in arabinoglucuronoxylans of cereals and etherification of their hydroxyl group to phenyl hydroxyls of lignin. Evidence of cross links by ferulic acid was found in the

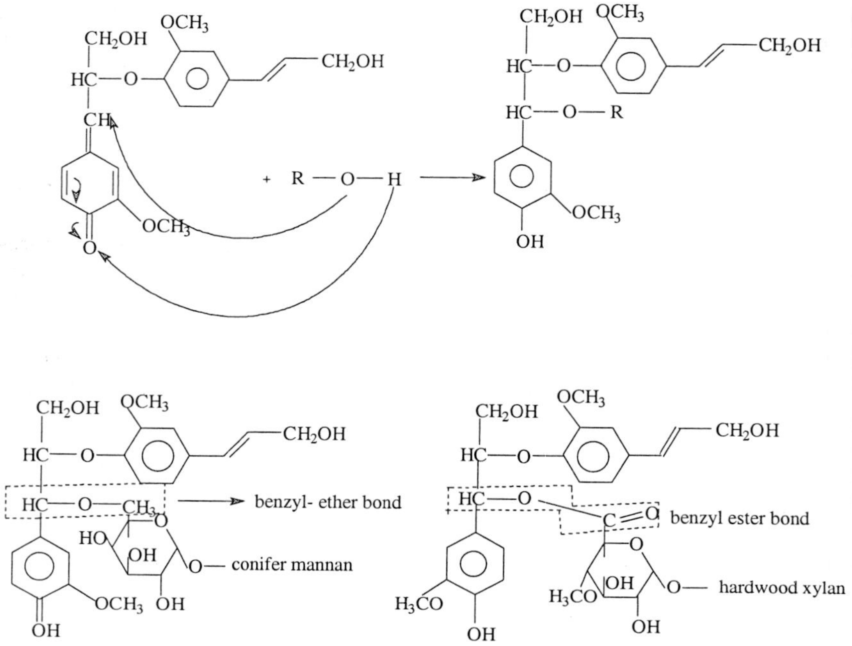

Fig. 5.4 Quinone methide of central importance for lignin chemistry and covalent bonds in LCC.

^{13}C NMR spectroscopy data (Scalbert *et al.*, 1985). Polysaccharide–polysaccharide cross linking is achieved by ferulate dimerisation either through a photochemical process or through radical coupling of ferulate–polysaccharide esters. Radical cross-coupling of polysaccharide linked ferulates with lignin monomers/oligomers results in lignin-polysaccharide cross-coupling. Dehidrodiferulates can also cross- link the cell wall polymers, resulting in extensive polysaccharide–polysaccharide– lignin cross-linking in cereal fiber (Monteon, Steinhard, 2000).

Polyphenols of Woody Plants Accompanying Lignins

The same "woody" phenolics that are found in a wide range of ferns, gymnosperms, monocotyledons and dicotyledons (in leaves in particular), which means in lignified vascular plants, are not found in major part of mosses, algae, lichens and fungi (the same was true for lignins). These phenolics are concerned with the development of vascular system in plants and therefore they are found in lignified tissues. They all derive from the same general phenylpropanoid biosynthetic pathway, which they share with lignins (see Fig. 5.1). The most regular occurrence of plant polyphenols is in leaves and it has been proven that the phenolic profile found in leaf is a good indication of the kind of phenolics that are likely to be present in other organs of the same plant, including flower, fruit, seed, stem and root (Harborne, 1997).

Three types of polyphenols predominate in the leaves of vascular plants, their structural pattern reflecting the pattern of polyphenols in other plants' parts.

- Leucoanthocyanidins (proanthocyanidins), to which the condensed tannins are related.
- Flavonol glycosides, principal ones being the glycosides of kaempherol (with phenolic B-ring), quercetin (with catechol B-ring) and myricetin (with pyrogallol B- ring). The A-rings of all aforementioned flavonols are of phloroglucinol type. The kaemferol is thus related to propelargonidins, the quercetin to procyanidins, while the myricetin is related to prodelphinidins.
- Esters, glycosides and amides of the various cinnamic acid derivatives. Esters of p- coumaric, caffeic, ferulic and sinapic acid, the acids which are important for monolignols biosynthesis, occur most frequently. One of the most familiar examples of this class of phenolic metabolites is the chlorogenic acid, 5-O-caffeoyl ester of quinic acid, which is directly related to lignin biosynthesis intermediates. Phytochemical surveys indicate that chlorogenic acid is one of the most abundant phenylpropanoid esters in vascular plants, both taxonomically and compositionally. The health benefits of dietary polyphenols, such as chlorogenic acid, have recently been related to their antioxidant power (Grace *et al.,* 1999).

The biosynthetic routes for lignins and for flavonoid polyphenols diverge in regard to the following features:

- Sequential hydroxylation and methoxylation of aromatic rings leading to cinnamic acid derivatives formation, which by successive reduction yield cinnamic aldehyde and ultimately cinnamic alcohol derivatives is important for the biosynthesis of lignins only.
- Two types of enzymes are involved in biosynthesis of flavonoids: general phenylpropanoid biosynthesis enzymes (phenylalanine ammonia lyase, PAL) until the formation of cinnamic acid derivatives, while the chalcone synthase is the key enzyme for chalcone intermediate formation by cyclisation reaction between CoA of p-coumaric acid and malonyl-CoA, which is the reaction proper to the biosynthesis of flavonoids.

The flavonoids are one of the major groups of polyphenols. Their chemical structure is based on $C_6C_3C_6$ skeletons, in which the B-ring and carbons 2, 3 and 4 in chroman (C-ring) originate from a phenylpropanoid metabolism, i.e. L-phenylalanine and that of ring A from acetate (malonate) metabolism. This was established by standard isotopic tracer methods. The central C15 metabolic intermediate in formation of flavonoids is the chalcone or its isomeric flavanone and the chalcone synthase is therefore the key enzyme for flavonoid biosynthesis, as presented in Fig. 5.5.

Pathways to the various classes of flavonoids starting from naringenin (flavanone isomeric to naringenin chalcone with 4',5,7-hydroxylation pattern) involve several enzymes, as presented in Fig. 5.6. The simpler flavonoids, such as flavones, flavanones, dihydroflavones and anthocyanidins are accompany by rule found in tannin rich hot water extracts of lignified materials.

The important roles the polyphenols play in the environment are related to the general capability of polyphenols to complex with metal ions such as transition metals: iron, vanadium, copper or manganese or metals like Ca and Al, to their antioxidant and radical scavenging properties and to their ability to participate various reactions with proteins and polysaccharides.

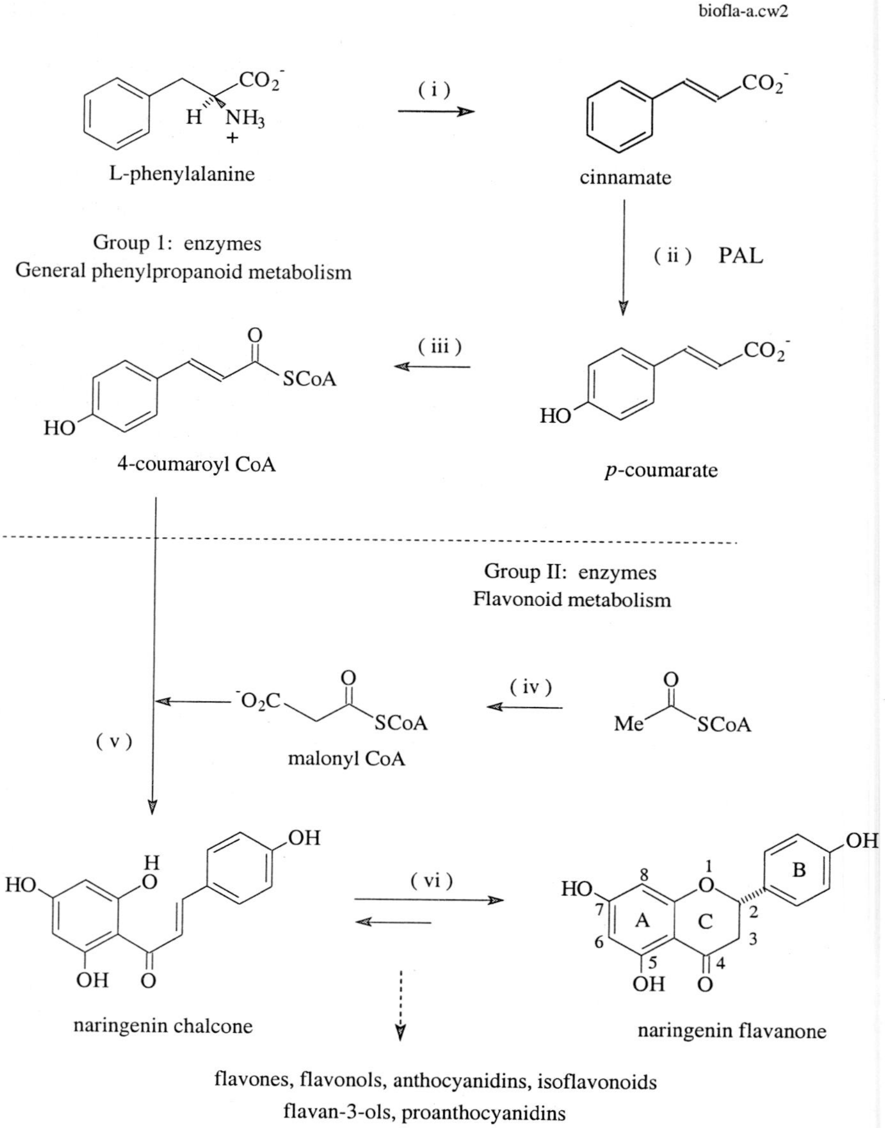

biofla-a.cw2

Fig. 5.5 Biosynthesis of chalcone, the central intermediate in biosynthesis of flavonoids (adopted from Haslam, 1998) (i) L-phenylalanine -ammonia lyase; (ii) hydroxylase; (iii) CoA- lygase; (iv)- acetyl-CoA carboxylase; (v) chalcone synthase; (vi) chalcone-flavanone isomerase.

Tannins (Plant Polyphenols)

Vegetable tannins occupy the borderland of science between botany and chemistry (Haslam, 1998). Virtually all of the tannins found in land plants consist of catechol or pyrogallol polyphenolic residues rather than simple phenolic residues. This makes them excellent chelators (complexing agents) for metal ions. This property represents the basis of various technological applications of tannins. The iron chélation of tannins has been studied recently in the context of tannin complexation importance for biology: for plant defence and in human nutrition (Scalbert *et al.*, 1999). The high affinity of tannins for iron cations may limit the iron availability to pathogens or decay micro-organisms, contributing in such a way to plant defence. This does not deny, however, the well established role tannins play in complexation of proteins.

The capacity of tannins to precipitate alkaloids, gelatine and other proteins from solution through complexation reactions is of importance for their molecular recognition, for their biological function, but also for a variety of applications of tannins as the major relatively soluble plant polyphenols. Bate-Smith (1973) has formulated a definition of vegetable tannins which succinctly stated the role of these polyphenols in plant defence: "From the biological point of view the importance of tannins in plants lies in their effectiveness as repellents to predators, whether animal or microbial. In either case the relevant property is *astringency* rendering the tissues unpalatable by precipitating proteins or by immobilising enzymes, impeding invasion of the host by the parasite". The most prominent property of plant polyphenols is therefore their affinity for proteins.

There are several physico-chemical properties that define plant polyphenols (vegetable tannins) as natural polymers. These polymers are difficult to dissolve in water as pure forms, even though they do present certain (even though minimal) water solubility when co-extracted from plant materials with lower molar mass polyphenols which modify (enhance) their solubility.

Molar masses of vegetable tannins (plant polyphenols) range between 500 and 4000 daltons. Their polyphenolic character is well reflected in the average phenolic hydroxyl group determined for these materials to range between 12 and 15 phenolic groups per $Mr = 1000$, which means per 5-7 aromatic rings (Haslam, 1998). These plant polyphenols are responsible for the common phenolic reactions, such as precipitation of some alkaloids, gelatine and other proteins from the solution.

There are three types of plant polyphenols which comply with the definition of vegetable tannins. The first two cited below are the major classes, with wide distribution in land plants, while the third is a minor class which is restricted to brown algae. The condensed tannins are common constituents of all wood barks and heartwood of many wood species, while the hydrolysable tannins are found only in hardwoods. More precisely, hydrolysable tannins are encountered only in 15 of 40 orders of dicotyledons (Harborne, 1997).

Three types of vegetable tannins are as follows:

- *condensed proanthocyanidins* with flavan-3-ols as fundamental units (catechin nucleus). Condensed proanthocyanidins exist as oligomeric forms (soluble) consisting of 2-5 or six catechin units and polymers which are insoluble. The flavan-3-ols are linked primarily through 4 and 8 positions by C-C bonds. The principal flavan-3ols are procyanidins and prodelpinidins (with pyrocatechol and pyrogallol B-ring hydroxylation patterns, respectively. The propelargonidin type condensed proanthocyanidins (presented in Fig. 5 and 6), which have B-rings derived directly from p-coumaric acid, are the least distributed proanthocyanidins in plants and are restricted only to the more primitive ferns (Scalbert *et al.*, 1999). The proanthocyanidins with catechol

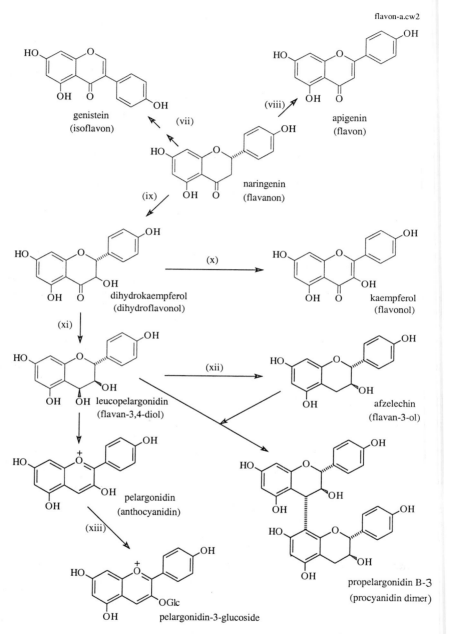

Fig. 5.6 Biosynthesis of various classes of flavonoids starting from chalcone intermediate (adopted from Haslam, 1998). (vii) (hydroxy)isoflavone synthase; (viii) flavone synthase; (ix) flavanone 3- hydroxylase; (x) flavonol synthase; (xi) dihidroflavonol reductase; (xii) flavan 3,4-diol reductase; (xiii) O-glucosyl transferase.

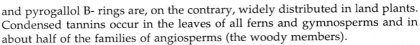

and pyrogallol B- rings are, on the contrary, widely distributed in land plants. Condensed tannins occur in the leaves of all ferns and gymnosperms and in about half of the families of angiosperms (the woody members).

Oligomeric proanthocyanidins are responsible for those plant properties which are attributed to condensed tannins. Condensed proanthocyanidins are further subdivided into those that are: (i) insoluble in water and usual organic solvents- polymers; (ii) readily soluble in water and not extractable therefrom by ethyl acetate; and (iii) readily soluble in water and extractable therefrom by ethyl acetate. The latter, comprising monomers, dimers and lower oligomers, represents just "the tip of the iceberg", as it is now well known that the first two categories predominate over the soluble ones, particularly in woody plants (Haslam, 1998). The soluble condensed tannins occur in ferns and fruits, as well as in gums and exudates.

- *galloyl and hexahydroxydiphenoyl esters* and their derivatives are usually found as multiple esters with D-glucose, and otherwise designated as *hydrolysable tannins*. The majority of known hydrolysable tannins can be regarded as derived from the key intermediate in biosynthesis of hydrolysable tannins: β-1,2,3,4,6- penta- O-galloyl -D-glucose.

Acid hydrolysis of hexahydroxydiphenoyl esters produces *bis*-lactone of hexahydroxydiphenoyl acid, the planar and virtually insoluble ellagic acid, from which the nomenclature of ellagitannins is derived.

Gallic acid is most frequently present in form of esters, which can be either

(1) simple esters,
(2) depside metabolites (synonym gallotannins)
(3) hexahydroxydiphenoyl and dehydrohexahydroxydiphenoyl esters forms (synonym ellagitannins) which are based upon (a) $^4C_1$ conformation of D-glucose, (b) $^1C_4$ conformation of D- glucose and (c) "open" chain derivatives of glucose.
(4) "dimeric" and "oligomeric" forms of hydrolysable tannins, formed by oxidative coupling of monomers, principally of the type (3).

The third group of vegetable tannins is represented by:

- *phlorotannins*, a minor group recognized in several genera of red-brown algae. Their structures are composed entirely from phloroglucinol sub-units, linked by C-C and C-O chemical bonds. (Haslam, 1998).

By far the greatest proportions of proanthocyanidins found in plants is invariably in form of higher oligomers and polymers. Proanthocyanidin higher oligomers are built in the same way as the dimers and trimers, by the successive addition of phenolic flavan-3-ol extension units to quinone methide intermediates, through C-4 to C-8 or C-4 to C-6 interflavan linkages and they represent recalcitrant substances in the environment.

Biosynthesis of Proanthocyanidin Oligomers, Soluble and Insoluble Polymers

Present status of knowledge on biosynthesis of oligomeric proanthocyanidins in plants suggests that it is intimately associated with formation of phenolic flavan-3-ols leading first to formation of a proanthocyanidin dimer, by a stereospecific nucleophilic capture at C-4 of the putative quinone methide intermediate in flavan-3-ol biosynthesis. It is interesting to remind here of the importance of quinone methides in lignin formation, which is the most important intermediate dimeric structure (see Fig. 5.4) for quinone methide structure is lignin biosynthesis. The dimer formed first in the course of proanthocyanidin biosynthesis, captures a further quinone methide intermediate (or its protonated carbocation equivalent) from flavan-3-ol biosynthesis to generate a trimer and so progressively by the capture of further quinone methides (or protonated carbocation equivalents) oligomers which finally lead to polymers, as presented in Fig. 5.7.

Quinone methides are therefore central intermediates in polymer growth of proanthocyanidins, as they are in lignin biosynthesis. As in case of lignins, the quinone methides can be attacked not only by phenolic hydroxyls from flavan-3-ols, resulting thus in an increase of proanthocyanidin polymers growth, but also by

Fig. 5.7 Biosynthesis of oligomeric proanthocyanidins.

hydroxyls originating from sugar residues of polysaccharides already present in the cell wall. This explains the intimate association of proanthocyanidins with wood cell walls components, which implies that tannins cannot be regarded purely as wood extractives in the true sense.

The fraction of tannins extractable with organic solvents is only a small proportion of total proanthocyanidins present in lignified materials. This is exemplified for some plant materials which even after repeated extractions with methanol still contain substantial amount of substances which give positive response to various colour tests for proanthocyanidins. One interpretation of these observations is that these forms of plant proanthocyanidins are covalently bound to an insoluble polysaccharide matrix within the plant cell wall, in much the same way as lignins are bonded to hemicelluloses, which impedes their complete solubilization. Covalent bonding between proanthocyanidins and lignins are also very likely to occur.

Just as in case of lignins, the *in vitro* synthesis of proanthocyanidins (which mimics the proposed biosynthetic pathways) provided the soluble products which qualitatively and quantitatively resembled the procyanidins isolated from the particular plant tissues (Haslam, 1998).

Different bark materials are important sources of condensed tannins (proanthocyanidins). Branches and twigs used in this study are bark rich materials in which the presence of proanthocyanidins biosynthetically related to lignins is very important. As fragmented forest residues applied in our project are bark rich materials, a study of proanthocyanidins accompanying lignins is of importance for understanding the process of biotransformation of these materials in nature leading to formation of humic substances.

Plant Polyphenols and Their Relation to Humic Substances in Forest Soils

Organic residues rich in lignins and other polyphenols decompose more slowly and contribute much more significantly to recalcitrant soil organic matter (humus) than proteins and sugars, for example. The theory that considers lignin, but also the polymeric polyphenols such as proanthocyanidins, as preserved plant polymers bonded to soil matrix as only partly degraded and chemically modified polymers, can therefore be regarded as the most probable hypothesis when forest soils are considered. Both degradation and condensation reactions are likely to occur during humification process. Tentative structural formula for humic acids deriving from forest soils should probably include more lignin related substructures and inter-units linkages, as lignin is the most abundant polyphenolic polymer in nature, its major source being the forests of the world. This hypothesis can be verified through the comparative study of lignins and humic acids.

Conceptually, humic substances can be visualized as combinations of biodegraded biopolymers from plant and animal residues, where lignins play a paramount role because of their abundance when forest residues are considered. The structure of the resulting humus is therefore dependent on both the plant components and on the environmental conditions in which their biodegradation take place.

Lignin biodegradation is characterized by oxidation, side chain and ring cleavage, secondary condensation and hydroxylation (Chen, Chang, 1985). The extent of these transformations is dependent on the initial lignin structure and on the nature of the microbes present in soil. The heterogeneity discussed previously in relation to lignin origin has to be taken in consideration when discussing lignin biotransformation in the environment. The recalcitrant compounds, such as various polyphenols (lignans various flavonoids and tannins) and triterpenes, for instance, are also to be considered along with lignin as they are also present in wood and bark.

The elementary analyses (C, H, O and methoxyl content) of lignin isolated from sound wood and that of soil humic acids demonstrate better resemblance than do the results for lignin isolated from biodegraded wood and humic acids. The methoxyl content is steadily decreasing from MWL through biodegraded wood lignin and humic acids, as can be seen from the data presented in Table 5.1 (Chen and Chang, 1985; Schnitzer, 1994).

Table 5.1. Elementary Composition and Methoxyl Content of Fungus- Degraded Lignins and Milled Wood Lignin from Spruce and Birch (Polymeric part of lignin is extracted from decayed wood by methanol).

Source	C%	H%	O%	OCH_3, %
MWL Spruce American	62.85	6.08	31.07	15.11
MWL Spruce European	62.81	5.88	31.31	15.24
Average Spruce liginin, Freudenberg	65.08	5.90	29.02	15.60
Lignin decayed by *P. anceps*	58.71	4.99	36.30	11.21
Coriolus versicolor	57.97	4.70	37.23	11.33
Poria subacida	58.54	5.23	36.22	11.75
MWL birch B. papyrifera	57.47	6.22	36.31	20.52
birch B. papyrifera decayed by P. *chrysosporium*	55.29	5.87	38.84	12.50
Humic acid * Haplaquodl	58.2	5.4	32.6	N– 3.1%; S– 0.7%; OCH_3 –0.93%
Humic acid Udic Boroll	56.4	5.5	32.9	N– 4.1%; S– 1.1%; OCH_3 –1.24%
Humic acids corrected for carbohydrates	61.8	5.9	29.8	N – 2.5% OCH_3 –0.93%

* humic acids data from Schnitzer (1994).

Study of plant polyphenols: lignins and polymeric proanthocyanidins in ramial chipped wood RCW (branches with diameter less than 7 cm and twigs) is fundamental for understanding and optimizing the transformation of these materials into humus constituents which are essential for soil quality. Polymeric proanthocyanidins are studied along with lignins as they are co-components of the lignified tissues present in RCW. Small branches and twigs consist of high proportions of bark, which is a rich source of polymeric proanthocyanidins (condensed tannins), and of juvenile wood which is characterised by different lignin structures from those found in lignin of mature wood.

Our preliminary studies of spectroscopic characteristics of humic acids isolated in Quebec from a soil amended with RCW of a single wood species (yellow birch, *Betula alleghaniensis*) suggest the occurrence of lignin related structures in humic acid macromolecules isolated from treated soil (D'Orazio *et al.*, 2000). The comparisons were made with milled wood lignin (MWL) isolated by the modified procedure which is usually applied to wood from the chipped wood of branches of yellow birch, which corresponds to the raw material used in soil amendment trials (Schunemann *et al.*, 2001).

Importance of Polyphenols for Humus Formation (Polyphenol Theory)

Conceptually, humic substances can be visualised as combinations of thoroughly and partially biodegraded biopolymers of plant and animal residues. The structure and combinations of these blocks depend on the environmental conditions of bio-degradation and the nature of the biopolymers involved.

Shevchenko and Bailey (1996) proposed that the identification of basic structural blocks in humic acids rather than the tentative structural formulae for humus would better contribute to the elucidation of structure of humic acids of different origin. Humic acids are usually defined as heterogeneous macromolecules that consist of highly condensed aromatic nuclei branched by peripheral aliphatic chains.

Typical bonds between phenylpropane units in lignin structure, the β-O-4, diaryl ether, biphenyl, phenylcoumarane and pinoresinol have good chances for survival from living plants to fossils. The amount of methoxyl and phenolic hydroxyl groups depends on the plant source. It can reasonably be assumed that some methoxyl groups are preserved and subsequently demethylated leading to formation of hydroquinones and finally to o-quinones, to which the deep colour of humic substances is ascribed.

Lignin certainly contributes significantly to the structure of humic and fulvic acids, but these are not the only contributors: cutins, suberins, triterpenes and extractable (soluble polyphenols), but also, and particularly so, the insoluble polyphenols, are all likely participants in formation of polycondensed structures in humic acids. The proanthocyanidins are widely distributed in plant kingdom and as a rule accompany the lignified tissues. Information on structure and reactivity of condensed tannins is critical to the development of any type of their application and understanding of their interactions with the environment. Lignins were proposed to contribute to the core structure of the humic substances, the structure of which is formed over a wide time range.

The forest residues dealt with in this study are extremely lignin rich materials and therefrom our interest for lignins and accompanying polyphenols and their role in forest humus formation. One should bear in mind that not only forest soils are dealt with here, but also the forest humus of well defined history (treatment with forest residues of known composition).

How the Polyphenols (Lignins and Polyphenolic Extractives) Contribute to Forest Humus Formation

A comparative study of lignins and proanthocyanidins from defined wood species with humic (and fulvic) acids isolated from humus of related origin could widely contribute to better understanding of the transformations of lignins (and other polyphenols) leading to major structural fragments of humic acids and their reactivities in soils. Additional information about humus formation processes could help optimize the process of RCW transformation into humus and enable its more efficient application for upgrading and remediation of arid soils in different regions of the world.

The polyphenols accompanying lignins could play important role in the environment in two different ways: they can act as inhibitors of decomposition phenomena of forest residues by reducing palatability, inactivating the enzymes and precipitating proteins, thus influencing the ecology of soil fauna and flora.

On the other hand, the proanthocyanidins intimately associated with cell wall components (the insoluble polyphenols) could contribute to humus formation by direct incorporation, in much the same way as was proposed for lignins. Due to its complex aromatic structure and specific features of its biodegradation, lignin is of great importance for the bio geo-chemical cycle of carbon.

The polyphenol theory of humus formation usually assumes the initial complete depolymerization of plant polymers into simple phenols, which are then resynthesised and repolymerised to form humic substances.

There are two major objections to the often proposed hypothesis of humus formation based on repolymerisation of phenols released from biologically (enzymatically) degraded macromolecular polyphenols. The high mobility of such monomeric phenols and their low concentration upon such potential depolimerisation, would make it quite difficult to visualize their intensive condensation. The other serious objection to the hypothesis of complete lignin depolimerisation is the experimental evidence indicating that only a small portion of lignin is degraded by white-rot fungi, otherwise the most efficient ligninolytic micro-organisms. The data presented in Table 5.1 comply also with this view (Chen, Chang, 1985).

Therefore, the concept which considers lignin as a preserved part of plants which is bonded to soil matrix as only partly degraded and chemically modified macromolecule seems to be acceptable at least when forest soils humus is considered. Both degradation and condensation reactions of lignins could contribute to humification processes in forest soils, the participation of polyphenols, but also of other recalcitrant materials should also be taken into account.

In the classic alteration scheme, increasing degradation first leads to a progressive evolution of humin, then the more soluble humic acids and finally the most soluble fulvic acids are formed, by redox and condensation reactions. The fact that lignins are natural polymers with high degree of polydispersity and structural variability leads to the assumption that humic acids and fulvic acids from forest soils originate mainly from lignin, while humin should be related to other organic substances (Shevchenko, Bailey, 1996).

Crucial Role of Lignin Bio-transformations in Forest Humus Formation (Environmental Aspect of Lignin)

Fungi and bacteria decompose carbohydrates more rapidly and more extensively than lignins. Lignin serves as a biological shield for polysaccharides in plant cell walls, according to its previously discussed biological roles. However, certain carbohydrates can also be preserved during the biodegradation, as they are covalently bonded to the lignin aromatic core. The strong linkages are in fact responsible for residual lignin structures in pulps and are hardly broken without producing extensive degradation of the remaining polysaccharides, which are the problems in focus of modern studies of pulp bleaching. The results of these studies confirm the importance and persistence of covalent bonds in lignin-carbohydrate complexes (LCC).

Some fungi and bacteria modify functional groups in lignins. Brown-rot fungi split the methoxyl groups, other are capable to split inter-unit linkages. Extensive depolymerisation is not occurring, however, as it requires the co-operation of depolymerases and oxidases.

The extent of the transformation is dependent on the initial lignin structure and on the nature of the microbial consortium present in soil. It has been confirmed that lignins are not completely degraded into phenolic units, as the lignin structure is altered before any extensive depolymerisation and elimination of carbon dioxide occurs.

Significant alteration of structure of lignin during humification is related to different chemical resistance of different types of lignins to biodegradation, but also between different types of substructures within the same lignin. Syringyl structures split more easily than the guaiacyl ones, for instance.

Under aerobic conditions lignin can be randomly biodegraded by white-rot fungi, otherwise the most efficient lignin degrading micro-organisms, into lower-molar mass fragments, but not necessarily to monomers, and incorporated in humus as thus transformed blocks. Under anaerobic conditions, less biodegradation is expected and transformation of the initial lignin skeleton through humus and peat to coal is taking place. It is of interest to mention here that the lignin- related guaiacyl structures have been identified in coals. The predominant anaerobic process otherwise is degradation of polysaccharides debris.

Lignin transformation in soil defines therefore the restructuring of the polymer due to : (1)- oxidation (2) side chain and ring cleavage, (3) secondary condensation and (4) hydroxylation.

The initial oxidative transformation of side chains renders easier the processes of adsorption on clay, complexation with metal ions and incorporation in humus matrix. Enhanced oxidation also increases solubility of lignin fragments.

Chemical properties of plant materials are affected greatly by supramolecular organization (ultrastructure). At the micrometer scale lignins are reported to be granular (aggregates of spherical granulae) while the LCCs are reported to be homogeneous. Based on X-ray diffraction data described recently, it was assumed that the supramolecular organisation of humic acids is a spatial lattice that include several layers of condensed aromatic systems with chain fragments of different lengths and degrees of order (Shevchenko et al., 1998). Humic and fulvic acids are reported to have mostly chain and filamentous structures. Certain similarities between LCCs and humic substances based on AFM (atomic force microscopy) data. It was an important observation that fractal dimensions of stream sediment humic acid and dioxane lignin (a common lignin preparation) coincided. The fractal dimension of dissolved humic acid was slightly larger, assuming a less compact, more open form. The same tendency was observed when going from native lignin to water soluble lignosulfonates (Shevchenko *et al.*, 1998).

The large quantity of lignin in forest residues and its discussed heterogeneity make the study of lignin contribution to forest humus formation an important issue which is in focus of our research on forest residues application to soil rehabilitation.

Conclusions

The increasing requirements of restoring soil organic matter content and its fertility functions have promoted in the last decades the recycling of waste materials in the form of organic soil amendments. In this context, the forest residues such as small branches and twigs with the foliage they contain, being available in large amounts following the forest operations, have been applied for soil amendments in chipped form (RCW) in field trials in Quebec.

Lignins as major polyphenol constituents of such materials are the biopolymers with high resistance to microbial degradation. Their structure and environmental conditions in which their biodegradation is taking place are at the basis of the structure of the final products of their bio-transformation. The polyphenols structurally related to lignins, such as proanthocyanidins, are also of interest when organic matter constituents of forest soils are studied. The preliminary spectroscopic studies of lignins isolated from yellow birch (*Betula alleghaniensis*) and humic acids isolated from the soil 15 years after it was treated with the RCW of the same species indicate to the persistence of lignin structures in humic substances. Therefore, the application of forest residues for soil amendments is a venue to the reconstruction of the humus layer. The humic acids constituents of that humus are likely to incorporate some lignin related structure. The optimization of soil treatment technology could consist of application of defined mixtures of wood species, as the synergistic effects have been reported from the related studies with RCW.

The global importance of lignins for bio-geochemical carbon cycle is strongly confirmed through this research. This investigation exemplifies yet another form of lignin utilisation through an extremely important application of lignin and related polyphenols rich forest materials for the soil humus reconstitution. Despite the fact that the soils feed our world there still remain many aspects of soil chemistry and biology to learn about.

Keywords: lignin, polyphenols, phenylpropanoids, biosynthesis, biodegradation, soil, humus, fertility, hardwood, softwood, forest residues, soil amendment, ecology

REFERENCES

1. Bate-Smith, E.C. (1973): Haemanalysis of tannins- the concept of relative astringency. *Phytochemistry*, 12, 907.

2. Beauchemin, S., N'Dayegamiye, A., Laverdière, M.R. (1990): Effet d'apport d'amendement ligneux frais et humifiés sur la production de pomme de terre et sur la disponibilité de l'azote en sol sableux. *Can. J. Soil Sci.*, 70: 555-564.

3. Chen, C.-L., Chang, H.-M. (1985): Chemistry of Lignin Biodegradation. In : *Biosynthesis and Biodegradation of Wood Components*, T. Higuchi, ed., Academic Press, Orlando, New York, London, p. 535- 556.

4. Davin, L.B., Wang, H.-B., Crowll, A.I., Bedgar, D.I., Martin, D.M., Sarkanen, S., Lewis, N.G. (1997): Stereoselective bimolecular phenoxy radical coupling by an auxiliary (dirigent) protein without an active center. *Science*, 276, 362-366.

5. Davin, L.B., Lewis, N.G. (2000): Dirigent proteins and dirigent sites explain the mystery of specificity of radical precursor coupling in lignan and lignin biosynthesis. *Plant Physiol.*, 123, 453- 461.

6. D'Orazio, V., Cocozza, C., Stevanovic Janezic, T., Miano, T., Senesi, N. (2000): Effects of application of forest residues on spectroscopic properties of soil humic acids. *Proceedings of the 10th International Meeting of the International Humic Substances Society.* Vol. 2, p. 703-706.

7. Förster, H., Pommer, U., Savidge, R.A. (1999): Metabolic activity of uridine 5-'diphosphoglucose cinnamyl alcohol glucosyltransferase as an intrinsic indicator of cambial growth in conifers. In: *Plant Polyphenols* 2. G. Gross and Kluwer Academic/Plenum Publishers, New York. p. 371-391.

8. Freudenberg, K., Neish, A.C. (1968): Constitution and Biosynthesis of lignin. *Molecu. Biol. Biochem. Biophys.* Series 2. Springer Verlag, New York, p. 3-43.

9. Grace, S.C., Yamasaki, H., Pryor, W.A. (1999): Spin stabilising approach to radical characterization of phenylpropanoid antioxidants: an ESR study of chlorogenic acid oxidation in the horseradish peroxidase, tyrosinase and ferrylmyoglobin protein radical systems. In: *Plant Polyphenols* 2. G. Gross and Kluwer Academic/Plenum Publishers, New York. p. 435- 450.

10. Gross, G.G. (1985): Biosynthesis and Metabolism of Phenolic Acids and Monolignols. In T. Higuchi ed. *Biosynthesis and Biodegradation of Wood Components,* Academic Press, Orlando, p. 229-271.

11. Harborne, J.B. (1997): Role of phenolic secondary metabolites in plants and their degradation in nature. In: *Driven by Nature. Plant Litter Quality and Decomposition.* G.Cadisch and K.E. Giller, eds., CAB International. Ch 4. p. 67-74.

12. Haslam, E. (1998): *Practical Polyphenolics. From Structure to Molecular Recognition and Physiological Action.* Cambridge Univ. Press., Cambridge.

13. Higuchi, T. (1957): Biochemical studies of lignin formation. *Physiol. Plant.* 10, 356-372.

14. Kwon, M., Burlat, V., Davin, L., Lewis, N. (1999): Localisation of dirigent protein involved in lignan biosynthesis: Implications for lignification at the tissue and subcellular level. In: *Plant Polyphenols* 2. G. Gross and Kluwer Academic/Plenum Publishers, New York. P. 393-411.

15. Larochelle, L. (1994): L'impact du bois raméal fragmenté sur la dynamique de la mésofaune du sol. Mémoire de maîtrise. Université Laval.

16. Lemieux, G. (1986): Le bois raméal et les mécanismes de fertilité du sol. Ministère de l'Énergie et des ressources naturelles and la Faculté de foresterie et de géomatique de l'Université Laval, Québec, 20 p.

17. Lemieux, G. (1990): Le bois raméal et la pédogenèse: une influence forestière et agricole directe. Ministère de l'Énergie et des ressources naturelles and la Faculté de foresterie et de géomatique de l'Université Laval, Quebec, p.34.

18. Lemieux, G. (1998): A new forested technology for agricultural purposes: the RCW technology. Publ. No. 95; Published by Faculty of Forestry and Geomatics, Laval University, p.9.

19. Li, L., Popko, J.L., Umezawa, T, Chiang, V.L. (2000): 5-Hydroxyconiferyl Aldehyde Modulates Enzymatic Methylation for Syringyl Monolignol Formation, a New View of Monolignol Biosynthesis in Angiosperms. *J. of Biol. Chem.,* 275 (9), 6537-6545.

20. Monteon, G. Steinhard, H. (2000): Difeulate analysis: new difeulates.

21. N'Dayegamiye, A., Dubé, A. (1986): L'effet de l'incorporation de matières ligneuses sur l'évolution des propriétés chimiques du sol et sur la croissance des plantes. *Can. J. Soil Sci.,* 66, 623-631.

22. Pettigrew, D. (1998): Perte de masse anhydre et dynamique des éléments chimique du bois raméal fragmenté de tremble (Mass loss and dynamics of chemical elemnts of rameal chipped wood of aspen). Mémoire de maîtrise (M.Sci. thesis). Université Laval, Ste-Foy, Québec, p 94.

23. Ralph, J., Bunzel, M., Marita, J.M., Hatfield, R.D., Lu, F., Kim, H., Grabber, J.H., Ralph, S.A., Jiminez-Monteon, G., Steinhardt, H. (2000): Diferulate analysis: new diferulates and disinapates in insoluble cereal fiber. *Polyphénols actualités*, 19, 13-17.

24. Saiz-Jimenez, C. (1996): The Chemical Structure of Humic Substances: Recent Advances. In: *Humic Substances in Terrestrial Ecosystems*, ed., Alessandro Piccolo, Elsevier, Amsterdam, Lausanne. p. 1-44.

25. Saka, S., Thomas, R.J. (1982): Evolution of the quantitative assay of lignin distribution by SEM-EDXA technique. *Wood Sci. Technol.*, 16, 1-18.

26. Saka, S., Goring, D.A.I. (1985): Localisation of lignins in wood cell walls. In: *Biosynthesis and Biodegradation of Wood Components*, T. Higuchi ed. Academic Press, Orlando, FI, p. 51-61.

27. Savidge, R.A., Udagama,-Randeniya, P. (1992): Cell-wall bound coniferyl alcohol oxidase associated with lignification with conifers. *Phytochemistry*, 31, 2959-2966.

28. Scalbert, A., Monties, B., Lallemand, J.-Y., Guittet, E., Rolando, C. (1985): Ether linkage between phenolic acids and lignin fractions from wheat straw. *Phytochemistry*, 24, 1359.

29. Scalbert, A., Mila, I., Expert, D. (1999): Polyphenols, metal ion complexation and biological consequences. In: *Plant Polyphenols* 2. G. Gross and Kluwer Academic/Plenum Publishers, New York. p. 545- 555.

30. Schnitzer, M. (1994): A chemical structure for humic acid. Chemical, ^{13}C NMR, colloid chemical and electron microscopic evidence. In: *Humic Substances in the Global Environment and Implications on Human Health*, N. Senesi and T.M. Miano eds. Elsevier Science B.V., p. 57-69.

31. *Schünemann, K.*, Stevanovic Janezic, T., Riedl, B., Bley, T. (2001): Isolation and characterisation of bark lignins from yellow birch (*Betula alleghaniensis*), *Grenoble Workshop*, Grenoble June 2001, (8-19), p.

32. Senesi, N., Loffredo, E. (1998): The Chemistry of Soil Organic Matter. In: *Soil Physical Chemistry*, ed., D.L. Sparks, CRS Press, Boca Raton, p. 239- 370.

33. Shevchenko, S.M. Bailey, G.W. (1996): Life after death: Lignin - humic relationship re-examined. *Critical Reviews in Envir. Sci. and Technol.*, 26: 95-153.

34. Shevchenko, S.M., Yu, Y.S., Akim, L.G., Bailey, G.W. (1998) : Comparative Surface Morphology of Lignin- Carbohydrate Complex and Humic Substances: AFM/VR Approach. *Holzforschung*, 52, 149 – 156.

35. St-Amand, D., Lauzon, M. (1995): Fragmentation en Bois Raméaux Fragmentés de biosurplus forestiers. Projet no. 1051. Essais, expérimentations et transfert technologique en foresterie. Service canadien des forêts, p 61.

36. Stevenson, F.J. (1982): *Humus Chemistry. Genesis, Composition, Reactions*. Wiley-Interscience, New York.

37. Timell, T.E. (1986): *Compression Wood in Gymnosperms*. Volumes 1-3. Springer Verlag, Berlin.

38. Udagama,-Randeniya, P., Savidge, R.A. (1994): Electrophoretic analysis of coniferyl alcohol oxidase and related laccases. *Electrophoresis*, 15, 1072-1077.

39. Tissaux, J.-C. (2000): *Caractérisation des bois raméaux fragmentés et indices de décomposition. Mémoire de maîtrise*, Université Laval, p. 114.

40. Tremblay, J., Beauchamp, C.J. (1998): Fractionnement de la fertilisation azotée d'appoint à la suite de l'incorporation au sol de bois raméaux fragmentés: modifications de certains propriétés biologiques et chimiques d'un sol cultivé en pommes de terre. *Can. J. Soil Sci.*, 78: 275-282.

41. Zech, W., Senesi, N., Guggenberger, G., Kaiser, K., Lehmann, J., Miano, T.M., Miltner, A., Schroth, G. (1997): Factors controlling humification and mineralization of soil organic matter in the tropics. *Geoderma*, 79 , 117 – 161.

6

Bioreactors for Wastes Fermentation and Fuels Production—Relevant Thermodynamic and Process Parameters, Knowledge and Engineering Data

Marija S. Todorovic, Franc Kosi and Ljiljana Simic

Laboratory for Thermodynamics and Thermotechnics of The Division for Energy Efficiency and Renewable Energy Sources, Agricultural Faculty[1], University of Belgrade

Introduction

Broader biowaste processing for gaseous and liquid fuels production, design and engineering applications as well as spreading of effective and practical implementation of biofuels production require: reliable operational results of specific pilot plants, and approved engineering data and calculation procedures. To be able to design and perform efficient processing control as well as approach to processing optimization, predictive analytical modeling is crucially necessary.

Significant R&D progress has been made in the area of wastes processing and related systems engineering including anaerobic bioreactor systems for integrated fermentation and products separation. Bioprocessing engineering applications and modern utilization of biowastes and/or co-utilization of bioresidues mixed with wastes of other, municipal or industrial origin, for fuels and chemicals production require, critical evaluation of available R&D results and data within an organized engineering database, and further development of more efficient and methodologically clear analytical and experimental methods (Todorovic, M., and E. Boyce, 1995). Simultaneous fermentation and separation of an inhibitory fermentation product, in a biparticle fluidized bed reactor containing immobilized cells, which increase productivity by easy removal of potentially-inhibitory products, as well as a few other innovative processing routes have been studied. It has been shown that predictive, analytical reactor modeling has been still hampered by limited data on biophysical and thermophysical properties, including reactor kinetics and transport phenomena (Todorovic, M., *et al.*, 1996).

---

[1] 11080 Belgrade, P.O. Box 127, Yugoslavia.

Thermodynamical and rheological properties of agricultural effluents relevant for mechanical and biochemical processing have been obtained experimentally and the intrinsic phenomenological and theoretical aspects of their determination have been given. The influence of the different growth stages of the immobilized biomass on bioreactor phenomena and bioreactor's content transport properties, and likewise the influence of transport phenomena on microbial growth and activity has to be furthermore systematically studied and experimentally more clearly characterized.

Aimed at improving understanding, biophysics of synergetic microbiological and physicochemical processes have been analyzed. Relevant phenomena occurrence and interaction have been analytically described and parametrically investigated. Reactor phenomena are being studied on both micro (i.e., pore- and particle-size) and macro (i.e., reactor) levels, to describe synergetics between biodynamics and physicochemical dynamics as well as their intrinsic micro-macro relations. Different formulations of fluid flow, mass and heat transfer in two or multiphase reactor media—content, which exist in the literature as a result of the importance of this topic to bioengineering and bioprocessing—including aerobic and anaerobic systems, have been reviewed.

Current approach to kinetics study aimed at advancing basic knowledge and parametric data determination is a basis for development of physically sound modeling. It encompass the study of biochemical mechanisms of microbial growth and activity, and the study of mechanisms of physical processes. To describe the occurrence and correspondence of micro- and macro-phenomena and synergetism of microbiodynamic and dynamical changes in physicochemical fields, a mathematical formulation based on a hierarchical volume averaging method, has been applied in defining system of governing equations. A path of analysis that originates with the continuum axioms for the mass and momentum of multicomponent systems, has been presented. Corresponding mathematical methods of closure, which are in development, have been reviewed.

Regarding the interwoven nature of wastes as one of the most critical environmental problems, as well as the number of possible processing routes and technologies, development of new more efficient multiparametric system approach is discussed. Even in the phase of fundamental and applied research related to technology development, the transition to a system approach, as an evolutionary process of integrating all energy and environment relevant criteria and constraints is needed. Thus, this paper describes projects and results of interactively connected, integrated research aimed to advance fundamental knowledge and a technologically intrinsic system approach.

Biowastes for Biofuels

A number of biowastes can be converted to fuels. The lignocellulosic biomass wastes, cellulose and hemicellulose, can be broken down into sugars that can be fermented into ethanol or can be gasified to a mixture of carbon monoxide and hydrogen for catalytic conversion into methanol. By anaerobic digestion, a mixture of bacteria can break down lignocellulosic biomass and produce a gas that can be cleaned up for pipeline—quality methane. Thus, the energy content of biomass can be biologically or thermally transformed to liquid or gaseous fuels that can be integrated within the existing fuel distribution and use infrastructures (Benefild L.D., W.R. Clifford, 1980).

Several processes have been used for conversion of lignocellulosic biomass to ethanol catalyzed by dilute acid, concentrated acid, or enzymes known as cellulases. Enzymes are also biodegradable and environmentally benign. Enzyme-catalyzed processes provide the high yields of ethanol necessary for economic viability, under mild conditions, with low concentrations of enzymes and in addition these processes have still tremendous potential for technology improvements: improvement of glucose and xylose yields from pretreatment; further increase of ethanol yields and concentrations; decrease of stirring and pretreatment energy use; better productivities via continuous processing; reduction of bioreactor's fermentation times and lowering of production costs of octane enhancers or chemicals from lignin (Wyman, E. C., 1994). The simultaneous sacharification and fermentation (SSF) is one of the enzyme-based processes which has became a favoured route to achieve cost-effective ethanol production.

Biogas is a medium-energy-content gas, mixture of methane and CO_2. By anaerobic digestion complex organic compounds are decomposed by microorganisms in bioreactors. In the anaerobic digestion process, one group of bacteria enzymatically breaks down cellulose and other complex molecules into simple sugars and other monomers. Then, other types of bacteria digest these products, producing organic acids that are broken down to form still smaller molecules of acetate, formate, hydrogen, and CO_2, finally, to be, by the specialized bacteria called methanogens, used to produce methane and CO_2. The methane-rich gas, after and CO_2 removal, is a high-energy-content gas, excellent substitute for the natural gas. If MSW is employed as the substrate, anaerobic digestion also provides an environmentally sound processing method. In the anaerobic digestion of MSW, the solid waste is shredded, and ferrous materials are removed. Generally, it is also necessary to separate the extra fine and oversized materials for landfill disposal. The mixture is fed into digesters, and the microbial process converts about half the solid waste into biogas. MSW can be also co-digested with agricultural or industrial wastes (Todorovic, M., and F. Kosi, 2000).

In sanitary landfills, naturally occurring anaerobic bacteria break down the biodegradable fraction of the MSW buried there to form biogas. Gas production rates and yields vary widely and only a small percentage of that gas is now being effectively and economically recovered. Anaerobic digestion is also used to remove soluble wastes from chemical plant effluents. However, although the anaerobic digestion processes are effective in meeting wastes disposal requirements, current systems are still not well developed enough for the utilization of solid waste lignocelullosic biomass. Work is also needed on landfill gas recovery to understand the effect of atmospheric conditions on gas flow, evaluate microbial populations that produce landfill gas, improving gas generation and capturing methods (Wyman E. C., 1995).

Knowledge Status

Immobilized-Cell Reactor's Technology

Fermentation—the biological production of organic liquid and gaseous fuels and solvents represents tremendous potential for microbiological production of energy-rich biogases, ethanol, methanol and butanol, volatile fatty acids used as acidulents, preservatives, flavoring agents or chemical feedstocks. Both, pilot- and large-scale bioreactor systems operate worldwide, but the potential bioprocessing efficiency and productivity of bioreactors is still far greater than that actually achieved. Functional imperfections and uncertainty in continuous process control (Todorovic, M., and E. Boyce, 1995 and Todorovic *et al.*, 1993)—both features of current bioreactor hardware—have often precluded maintaining sufficient long-term productivity rates. This can be attributed to present limitations in knowledge of microbiological, thermophysical and rheological properties of biomass and fermentation substrates, and to a lack of parametric data on process kinetics and the synergism between biochemical-, mass-, and energy-transport processes. Clearly, a better understanding of basic bioreactor phenomena, plus supplementation of the bioengineering knowledge database, are essential for making bioreactors more economically feasible.

A key element in developing systems capable of very high rates of substrate conversion, is retention of high concentrations of biocatalyst (i.e., high cell densities of active microorganism) within the bioreactor, plus establishment of intimate contact between substrate and biocatalyst. One extremely effective method of achieving both goals is immobilization of microorganisms within a well-defined matrix, such as a permeable gel. Biocatalyst immobilization in reaction environments conducive to effective heat- and mass-transport can result in highly productive bioreactor systems. Moreover, immobilization enhances process stabiliity and facilitates removal of soluble products such as organic acids, which can otherwise reach concentrations that are inhibitory or even lethal to the cells producing them.

The best-developed methods of biocatalyst immobilization include entrapment of microbial cells in polymer gels, composite beads, or hollow fibers, and microencapsulation in polymer membranes (Todorovic, M., and E. Boyce, 1995). The economic potential of many systems (e.g., hollow-fiber reactors) is limited by high start-up costs, and productivity of many immobilized cell reactors is frequently limited by diffusion rates of substrates and/or products, unsatisfactory cell viability, and leakage of cells from the immobilization matrix. **Biofilm** reactors, which employ a second microorganism (the biofilm-producer) to immobilize the biocatalytic organism by adhering both to it and to a solid support, have shown considerable promise in addressing some of the limitations of other types of bioreactors (Todorovic, M., and E. Boyce, 1995 and Todorovic, M., *et al.*, 1993).

Superior reactor performance and high productivity achieved with **fluidized-bed** reactors (FBR) containing immobilized *Propionibacterium acidipropionici* (Todorovic, M., and E. Boyce, 1995) were attributed to high cell density in reactor. Scanning electron micrographs indicated that cells were immobilized both by attachment through surface colonization and by entrapment of large cell clumps within the fibrous matrix. This immobilization scheme caused cells to be swept from the matrix but also to be constantly renewed, allowing the reactor to maintain a stable, long-term productivity without degeneration due to cell aging.

Use of FBR containing immobilized cells increases productivity because potentially-inhibitory products can be easily removed from the reactor system. Combining the benefits of cell immobilization and *in situ* product separation, Davidson, Thompson, Kaufman and Cooper (Todorovic, M., and E. Boyce, 1995) demonstrated simultaneous fermentation and separation of an inhibitory fermentation product, lactic acid, in a **biparticle** fluidized bed reactor (BFBR). Their reactor included both biocatalyst particles (i.e., active cells immobilized in carrier beads) which were fluidized by the liquid medium, and other particles which served as product adsorbents. Integrating fermentation and separation processes in a single reactor—and its modified biochemical processing helps to minimize byproduct formation and to reduce reactor waste production at its source.

Conducted, comprehensive review (Todorovic, M., and E. Boyce, 1995) of recent engineering, biotechnology, and microbiology literature on anaerobic fermentation and bioprocessing reveals numerous papers analyzing relevant process kinetics and determining productivities of bioreactor systems. For the most part, comparison and evaluation of these can only be made with regard to types of bioreactors described, origins and identities of substrates, and identities of desired products—in fact, the majority of studies focused on production of relatively few feedstock chemicals and fuels, including methane, ethanol, butanol and lactic, propionic and acetic acids.

Published characterizations of bioreactor performance are almost always described in terms of two well-known quantities: **volumetric productivity** and **specific mass productivity**. This severely limits or negates outside observers' abilities to directly compare and evaluate results between studies, or to use the results for bioprocess engineering predictions or for process optimization.

One major difficulty in describing specific mass productivity is determining what proportion of immobilized biomass is indeed active at a given time; this is technically difficult to determine, and in addition, may vary over time due to decreased cell viability and reproductivity, and to changes in metabolic activity levels—intrinsic to batch processing (time-related fluctuations can occur in continuous flow processes as well). Volumetric productivity is parametrically dependent on reactor shape and dimensions. As an integral quantity, it is dependent on spatial and temporal uniformities of substrates, on viability and metabolic state of immobilized biomass, and on relevant biophysical properties and transport forces.

Mathematical Modeling of Biofuel's Reactors

Although the anaerobic fermentation has significant advantages over other methods of organic wastes processing, it is not implemented world wide, and not adequately well recognized as an efficient and reliable technology because of its still relatively limited record with respect to effective and maintainable process control strategies.

Andrews (Andrews J. F., 1997) has developed a dynamic model which can be used to predict the dynamic response of the five variables most commonly used for monitoring process stability: pH, volatile acids concentration, alkalinity, gas composition, and gas flow rate. The model indicates that there are strong interactions between the phases and uses equilibrium relationships, kinetic expressions, stoichiometric coefficients, change balances and mass transfer equations to reflect these interactions.

Comprehensive, more recent review of engineering, biotechnology, and microbiology literature on fermentation processing reveals numerous papers analyzing relevant process kinetics and determining productivities of bioreactor systems (Todorovic, M., and E. Boyce, 1995). For the most part, comparison and evaluation

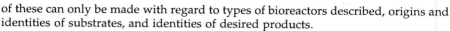

of these can only be made with regard to types of bioreactors described, origins and identities of substrates, and identities of desired products.

Re-examination of bioreactor modeling reveals that greatest emphasis has been placed on empirical relationships within reactors. Development of predictive, analytical models has been retarded by poor applicability of mathematical formulations describing kinetics, transport and constitutive relationships (Todorovic, M., and E. Boyce, 1995). This can be directly attributed to limited availability of experimental data on bio- and thermophysical properties of heterogeneous reactor contents, and to incomplete descriptions of the synergetics between microbiological and physicochemical phenomena. For predicting gas and liquid as well as variable biomass volume of bioreactor's content fractions, widely used correlations are providing parameters and their ranges of values with limited generic applicability, since they are defined by the specific physical and microbiological properties of individual systems.

Different formulations of fluid flow, mass and heat transfer in two or multiphase bioreactor media—content, exist in literature as a result of the importance of this topic to bioengineering and bioprocessing—including aerobic and anaerobic systems. In view of the increasing research activity in this area, a critical review and summary of the most prevalent formulations is of timely interest. The objective of this work is to examine each formulation in terms of the availability of accurate transport coefficients, the order and types of physical phenomena which are described and intrinsic adherence to relevant constraints. It is concluded that most approaches require knowledge of numerous coefficients which are difficult to obtain experimentally. These difficulties presumably arise from the definition of coefficients which do not have clear physical nor biophysical meaning. Predictive, analytical reactor modelling has been hampered by limited data on biophysical and thermophysical properties, including reactor kinetics and transport phenomena, and by incomplete understanding of the biophysics of synergetic microbiological and physicochemical processes (Todorovic, M., and E. Boyce, 1995). In this study, the current research results of different phenomena in various bioreactor types and its working media—content, are analyzed and re-examined within the context of a more generalized formulation. Models which have been described so far have been incomplete or deficient for several possible reasons: it may have ignored certain relevant features of specific bioreactors (e.g., axially varied gas production during fermentation, and its effect on concentration profiles; effects of varying diameters in tapered columns; non-steady-state growth conditions leading to changes in effective biomass concentration); **it failed** to account for all relevant phenomena (e.g., solid holdup due to incomplete mixing; diffusion and reaction occurring within reactor particles); **it made erroneous assumptions** about reactor conditions (e.g., that axial dispersion coefficient is constant).

Both gas and liquid superficial velocities have been assumed by some researchers (Todorovic, M., and E. Boyce, 1995) to be functions of **axial** position within the bioreactor, and to be uniform with respect to **radial** position. The same researchers assumed that temperature within the whole volume of the bioreactor was uniform. Researchers at the Oak Ridge National Laboratory and Washington State University performed interactive experimental and analytical studies (Todorovic, M., and E. Boyce, 1995) which have resulted in steady-state mathematical description applicable to immobilized cell, three-phase fluidized bed bioreactors—describing effects of the tapered bed, of variable solid holdup, and of liquid dispersion with respect to axial position of biocatalyst particles, upon concentration profiles developed in the bed, taking in account for concentration profile development inside the biocatalyst bead, and a variable dispersion coefficient affecting concentration profile within the fluidized bed.

Todorovic *et al.*, (1993) have conducted experimental investigations of hydrodynamically and thermally fully-developed, forced convection in bioreactor models, determining heat conduction in a stationary regime without fluid flow, and thermal instability onset in a bioreactor's porous system model (confirming the existence of different axial and radial superficial velocity and temperature profiles, identifying the boundary layer domain of the non-darcian flow effects. In the **radial** direction, in the zone near the reactor wall, molecular diffusion was found to be the dominant phenomenon, whereas in the **core** zone of the porous bed, dispersion was found to control transport.

Hydrodynamics of **biparticle** fluidized bed reactors, which is even more complex, has been investigated (Todorovic, M., and E. Boyce, 1995) along with kinetics of the adsorption and desorption processes occurring in adsorbent particles.

For predicting **gas, liquid and solid holdup,** widely used correlation of Begovich and Watson (Todorovic, M., and E. Boyce, 1995) is providing parameters and their ranges of values with limited generic applicability, since they are defined by the specific physical and microbiological properties of individual systems.

To describe **liquid dispersion,** an axial dispersion coefficient (D_z) has been determined experimentally for three-phase fluidized beds containing glass particles and then implicitly expressed by the Peclet Number (P_e) equation (Todorovic, M., and E. Boyce, 1995), which is based on the isotropic turbulence theory, takes into account the interaction between column (D_c) and carrier diameters (d_p), and the effects of liquid and gas superficial velocities upon D_z at constant pressure. The bubble coalescing coefficient (C) is assigned a value depending on whether gas bubbles are coalescing or breaking up in the column. It is possible that C differs for bioreactor systems in which particles themselves are porous (so that immobilized cells have morphologically-changed surfaces and are surrounded by biofilm), compared to ordinary three-phase fluidized beds. This possibility was recognized in the experimental method (Todorovic, M., *et al.*, 1998), as was determination of D_z in real systems involving various bioreactor types. **Diffusion** is one of the most important phenomena in bioprocessing. Methods of measuring diffusion have been reviewed in (Todorovic, M., and E. Boyce, 1995).

Anaerobic Bioreactor's Kinetics

Bioreaction kinetics of reactors has been widely studied. Most researchers use relatively few types of mathematical descriptions of biological reactions employing Michaelis-Menton kinetics to account for substrate and product inhibition. These incorporate a threshold for inhibition onset and one for reaction cessation.

The anaerobic fermentation of biowastes involves two distinct phases. In the first phase, hydrolysis of complex waste components, including fats, proteins, and polysaccharides to their component subunits by a heterogeneous group of facultative and anaerobic bacteria, which then act on the products of hydrolysis to fermentations, β-oxidations, and other metabolic processes leading to the formation of simple organic compounds, mainly short-chain volatile acids and alcohols (reason to be often referred as acid fermentation). In the second phase the end products of the first phase are converted to gases (methane and carbon dioxide mainly and hence referred to as "methane fermentation") by different species of strictly anaerobic bacteria. These two phases occur simultaneously and in active sinergetics in a bioreactor. The bacteria causing acid fermentation are relatively tolerant to variations in pH and temperature and have much higher rate of growth than bacteria responsible for methane fermentaion. Hence, methane fermentation is the generally dominant kinetics-controlling biogas production process in anaerobic wastes processing.

In multiphase bioreaction, the kinetics of the slowest phase is controlling the kinetics/rate of occurring of the overall reaction—bioprocessing. There is strong experimental evidence that methane fermentation of short—and long-chain fatty acids is the rate-limiting phase in anaerobic biowastes fermentation.

Thus a relevant kinetics model is to be developed which should encompass formally—physically and mathematically all relevant process phases and subphases/subprocesses, and which should inherently be able to describe governing mechanisms and dominant subprocess in both special and temporal domain.

An approach aimed at developing a more biophysically sound and mathematically descriptive kinetics model shall provide a detailed understanding of a nonlinear interactive relationships between biomass (microorganisms) concentration/growth and productivity status/activity, substrate limitation and inhibition, as well as bioreactor productivity and product inhibition.

Thermodynamics and Process Parameters

Thermodynamical and rheological properties of agricultural liquid effluents relevant for mechanical and biochemical processing including methods of their experimental determination have been studied (Kosi F., and Todorovic M., 1991 and 1995), and a sample of obtained measurements results is presented here. The intrinsic phenomenological aspects of the relevant property determination need improvements of their phisico-chemical and measuring-technical characterization as well as corresponding analytical formulations. The intrinsic time and space dependent variability of the physical and chemical properties of the studied materials were confirmed. Their complex structure, distinct polyphasic heterogeneity and anisotropicity complicate material description and repeatable measuring characterization.

Specific heat capacity of samples of livestock waste slurries has been measured by the Bunsen "ice calorimeter". The specific heat capacity of the material is obtained through measuring of the enthalpy change by cooling of the sample until the temperature of thermal equilibrium of the system "sample-calorimeter" is achieved.

Experimental determination of heat conductivity has been performed using concentric-cylinder-measuring apparatus based on the one-dimensional steady state Fourier heat conduction law. Due to the heterogeneity of manure, specific problem in performing these measurements regarding the time lapse for the steady-state measurements and the tendency of solid particles suspended in the manure to settling and wall deposit. Relevant analysis has been made, the result of which shows that due to the specific rheological properties of manure and to the tendency of solid particles to form irregularly shaped clusters, in the narrow space between the cylindrical surfaces of the measuring device appears a "local" settling, i.e. clogging which prevents a vital jeopardising of the assumed homogeneity of the sample.

Materials specific density has been determined by the direct measurement of volume V of a certain mass m of the sample previously brought to a desired temperature, assuming that intensive mechanical mixing during the measurement procedure guaranties the uniformity of materials sample temperature and concentration fields and hence the relevance of obtained density values as $\rho = M/V$.

Rheological model characteristics have been obtained by the determination of the correlation between the measured fluid shearing stresses and shear rates on the wall surface. Concentric- cylinder rotary viscometer of the Brookfield Synchro-Lectric type with four spindles has been used for the measurements. The rheological behaviour of the pseudo-plastic fluids can be described by the Ostwald de Waele equation as follows (Kosi F., and Todorovic M., 1995):

$$\tau_w = K \cdot \left(\frac{\partial v_x}{\partial y}\right)^n \tag{1}$$

where K $(Pa\ s^n)$ is the rheological consistency index and n the rheological behaviour index. The second order regression analysis is used to relate measured thermophysical properties to the temperature:

$$\pi(t) = a + b \cdot t + c \cdot t^2 \tag{2}$$

where π is values of thermophysical property and a, b and c are the constant coefficients. The performed measurements of the rheological properties confirm that manure is a pseudoplastic fluid. A significant influence of the sludge volume fraction on the values of rheological parameters K and n is observed.

Microbiological and Physicochemical Synergetics

Simultaneous fermentation and separation of an inhibitory fermentation product has been studied (Wyman E. C., 1995). The performed investigation employs an immobilized-cell biparticle fluidized bed reactor for homolactic fermentation as a relatively simple model system from which to generate predictive analytical mathematical model, including relevant parameters and data description via microbiological and physicochemical synergetics. Here conceptualized bioreactor design procedure is given in Fig. 6.1.

The crucial part of this study is aimed at enabling description of the influence of different growth stages of immobilised cells on bioreactor phenomena, and likewise the influence of transport phenomena on microbial growth and activity. Corresponding experiments have been defined, with the purpose to obtaining more complete measurement data under controlled and repeatable conditions. Current search encompasses the influences of surface morphology of bed particles with immobilized cells and of the particles effective density/specific volume on interactions between interfacial forces (both liquid/solid and gas/solid) and superficial velocities of gases and liquids.

Development of a biophysically sound set of governing equations for bioreactors necessarily encompasses micro-scale bioreaction kinetics and diffusion. Crucial for

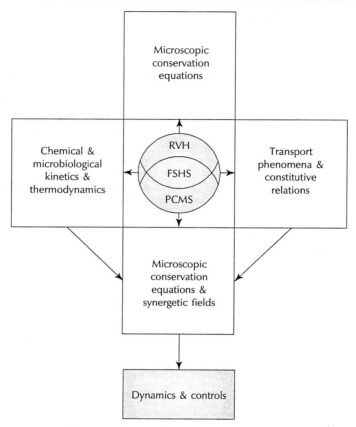

Fig. 6.1 Scheme of bioreactor design process
RVA—Representative Volume Averaging
PCMS—Physicochemical and Microbiological Synergetics
FSHS—Functional & Structural Hierarchy & Synergetics

its development is description of diffusion processes and establishment of relevant measurement methods to be performed at four levels: *pore-scale* (in pores within particles—i.e., mass transfer through capillary fluid and/or through capillaries partially filled with biomass); *particle-scale* (between particles consisting of uniform-sized or hierarchical-sized channels partially filled with fluid and biomass); *biofilm-scale* (i.e., mass transfer to immobilized cells through a diffusional boundary layer or through the biofilm formed on particles or on the membranes), and in *bulk flow* through the bioreactor.

At different scales of observation and resolution bioreactor porous system (packed bed or fluidized bed of porous particles with immobilized microbial) may appear as random or deterministic, homogeneous or inhomogeneous, and isotropic or anisotropic. Yet at a scale of observation and resolution immediately above or below this scale, the medium may appear to have completely different character.

Beside spatial distribution of porosity and heterogenity, intrinsic characteristic of bioreactor's porous systems is that they may change with time. Thus, bioreactor's substrate/particles/biomass-porous formation possesses structural hierarchy as well as a hierarchy of transport processes, i.e. functional hierarchy (Cushman J. H., 1992). However, it is often not obvious what the relation is between a hierarchical structure and the functional hierarchy. To determine the character of linkage, processes taking place within the formation have to be analysed as decomposed into subprocesses (Cushman J. H., 1992). If interacting subprocesses are successively nested then the formation is said to exhibit functional hierarchy. In addition, bioreactor's complexity is enriched on both micro (i.e., pore- and particle-size) and macro (i.e., bioreactor) levels, by the synergetics between biodynamics and physicochemical dynamics as well as their intrinsic micro-macro relations.

Closure of field equations requires the definition of constitutive variables that have meaning only with respect to an inverse problem which is defined in terms of measured field variables or rates of change of these variables. In a discrete hierarchical system closure requires instruments be designed to decouple processes between successively nested natural scales. Practically it means that the measurement process (averaging process) has to separate scales, to make possible the adequate balance laws be closed via local constitutive theory.

At the particle/immobilized biomas level the dominant problem is synergetics of the difussive transport of species to living cells and of their products to bioreactor medium and sorbent particles, as well as heat and microbial growth and microbiological and chemical reactions kinetics.

Hierarchical Volume Averaging

To describe the occurrence and correspondence of micro- and macro-phenomena and synergetism of microbiodynamic and dynamical changes in physicochemical fields, a mathematical formulation based on a hierarchical volume averaging method, has been applied in defining governing equations system.

The type of hierarchy of scales encountered in the derivation of the design equations for biparticle fluidized bed reactor is as follows. At the micro (pore and particle-size) level are defined two porous systems types: particles with immobilized bacteria—biointegrated solid/liquid and particles which are absorbing bioreactor product—solid/liquid system.

To describe hierarchical volume averaging method (Quintard M. and S. Whitaker, 1992 and Whitaker S., 1986), on which technically is based the predictive analytical modelling via microbiological and physicochemical synergetics, following development may be illustrative. The governing point equation of interest shall be (Todorovic M., et al., 1998):

$$\frac{\partial c_A}{\partial t} + \nabla \cdot N_A = R_A \tag{3}$$

The phase average of this equation is given by:

$$\frac{1}{V_A}\int_{v_{\alpha(t)}}\left(\frac{\partial c_A}{\partial t}\right)dV + \frac{1}{V_A}\int_{v_{\alpha(t)}} \nabla \cdot N_A dV = \langle R_A \rangle \tag{4}$$

General transport theorem is used to write:

$$\frac{d}{dt}\int_{V_{\alpha(t)}}C_A dV = \int_{V_{\alpha(t)}}\left(\frac{\partial c_A}{\partial t}\right)dV + \int_{A_{\alpha\omega}}C_A w.n_{\alpha\omega}dA \tag{5}$$

where w represents the velocity of the different phases interface.

Eq. (5) along with the averaging theorem and concerning the fact that the phase averaged concentration is associated with a fixed point in space following expression is obtained:

$$\frac{\partial\langle c_A\rangle}{\partial t} + \nabla\langle N_A\rangle + \frac{1}{V_A}\int_{A_{\alpha\omega}}c_A(v_A - w)n_{\alpha\omega}dA = \langle R_A\rangle \tag{6}$$

In this result the first term represents the obvious accumulation of species A, while the second term represents the convective transport. The third term represents the interfacial transport of species A and the right hand side represents the rate of production owing to chemical and microbiological reactions.

Using the divergence theorem, the averaging theorem, assumptions that convective effects at entrance and exit are large comparing to diffusive effects, with the normal decomposition of molar flux of species according to (Todorovic M., et al., 1998) Eq. (5) becomes:

$$\varepsilon_\alpha\frac{\partial\langle c_A\rangle^\alpha}{\partial t} + \langle c_\alpha\rangle^\alpha\frac{\partial\varepsilon_\alpha}{\partial t} + \nabla\cdot\left(\varepsilon_\alpha\langle c_\alpha\rangle^\alpha\langle v\rangle^\alpha\right) + \nabla\cdot\langle\bar{c}_A\bar{v}\rangle$$

$$+ \nabla\cdot\langle J_A^*\rangle + \frac{1}{V_A}\int_{A_{\alpha\omega}}C_A(v_A - w)\cdot n_{\alpha\omega}dA = \varepsilon_\alpha\langle R_A\rangle^\sigma \tag{7}$$

In given equations are c_A is species A concentration, t is time, N_A is molar flux of species A, R_A is molar rate of production per unit volume of species A owing to chemical reaction, V_A is averaging volume, V is volume, v_A is species A velocity, $\bar{c}_A$ is equal to $c_A - \langle c_A\rangle^\eta$ —concentration deviation.

This result shall be thought of as a design equation for the defined bioreactor system. To use it the volume fraction has to be known. Assuming that the solid phase is rigid, we can determine the fraction of liquid-phase by experimental methods. The velocity field has to be determined in order to evaluate the convective transport and the dispersive term is to be modeled to produce reliable longitudinal and lateral dispersive fluxes. For steady laminar flow a closure scheme allows to express the dispersive and diffusive flux as:

$$\langle\bar{c}_A\bar{v}\rangle + \langle J_A^*\rangle = -\varepsilon_\alpha D^*\cdot\nabla\langle c_A\rangle^\alpha \tag{8}$$

Here D^* is referred to as the total synergetic dispersivity tensor. J_A^* is molar diffusive flux. For completeness, the diffusion flux in Eq. (8) needs to be represented as:

$$\langle J_A^*\rangle = -\varepsilon_A D_{eff}\nabla\langle c_A\rangle^\alpha \tag{9}$$

in which virtually and synergetic estimate of D_{eff} may be satisfactory when the convective and dispersive transport dominate. The interfacial flux term in Eq. (8) represents a crucial term in the design of multiphase bioreactors and it can be given as follows:

$$\frac{1}{V}\int_{A_{\alpha\omega}}c_A(v_A - w)\cdot n_{\alpha\omega}dA = a_{\alpha\omega}\cdot K^*\cdot\left(\langle c_A\rangle^\alpha - \langle c_A\rangle_{\alpha\omega}\right) \tag{10}$$

It is dependent on the sinergetic "biomass" transfer coefficient K^*, *which* is itself mutually dependent on the diffusion process occurring at the a -ω interface that is direct result of bacterial activity kinetics and synergetics of the interfacial transport phenomena.

Diffusion process is often a high concentration, multicomponent transport process. The rate of microbiological reaction (or absorption in the case of adsorbent particle) is commonly obtained experimentally.

The expressions relevant for assumed homogeneous reaction rate are to be obtained experimentally. This is valid for the term R_A – local molar rate of production per unit volume of species A owing to chemical reaction (related volume averaged value is denoted with $\langle R_A \rangle^\gamma$ in Eq. (8).

Current approach to kinetics study aimed at advancing basic knowledge and parametric data determination is necessary for development of biophysically sound modelling, has been searched and only preliminary results are obtained. It encompasses the study of biochemical mechanisms of microbial growth and activity, and the study of mechanisms of physical processes.

The growth of immobilized microorganisms can be described generally as a function of the biomass cells concentration X, concentration of the limiting substrate S and the concentration of inhibitor I:

$$\frac{dX}{dt} = f(X, S, I) \tag{11}$$

The local rate of reaction itself—product formation by the immobilized biomass in carrier is dependent on the biomass rate of growth, its concentration and its life and activity status. Thus its volume averaged value $\langle R_A \rangle^\gamma$ can be expressed as follows:

$$\langle R_A \rangle^\gamma = f\left(\alpha_{CS}, \langle X \rangle, \left\langle \frac{dX}{dt} \right\rangle^\gamma \right) \cdot \eta \tag{12}$$

where α_{CS} = biokinetics intrinsic coefficient

$\langle X \rangle^\gamma$ = volume averaged biomass concentration

$\langle dX/dt \rangle^\gamma$ = volume averaged rate of growth of immobilized biomass.

Dynamic Programming Optimization of the Bioprocessing

According to the dynamic programming optimization technique described in (Bellman R., 1979), a particular system for which an optimum is sought, is broken down to a number of smaller series of stages and an optimal solution is obtained by making a sequence of interrelated decision processes at different stages. For the complex biochemical system of four stages with two backward junction lines, a new system of equations based on the Bellman optimality principle of dynamic programming expressed in the form (Bellman R., 1979):

$$f_n(\beta_{n+1}) = \max(g_n (\beta_{n+1}, u_n) + f_{n-1} (\beta_n)) \text{ for } u_n \in U_n, \tag{13}$$

for execution of the optimization procedure is founded.

In Eq. (13), associated with nth stage, variables β_n are defined in the domain given by non-equations of the following type:

$$B_n : \Phi_n (B_{n,1}) \leq 0 \text{ for } n = 1,2, \ldots, N \text{ and } 1 = 1,2, \ldots, L_N \tag{14}$$

are state variables of the nth stage. These variables, which, generally are multidimensional vectors with scalar physical values as "coordinates", present characteristics of the mass, energy and information flows between the stages. State variables undergo some changes in value at each stage defined by relationships:

$$\beta_n = T_n (\beta_{n+1}, u_n) \tag{15}$$

which are state-transformation functions T_n. The decision (control) variables U_n, $n = 1, 2, \ldots, N$, in the optimal control theory, the problem of prediction of the decision variables is solved by definition of the functions U_n in the domain U given by non-equations system of the following type:

$$U_n : \psi_n (u_{n,j}) \leq 0 \text{ for } n = 1, 2, \ldots, N \text{ and } j = 1,2, \ldots, J_N \tag{16}$$

Assuming the continuous bioprocessing and non-Newtonian (power-law) fluids, the relevant equations of state variable transformations affected by decision variables are adjusted to the Bellman's procedure by parameters discretisation. For the energetic use of gas produced by biochemical reaction, the overall objective function of the optimization model is defined by "the present-worth" method.

The optimization of a mesophilic anaerobic fermentation system is performed and related values of reaction temperature, retention time, overall heat transfer coefficients of the bioreactor wall, fluid velocity, dimensions of heat exchangers, gas—storage volume and total solid concentration of the treated material are determined.

The first step in applying the dynamic programming technique for optimization is suitable structuring of the process as a serial sequential process. The general multistage biochemical processing analyzed is structured as shown in Fig. 6.2.

Problem formulation of optimal value calculation (β_n, u_n, $n = 1, 2, \ldots, N$) of the process displayed in Fig. 6.2 is defined by formal system of the state variable transformation functions:

$$\beta_1 = T_1 (\beta_2, \beta_5, \beta_6, u_1) \tag{17}$$
$$\beta_2 = T_2 (\beta_3, u_2) \tag{18}$$
$$\beta_3 = T_3 (\beta_4, \beta_5, u_3) \tag{19}$$
$$\beta_4 = T_4 (\beta_7, \beta_6, u_4) \tag{20}$$
$$\beta_6 = T_5 (\beta_3, u_3) \tag{21}$$
$$\beta_7 = T_6 (\beta_7, u_4) \tag{22}$$
$$\beta_m \in B_m, m = 1, 2, \ldots, 7 \tag{23}$$
$$u_n \in Un, n = 1, 2, 3, 4 \tag{24}$$

Practical realization of the solving procedure has two phases with five steps in each of them.

The first phase is development of recurrent equations systems based on the optimality principle of dynamic programming (Eqn. 13) and solving of associated numerical problems of finding of the Bellman's functions for possible variables β_m, $m = 1, 2, \ldots, 7$ values, (because optimal values β_m are not known in advance). The

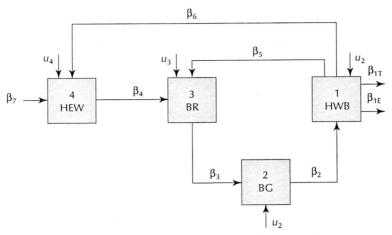

Fig. 6.2 The general structure of multistage biochemical processing
BR: Biochemical reactor, stage 3
HWB: Co-generation system, stage 1
BG: Gas low-pressure reservoir, stage 2
HEW: Heat exchanger, stage 4
β_{1T}: Thermal energy
β_{1E}: Electricity

second phase of predictions ("final calculation") consists of a direct determination of optimal values u_n, $n = 1, 2, 3, 4$ and state variables β_m, $m = 1, 2, \ldots, 7$ according to previously obtained values f_n, $n = 1, 2, 3, 4$ and equations (17) to (22) "going" in calculations from $m = 7$ to $m = 1$. Proceeding from the definition

$$f_0 (\beta_1) = 0 \qquad (25)$$

and successively applying the optimality principle (13) to different stages of the system, it is obtained the system of equations for calculating Bellman's functions:

$$f_1 (\beta_2, \beta_5, \beta_6) = \max (g_1(\beta_2, \beta_5, \beta_6, u_1)), u_1 \in U_1 \qquad (26)$$

$$f_2 (\beta_3, \beta_4, \beta_7) = \max (g_2 (\beta_2, u_2) + f_1 (\beta_2, \beta_5, \beta_6)), u_2 \in U_2 \qquad (27)$$

$$f_3 (\beta_4, \beta_7) = \max (g_3 (\beta_4, \beta_5, u_3) + f_2 (\beta_3, \beta_4, \beta_7)), u_3 \in U_3 \qquad (28)$$

$$f_4 (\beta_7) = \max (g_4 (\beta_7, u_4) + f_3 (\beta_4, \beta_7)), u_4 \in U_4 \qquad (29)$$

While calculating Bellman's functions f_n, $n = 1, 2, 3, 4$, values of f_n, $n = 2, 3, 4$ previously obtained in a definite number of "nodes" of domains of relevant arguments β_m and u_n, are included in appropriate expressions (13) to (22). Since in general case, the presumed and obtained values of variables β_m in set of grid points of the domains (14 and 16) are not corresponding, variables f_n, $n = 1, 2, 3, 4$ included in procedure of finding the conditions that give maximum (or minimum) values of functions defined in equations (27) to (29) will be approximated by interpolation polynomials of first or second order.

Final prediction of optimal decision values u_n, $n = 1,2,3,4$ and associated state variables β_m, $m = 1, 2, \ldots, 7$ is performed through the system of Eqns. (17) to (22), including the "optimal" values, obtained at the previous stage of the calculation, in the next expression.

Engineering Data Base

Relevant knowledge and engineering data base, referent to the current status of knowledge in this field and current technology development needs, presented in this paper, shall enable the following:

I. Comparison and evaluation of results of bioprocessing research, which is essential to informed decision-making regarding future research directions and technology development. Currently, no general method exists for systematic analysis of wastes—bioprocessing data, because unified evaluation procedures have yet to be practically established and there is great need for precise definition of relevant reactor parameters, for defined criteria for recognizing specific biophysical states, phases and reactions within reactors, and for describing relevant transport phenomena in biophysical terms.

II. Further development of a process engineering knowledge database which is dependent on the following:
1. Determination of which bioreactor parameters are most relevant for desired effects, and precise, repeatable measurement of these parameters;
2. Development and application of sound biophysical theories in describing such parameters;
3. Development of analytical models, predictive of actual bioreactor performance including its dynamics. Such development should occur through both experimental observations of biophysically well defined systems and explicit satisfaction of the characteristic microbiological, biophysical and thermodynamic constrains.

III. Further development of viable mathematical descriptions and practical modeling applications of governing reactor's equations system requiring data on:
1. Thermophysical and biophysical properties including kinetics and transport coefficients of bioreactors constituents (elements and multi-phase systems) as: effective stagnant thermal conductivity and diffusivity, effective specific heat capacity, dispersion coefficients, diffusion coefficients, bioreaction's kinetics coefficients, as temporal functions of biomass life-status and spatial functions of position of biocatalyst particles within the bioreactor.
2. Occurrence and correspondence of micro- and macro-phenomena and synergetism of microbiodynamics and dynamical changes in physicochemical fields.

IV. Development of effective system's optimization methodology based on the synergetics—integration of micro/macro processing and energy efficiency and economics optimization as follows:
1. Whole system's energy productivity and raw – waste materials utilization/ conversion efficiency maximization method improvement.
2. Whole system's operational and economic optimization through the all subsystems and different scenarios of structuring technologies and technical units and sub-units simulation/investigation method development.

Need for Future Research

Further research is still needed to better identify the organisms present in anaerobic digesters and improve description of their complex interactions. Specific organisms can then be selected for generic breeding and manipulation to allow operation of anaerobic digesters at optimal conditions. Feedback and control mechanisms should be further advanced to maintain effective and efficient operation of bioreactors (Todorovic M. and E. Boyce, 1995 and Todorovic M. and F. Kosi, 2000). Further R&D and engineering efforts should be undertaken to develop lagre-scale reactor designs that can process high concentrations of solids to decrease the cost of biogas production and increase stability and reliability of bioreactor productivity.

The influence of the different growth stages of immobilized cells on bioreactor phenomena, and likewise the influence of transport phenomena on microbial growth and activity, require more complete description under controlled, repeatable conditions.

The proposed study of phenomena and relevant analytical correlations that are characteristic of fluidization will be critical to further development of more generally-applicable, sound biophysical models. Studies should also be made of the influences of surface morphology of bed particles with immobilized cells — and of the particles' effective density-specific volume—on interactions between interfacial forces (both liquid/solid and gas/solid) and superficial velocities of gases and liquids. Done on both experimental and theoretical levels, such studies may possibly describe a new, more generic form of correlation.

Development of a biophysically sound set of governing equations for bioreactors must necessarily encompass **micro**-scale bioreaction kinetics and diffusion. However, **diffusion** measurements should be performed at four levels: *pore-scale* (in pores <u>within</u> particles—i.e., mass transfer through capillary fluid and/or through capillaries partially filled with biomass); *particle-scale* (<u>between</u> particles, consisting of uniform-sized or hierarchical-sized channels partially filled with fluid and biomass); *biofilm-scale* (i.e., mass transfer to immobilized cells through a diffusional boundary layer or through the biofilm formed on particles or on membranes), and in *bulk flow* (i.e., mass transfer in bulk fluid flowing through bioreactor).

The objective of further **kinetics** study must be advancement of basic knowledge and parametric data determination necessary for development of biophysically sound modelling. This must encompass the study of biochemical mechanisms of microbial growth and activity, and the study of mechanisms of physical processes. Both types of process occur in well-defined fields of relevant bio- and chemico-physical properties, as initially approached by Williamson and McCarty and Moo -Young and coauthors (Todorovic M. and E. Boyce, 1995).

Summary and Conclusion

Simultaneous fermentation and separation of an inhibitory fermentation product, in a bipartcle fluidized bed reactor containing immobilized cells increases productivity because potentially-inhibitory products can be easily removed from the system. However, predictive, analytical reactor modeling has been hampered by limited data on biophysical and thermophysical properties, including reactor kinetics and transport phenomena, and by incomplete understanding of the biophysics of synergetic microbiological and physicochemical processes. In this paper, a data base frame with a structural scheme and content based on available data have been presented.

A path of analysis that originates with the continuum axioms for the mass and momentum of multicomponent systems, has been presented. Bioreactor porous system has been analysed in terms of synergetics of physicochemical and microbiological processes.

Describing the occurrence and correspondence of micro- and macro-phenomena and synergetism of microbiodynamic and dynamical changes in physico-chemical fields, a mathematical formulation based on a hierarchical volume averaging method, has been formulated and applied in defining governing equations.

In addition to this, an approach to kinetics study aimed at advancing basic knowledge and parametric data determination necessary for development of bio-physically sound modelling, has been searched and obtained results are given. It encompasses the study of biochemical mechanisms of microbial growth and activity, and the study of mechanisms of physical processes.

Methods of closure have not yet been fully developed. However, presented innovative concept of hierarchical volume averaging and synergetics shall serve as a tool to account more accurately for the relevant biophysics and to trace more appropriate way to physically sound experimenting and predictive modeling.

Finally, we can stress that permanent knowledge advances, knowledgeable managed shall help to successfully develop further and implement biowastes processing to fuels and energy. Optimally planning management and disposal of the growing quantities of waste materials and economically converting it to renewable fuels is a difficult but essential task to be performed. Searching and finding acceptable scenarios of wastes management and disposal, reducing waste "leaks", preventing toxic waste streams from entering the environment, increasing waste use for energy production we can minimise disposal needs, and we may believe that we shall be closer to the frontier of energy sustainable society.

REFERENCES

1. Andrews J. F. (1997), Dynamic Models and Control Strategies for Wastewater Treatment Plants – An Overview, *Proceedings of the International Federation of Automatic Control Symposium on Environmental Systems Planning,* Design and Control, Kyoto, Japan.

2. Bellman R. (1979), *Dynamic Programming,* Princeton University Press, Princeton, N.Y..

3. Benefild L.D., W.R. Clifford (1980), *Biological Process Design for Wastewater Treatment,* Prentice-Hall Series in Environmental Sciences.

4. Cushman J.H. (1992), Hierarchical Problems: Some Conceptual Difficulties in the Development of Transport Equations, *Heat and Mass Transfer in Porous Media,* Eds. M. Quintard and M. Todorovic, Elsevier Science Publ., p., 123-136, 99-110, Amsterdam.

5. Kosi F., Todorovic M. (1991), Experimental Investigation of Rheological Model Characteristics of Non-Newtonian Fluids, *Proceedings of the International Conference "Energy Efficiency 2000",* Belgrade, p. 253-264.

6. Kosi F., Todorovic M. (1995), Experimental Determination of Specific Heat Capacity of Livestock Waste Slurries, *Contemporary Agricultural Engineering,* No. 1, p. 9-14.

7. Petersen N.J[*]. and Davison H.B. (1991), Modelling of an Immobilised-cell Three Phase Fluidised -Bed reactor, *Applied Biochem. Biotechnol.* 28/29, p. 685-698.

8. Quintard M⋅., and Whitaker S. (1992), Transport Processes in Ordered and Disordered Porous Media, *Heat and Mass Transfer in Porous Media*, Editors M. Quintard and M. Todorovic, Elsevier Science Publ., p. 99-110, Amsterdam.

9. Todorovic M., Boyce E., Kosi F., Simic Lj. (1996), Anaerobic Bioreactors Producing Chemicals and Fuels—Innovations and Engineering Database, *Proceedings of the 9th European Bioenergy Conference*, Vol.3, p. 1560-1565, Copenhagen.

10. Todorovic M., F. Kosi, Grover Velma, B.K. Guha, W. Hogland, S. McRae (2000), *Solid Waste Management*, Special Indian Edition, Oxford & IBH Publishing Co. Pvt. Ltd, New Delhi, Calcutta.

11. Todorovic M., F. Kosi, Simic Lj. (1998), Bioreactors for Fuels and Chemical Production – Predictive Analytical Modelling via Microbiological and Psysicochemical Synergetics, *The 10th European Conference end Technology Exhibition—Biomass for Energy and Industry*, pp. 669-672, Wuerzburg, Germany.

12. Todorovic, M., and E. Boyce (1995). Anaerobic Bioreactors: Engineering Database and Mathematical Modeling of Bioreactors Producing Chemicals and Fuels. *School of Engineering and Pharmaceutical Chemistry & Center for Biomedical Research*, KU—Lawrence.

13. Todorovic, M., F. Kosi, M. Stojanovic, M. Niksic, and M. Terzic (1993). The Investigation of Bioengineering Fundamentals of Anaerobic Fermentations", *Research Project Final Report, Serbian Academy of Sciences and Arts*, Belgrade.

14. Whitaker S. (1986), Transport Processes with Heterogeneous Reaction, Department of Chemical Engineering, *Chemical Reactor Analysis: Concepts and Design*, Editor S. Whitaker and A.E. Cassano, Gordon and Breach, New York.

15. Wyman, E.C. (1994), Alternative Fuels from Biomass and their Impact on Cabron Dioxide Accumulation; *Applied Biochemistry and Biotechnology*, pp. 897-915, Vol. 45/46.

16. Wyman, E. C. (1995), Progress on Ethanol Production from Lignocellulisic Biomass, *Alternative Fuel Division, National Renewable Energy Laboratory*, Golden, CO 80401.

NOMENCLATURE

A	surface area of the species A body (m^2)
$A_{\alpha\omega}$	area of the α-ω interface (m^2)
$\mathbf{B}_n$	domain of variables β_n
c_A	species A concentration (moles/m^3)
$\langle c_A \rangle$	spatial average concentration or phase average concentration for the α-phase (moles/m^3)
$\langle c_A \rangle^\alpha$	intrinsic phase average concentration for the α-phase (moles/m^3)
$\bar{c}_A$	$c_A - \langle c_A \rangle^\eta$, concentration deviation (moles/m^3)
D_{eff}	effective diffusivity tensor (m^2/s)
$D^\bullet$	the total synergetic dispersively tensor
$J_A^\bullet$	molar diffusive flux (moles/m s)
K	rheological consistency index
$K^\bullet$	synergetic mass transfer coefficient
$n_{\alpha\omega}$	unit normal vector pointing from the α-phase into the ω-phase
N_A	molar flux of species A (moles/m^2 s)
R_A	molar rate of production per unit volume of species A owing to chemical reaction (moles/m^2 s)
T_n	state-transformation functions

t	time (s)
u_n	decision variables
U_n	domain of decision variables u_n
V	volume (m^3)
V_A	averaging volume (m^3)
V_α	averaging volume for the α-phase (m^3)
v_A	species A velocity (m/s)
$\bar{v}$	species velocity resulting from chemical reaction (m/s)
w	the velocity of the different phases interface

Greek Letters

β_n	state variables of the nth stage
ε_α	volume fraction of the α-phase
ε_A	species A density (kg/m^3)
π	values of thermophysical property
τ_w	species A viscous stress tensor (N/m^2)

7

Electricity from Biomass and Biological Wastes: European Scenario

Emmanuel G. Koukios and Kyriakos D. Panopoulos

Bioresource Technology Unit, Department of Chemical Engineering
National Technical University of Athens, Zografou Campus, Heroon
Polytexnou 9 GR-15700 Athens, Greece

Introduction

The White Paper sets 12 % of primary energy consumption in the EU as a target for renewable energy development. Translating this target to the part of electricity, means that in 2010, over 20% of the electricity produced in Europe has to be generated from renewable energy sources. According to the White Paper, biomass is going to contribute almost half of the renewable electricity in 2010 (8% of European electricity production or 230 TWh_e bioelectricity annually) [European Commission 1996]. Nevertheless, very few things have been implemented towards a fully commercial EU bioelectricity scene since the White Paper statement and the Kyoto Protocol agreement.

The lion's share of these 230 TWh_e is most probably going to be produced from so-called thermochemical biomass-to-energy technologies, i.e. combustion (including co-combustion with solid fossil fuels in existing large power stations), and gasification in combined cycle mode for larger installations or coupled with internal combustion engines for smaller units. Several demonstration projects have been commissioned to illustrate the most advanced technology available i.e. the I.G.C.C. which attracted the interest of utilities in Europe. More specifically these have been the EU funded I.G.C.C. project at Varnamo (Sweden) and the similar ones under construction: the ARBRE project in Yorkshire (UK), and in Pisa (Italy) [Maniatis and Millich 1998].

Combustion and gasification are proven technologies and resemble, more than the rest of the renewable energy technologies, the traditional electric utility practice. Apart from the utilities, the other key players in bioenergy are the agricultural population of a country, and the industry including the SME's (Small Medium Enterprices). Most of the bioelectricity installations are going to be fuelled by solid biofuels which derive from agro-industrial activity i.e. agricultural residues (AR), forestry residues (FR), new high yield energy crops (EC), and also the organic fraction of municipal wastes (MSW). The White Paper also estimates the utilization of some 15 Mtoe of landfill gas (LG). A large number of power stations are to be built by either SMEs, if the station's size is relatively limited and also by the large utilities, in the cases of larger power stations.

Availability of Biomass for Electricity in the EU

The main source for bioelectricity production derives from solid bio-fuels. Each country shows a different profile of available and possible bio-fuels. This depicts the country's economy based on agriculture, forest exploitation and industrialization. This potential is simply evaluated according to the following assumptions about the types of solid bio-fuels. These were taken rather unfavourably in order to have a view of the actual technically and possibly economically achievable potential.

- **Agricultural Residues.** (AR) from existing plantations of food and non-food crops. The technical potential can be directly estimated using an average factor of 2.5 t dry matter/ha of cultivated land. This average factor is used across the board, although in several cases cultivated land can give a lot more residues [Faaij et al. 1998] [Gomes et al., 1997] [Kadam et al., 2000]. The cultivated land in the EU member states is annually recorded in the Annual Statistical Reviews of EUROSTAT [Eurostat: Agriculture 1999]. An average energy content of agricultural residues is 14.5 GJ/t [Bussam, N. El. 1998].

- **Forestry Residues** (FR). These include residues in situ as well as wood processing residues. A very conservative average factor of 1 t dry matter/ha was used. This by far satisfies the need for sustainable utilization of forests. Again, the forestland in the EU member states is annually recorded by Eurostat [Kadam et al., 2000]. The average energy content of wood forestry residues is 17 GJ/t [Eurostat: Regions 1997].

- **Municipal Solid Wastes.** The amounts of MSW are recorded annually by Eurostat [Eurostat: Regions 1997]. The average energy content can be taken as 7 GJ/t i.e. considering only half of their content as organic material. This value as well as the amount of MSW exhibit great fluctuation from region to region and also according to the period of the year.

- Finally, the implementation of **energy crops** is taken into account as well. These crops can, on average, reach 10 t dry matter/ha. Examples of energy crops are giant reed, eucalyptus, miscanthus, sweet and fibre sorghum, switch-grass miscanthus, etc. Their average energy content is 17 GJ/t [Bussam, N. El. 1998]. As a first step these could be cultivated in present fallow lands. Following the Common Agricultural Policy in the EU incentives, the food crop's cultivation is going to be cut down in the years to come thus allowing even more non-food crop's implementation (one kind of which are Energy Crops).

These primary data are given in Table 7.1.

According to the above, the overall technically feasible bio-electricity potential for the EU thus estimated at 719 TWh$_e$ (for a clearer view the potential is translated into electricity terms using a conversion efficiency factor of 30%). The contribution of the different bio-fuels can be seen in Fig. 7.1 whereas Fig. 7.2 illustrates each EU-member state's profile.

It is clear from Fig. 7.1 that some countries should play a significant role in bio-electricity development within the EU. These countries are Spain, France, Germany, Italy and the UK.

Table 7.1. Primary data for the evaluation of biomass-to-energy potential in European countries.

	Agricultural Land* (1000 ha)	Forest Land* (1000 ha)	Available Land for Energy Crops* (1000 ha)	Available MSW** (1000 tonnes/yr)
AU	3337	3877	75	4783
BE	1362	620	13	4000
DK	2580	417	141	2532
FI	1984	20032	166	3100
FR	29301	15034	867	27000
GE	16639	10741	696	21615
GR	2998	3359	467	3200
IR	4193	570	132	1106
LX	126	89	1	190
IT	14185	6821	500	20033
NL	1836	334	12	9175
PO	3046	2875	921	3270
SP	25917	10662	3732	14296
SV	2954	24425	223	3180
UK	15825	2469	33	20000
EU15	**126283**	**102325**	**7979**	**137480**

*[Eurostat: Agriculture 1999]
**[Eurostat: Regions 1997]

The Electricity Generation Context (Year 1995). Focus on Biomass

In order to proceed to the distribution of biomass-to-electricity in the year 2010 among the EU member states, one should have in mind the picture of Electricity Generation in the Year 1995 [Energy in Europe 1999], and what is the penetration of biomass. Table 7.2 summarizes several important figures for electricity production in Europe.

The EU member states have different greenhouse emissions (mainly CO_2) reduction obligations within the EU. It can be seen that Germany, Italy and the UK are alone responsible for more than 50% of the CO_2 from electricity production in EU and their CO_2 reduction obligations are –21%, -6.5% and –12.5% respectively (for the year 2010 compared to 1990 figures). Nevertheless, from this group of countries only Germany seems to have enough megawatts of bio-electricity installed. Austria and Sweden, leaders within the EU in biomass-to-energy utilization are very strong in biomass for heat but exhibit little in electricity production. Finland on the other hand is the only country that utilizes substantial proportion of its potential for production of bio-electricity. Last but not least it is important to see that the smaller but more technologically advanced countries in the EU, such as Denmark, The Netherlands and Belgium have seized the opportunity to exploit some of their potential for bio-electricity. It is clear that countries that have already made an early start in bioenergy will most certainly continue to more towards the 2010 common target.

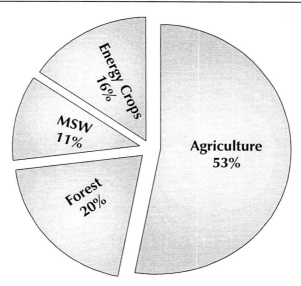

Fig. 7.1 Potential biomass-to-electricity distribution according to each source investigated in EU. Total Potential 719 TWH$_e$.

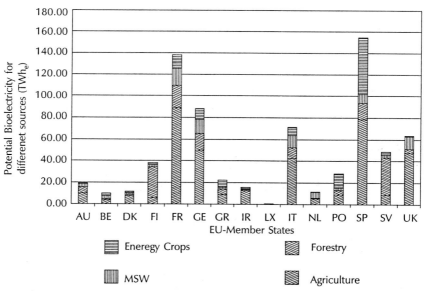

Fig. 7.2 Biomass-to-Electricity Potential of each EU member state.

Scenarios Investigated for Bio-Electricity in EU-2010

Four scenarios were investigated in two directions:
- The distribution of bio-electricity in each member state.

Table 7.2. The Electricity Generation Context (Year 1995). Focus on Biomass.

	Power Genera- tion 1995 (TWH$_e$)	Co$_2$ from Electri- city 1995 (Mtonnes)	% of European electricity CO$_2$	Biomass for power production (Mtoe)	Biomass in gross inland energy consumption (Mtoe)	Bio-electri- city % of power production	Utilization of bio-electri- city % of potential
AU	56.58	11.260	1.2	0.33	3.5	2.6	5.9
BE	74.42	22.880	2.5	0.5	0.6	3.0	23.2
DK	36.78	29.080	3.1	0.6	1.5	6.8	17.6
FI	63.87	20.630	2.2	1.19	5.7	8.5	10.8
FR	494.62	27.520	3.0	1.01	10.5	1.0	2.6
GE	555.24	318.070	34.4	1.3	5.9	1.0	5.2
GR	41.54	38.920	4.2	0.1	0.9	0.9	1.6
IR	17.86	13.440	1.5	0.1	0.2	2.6	2.2
LX	1.24	0.38	0.0	0.03	0.0	8.1	16.6
IT	241.44	125.860	13.6	0.27	6.7	0.5	1.3
NL	81.06	48.900	5.3	0.6	1.4	3.4	18.1
PO	33.26	19.140	2.1	0.15	2.4	2.0	1.9
SP	167.04	69.450	7.5	0.53	3.8	1.4	1.2
SV	148.32	6.090	0.7	1.56	7.5	3.2	11.2
UK	333.99	174.200	18.8	0.66	1.6	0.9	3.6
EU-15	**2347.26**	**925.820**	**100.0**	**8.93**	**52.3**	**1.7**	**4.3**

- The distribution of power station sizes and technology choice in each member state.

Each scenario favours one of the "Players" identified in the business of bio-electricity.

The main "Players" identified are:

- *Farmers and Foresters*. Favoured by large numbers of small-scale bio-electricity power stations. Big centralized power stations will force farmers to shoulder part of the logistical problems involved with the transportation and harvesting of biomass. Small-scale power stations are also in favour of residue's utilization rather than changing to energy crops. This seems to be easier for the farmer as they will not have to change their cultivation habits and move to an inexperienced field coupled with the initial costs related with new equipment. (**Scenario 1 Rural Development**).
- *SMEs*. Favoured by small extent of co-firing and medium sized power stations. In this scenario SMEs can do business by building and operating medium sized power stations. (**Scenario 3 Technological Development**).
- *Utility Companies*. Favoured by high co-combustion and large-scale dedicated biomass based power stations, probably I.G.C.C. (**Scenario 4 Utilities Centered Development**).

In addition a multi-player scenario was investigated (**Scenario 2 Balanced Development**).

The conversion technologies under investigation were:

1) Combustion steam cycle
2) Gasification IGCC
3) Gasification coupled with internal combustion engine
4) Co-combustion in large existing coal-fired power stations

The corresponding sizes, efficiencies of the above technologies (numbered 1 to 4) are shown in Table 7.3.

Table 7.3. Typical plant sizes and efficiencies for each technology

Typical plant sizes (MW_e):	Range (MW_e)	Efficiency	Conversion technology
1	0.5-2	20 %	[3]
5	2-10	27 %	[1], [2], [3]
20	10-30	30 %	[1], [2]
40	30-50	40 %	[2]

Furthermore, co-combustion in existing large coal-fired power station plants was estimated with efficiency:

$$\text{Efficiency} = 0.3\ ^x\ (\%\ \text{of biomass}) + 0.35\ ^x\ (\%\ \text{of coal})$$

in order to take into account the reduction in efficiency by the of use of biomass.

In each of the four scenarios, different minimum ratios of co-combustion, common for each country for the same scenario, were decided:

$$\frac{\text{biomass}}{\text{coal} + \text{biomass}}$$

The common target for each scenario is the total EU bio-electricity (including co-combustion) to be equal with 230 TWh_e (White Paper Target). The distribution of bio-electricity to the member states in 2010 is calculated proportional to the present (i.e. 1995) utilization of biomass-to-energy and the geometrical mean of the amount of CO_2 coming from power production in that country and the potential bio-electricity.

Finally, each scenario has a different arrangement for the distribution of technologies and power station sizes of dedicated power stations according to its favoured player. Figure 7.3 illustrates schematically the methodology followed in each scenario and Table 7.4 gives in detail the profile of each scenario.

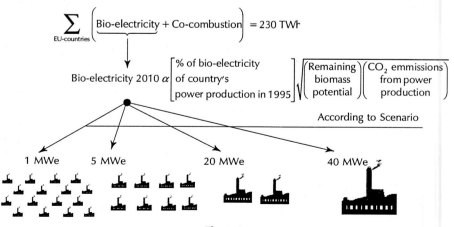

Fig. 7.3

The different scenarios are presented in Table 7.4.

Table 7.4. The profile of four scenarios investigated for bio-electricity in EU in 2010.

Scenario 1 Rural Development	Co-combustion 2.5% of coal based installed capacity. Distribution of power production by dedicated bio-electricity stations:			
	Plant Sizes (MW$_e$)	Range (MW$_e$)	efficiency %	Country's Bio- electricity %
	1	0.5-2	20	40
	5	2-10	27	20
	20	10-30	32	20
	40	30-50	40	20
	Emphasis on many small gasification systems with internal combustion engine.			
Scenario 2 Balanced Development	5	2-10	27	40
	20	10-30	32	20
	40	30-50	40	20
	Balanced emphasis on all conversion technologies.			
Scenario 3 Technological Development	Co-combustion 5% of coal based installed capacity. Distribution of power production by dedicated bio-electricity stations:			
	Plant Sizes (MW$_e$)	Range (MW$_e$)	efficiency %	Country's Bio- electricity %
	1	0.5-2	20	20
	5	2-10	27	20
	20	10-30	32	40
	40	30-50	40	20
	Emphasis on Combustion Steam Cycle technologies			
Scenario 4 Utilties Centered Development	Co-combustion 10% of coal based installed capacity. Distribution of power production by dedicated bio-electricity stations:			
	Plant Sizes (MW$_e$)	Range (MW$_e$)	efficiency %	Country's Bio- electricity %
	1	0.5-2	20	20
	5	2-10	27	20
	20	10-30	32	20
	40	30-50	40	40
	Emphasis on large IGCC power stations and co-combustion.			

5. Scenarios and Discussion

Scenario 1

This scenario is illustrated in Fig. 7.4. It is clear that due to the low efficiency operation of the smaller, and favoured here, power stations the utilization of the potential reaches for the total EU 36% of the total estimated for power production. More specifically this would mean that Germany, Finland, Italy and Spain would need to exploit 67%, 38%, 32% and 26% respectively of their biomass potential for power generation. Especially for Germany this would mean that around 10% of their own power production should derive from biomass. Under this scenario these countries would have to incorporate into their electrical grid system more than 2000, 600, 1000 and 1700 small 1 MWe power units respectively. This is clearly very difficult to implement as far as the technical difficulties that might arise to the stability and control of the electrical grid system because of so many smaller units connected to it. On the other hand this scenario could well fit into agricultural countries like Greece or Ireland where utilization of local residues could feed around 400 and 300 small units respectively in each of these two countries. Portugal on the other hand would not be favoured by this scenario because most of their potential sources are from energy crops implementation. Such schemes would normally require multi-MW systems surrounded by the energy plantation. Summing up, this scenario fits the implementation of bio-electricity in rural and remote areas where limited logistical margin is available.

Scenario 2

Going in to scenario 2, illustrated in Fig. 7.5, we can clearly see what is the effect of increased co-firing (7.5% of the total fuel input) in existing coal power stations, specially in countries which base there power production on this fuel. Germany and the UK would have to produce 22.5 TWhe and 11.06 TWhe by co-firing respectively. Spain, Italy as well as France, due to their small present dependence on coal have small co-firing capabilities. They would have to rely mostly on power stations of around 5 MWe i.e. more than 280, 160, and 130 respectively. In cases of countries like France, which has already based its electricity production in nuclear power and has a surplus, it is not normal to expect the planning of 150 power stations of that size for the next 10 years. This could partially be possible if the electricity market is opened completely and bio-electricity from France could, for example, be sold to other member states as part of the Kyoto emission trading methods.

Scenario 3

This scenario is clearly one of the most realistic as far as power station numbers and sizes are concerned. Almost 100 power stations of around 20 MWe and 26 of 40 MWe are necessary in Germany to cover most of the country's bio-electricity. (From Fig. 7.6 it is clear that other countries that should play a major role in bio-electricity implementation include Spain and the UK.)

Scenario 4

This scenario clearly favours large utilities. Germany and the UK, due to their large dependence on coal-fired power stations overtake by far the other countries in production of bio-electricity within the EU, producing together almost 100 TWhe from which 35 TWhe comes from co-combustion in existing coal fired power stations. This scenario favouring utilities foresees that 20% of Germany's electricity production should come from biomass. Spain would have to cover with bio-electricity around 15% of its electricity demand in 2010. Considerable amounts of bio-electricity is going to be produced by co-firing in the UK and lesser but significant (because of their coal based production) in, Belgium, Greece, the Netherlands, and Denmark.

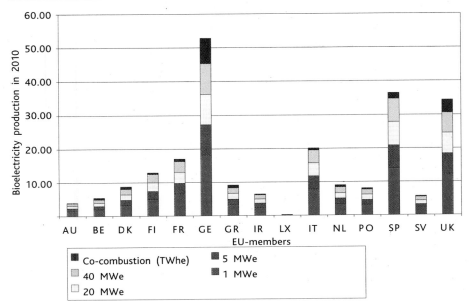

Fig. 7.4 Scenario 1. Bio-electricity production by different size power stations in the EU-member countries.

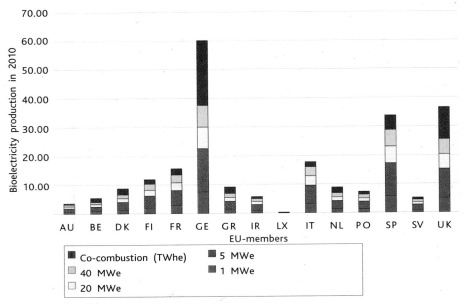

Fig. 7.5 Scenario 2. Bio-electricity production by different size power stations in the EU-member countries.

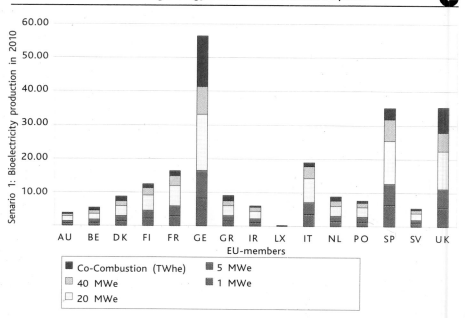

Fig. 7.6 Scenario 3. Bio-electricity production by different size power stations in the EU-member countries.

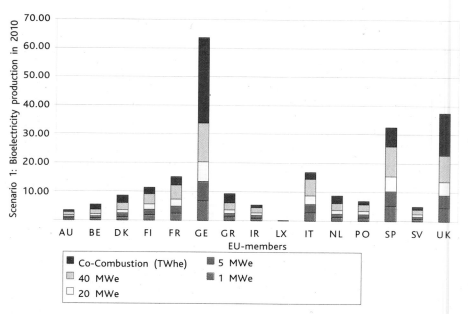

Fig. 7.7 Scenario 4. Bio-electricity by different size power stations in the EU-member countries.

Discussion

All four scenarios show clearly that several countries due to their larger CO_2 emissions and their potential bioenergy should take over more of the suggested 230 TWhe bioelectricity within Europe. The two leading countries (according to the scenarios) Germany and Spain have already adopted the most favourable renewable energy legislation amongst European countries. In the UK the large-scale utilization of natural gas and the free market mechanism adopted so far is rather difficult to help towards the deployment of bio-electricity in such extent proposed by this work. Austria, Sweden, Finland and Denmark are probably going to be the countries with the most experience in bioenergy utilization and are already playing a major role in spreading and providing the necessary technology. France although it has a considerable potential would have to sell the bio-electricity to other European countries, as it already exhibits surplus production. Through energy market liberisation there could be a way for Europe to take advantage of the large bio-electric capabilities of France. Portugal's largest share of potential lies in the implementation of energy crops. The country has a lot of experience in eucalyptus plantations and could therefore proceed more easily than others.

Looking at Europe as a whole, the White Paper target suggests (according to Scenario 4 for example) that almost 200 bio-electricity stations should be built in the size of around 40 MWe and another 200 of 20 MWe. This is clearly a target very difficult to implement and would require a rapid boost of the most advanced technology in bioelectricity. So far, i.e. year 2000, this is not the case. This boost has to take place in the following ten years otherwise the optimistic EU White Paper targets for renewable energy (based in large proportion to bionenergy) is going to be fulfilled in a small fraction only.

REFERENCES

1. Bassam, N. El. (1998). *Energy Plant Species*, James and James, UK, London.
2. European Commission. (1996). *An Energy Policy for the European Union, White Paper of the European Commission*.
3. Eurostat, Regions: *Statistical Yearbook 1997*, European Commission.
4. Eurostat. (1999). *Agriculture: Statistical Yearbook 1999*. European Commission.
5. *Energy in Europe: 1999 – Annual Energy Review*, Directorate General for Energy, European Commission.
6. Faaij, A., Steetskamp, I., Wick, A.V., Turkenburg, W. (1998). Exploration of the land potential for the production of biomass for energy in the Netherlands, *Biomass and Bioenergy*, **15**: 439-456.
7. Gomes, R., Wilson, P., Coates W., Fox, R. (1997). Cotton (*Gossypium*) plant residue for industrial fuel. An economic assessment, *Industrial Crops and Products*, **7**: 1-8.
8. Kadam, K., Forrest L., Jacobson A. (2000). Rice straw as a lignocellulosic resource: collection, processing, transportation, and environmental aspects, *Biomass and Bioenergy* **18**: 369-389.
9. Maniatis, K. and Millich, E. (1998). Energy from biomass and waste: the contribution of utility scale biomass gasification plants, *Biomass and Bioenergy*, **15**: 195-200.

8

Electrokinetic Soil Processing for Energy from Waste

Kyoung-Woong Kim and Soon-Oh Kim

Department of Environmental Science and Engineering, Kwangju Institute of Science and Technology (K-JIST), 1 Oryong-dong, Puk-gu, Kwangju 500-712, Republic of Korea

Soils can be contaminated with heavy metals derived from various sources including abandoned mining wastes, improper treatment of industrial wastes, incomplete collection of used batteries, leakage of landfill leachate, accidental spills and military activities (Adriano, 1986). The contamination often affects a large volume of soil underlying several acres of the surface area. There are also various types of contaminated lands such as paddy fields, farms, factory sites, mine fields and residential districts. Contaminants migrating from these sources threaten the human health and groundwater supply in the local area. However, technologies for decontaminating these sites have not been well developed. In addition, it has been recently reported that soil contamination is increasing in various sites such as residential areas near industrial complexes and reservoirs of drinking water.

Decontamination of hazardous waste sites is one of the most important technological challenges, and newly developed techniques for remedying soils can be generally classified into two groups. The first one is the biological remediation which has been mainly used to detoxify organic contaminants. The other is the physico-chemical decontamination that has been usually applied to remove inorganic contaminants (heavy metals). This includes excavation, soil washing and flushing, solidification and stabilization, electrokinetic soil processing (Chambers *et al.*, 1991). Although many techniques have been proposed for removing contaminants from waste sites, most suffer from several technical or economic disadvantages (Probstein and Renaud, 1986; Shapiro, 1990). A major limitation of the most successful remediation technologies, such as vacuum extraction and soil flushing, is that they are restricted to soils with high hydraulic conductivity and hence, cannot be used for fine-grained deposits. Furthermore, they are not specifically effective in removing contaminants adsorbed on the soil particles (such as pump-and-treat). Such adsorption may pose a threat for ground-water and plant contamination (Alshawabkeh, 1994).

Electrokinetic soil processing is also called electrokinetic remediation, electroreclamation, electrochemical decontamination. It needs low-level direct current in the order of mA/cm^2 of cross-sectional area between the electrodes to remove contaminants from soils. The low-level direct current results in physico-chemical and hydrological changes in the soil mass, leading to species transport by coupled mechanisms. Electrokinetic soil processing is an effective technology for removing contaminants in low-permeability soils ranging from clay to clayey sand. The advantages of this technology are its low cost of operation and its potential applicability to a wide range of contaminants (Pamukcu and Wittle, 1994). Electrokinetic soil processing is envisioned for the removal/separation of organic and inorganic contaminants and radionuclides. The potential of the technique for waste remediation has resulted in several studies (Lageman, 1993; Acar *et al.*, 1995; Hsu, 1997; Alshawabkeh *et al.*, 1999; Kim *et al.*, 2000a; Kim *et al.*, 2000b; Kim *et al.*, 2000c). Electrokinetic remediation technology has recently made significant strides and has been tested for commercial application in the United States and the Netherlands. The company named Geokinetics (The Netherlands) has successfully completed several field studies (Lageman, 1993) ; Electrokinetics Inc. (Baton Rouge, LA) has completed several large-scale pilot studies using 2-4 ton soil specimens (Acar and Alshawabkeh, 1993).

Most metal contaminants are positive ionic compounds in soil-water-electrolyte system, and they migrate towards cathode when the electric field is applied on the system. Furthermore, these metal contaminants are removed through the cathode effluent solutions from the contaminated soils. The cathode effluent solutions should be properly treated in order to recover metals contained in these solutions. Several technologies have been proposed to clean wastewater and recover metals from effluent solutions. Recently, ion exchange and electrochemical technologies, such as electrolysis, electrodialysis, and electrowinning, are widely used. Despite the fact that these technologies have significantly become available, the extent of application of these technologies remains limited in metal recovery applications, largely because of the fundamental limitations of each individual process. By combining the different processes, these restrictions can sometimes be overcome and the range of application considerably extended.

This chapter contains two sections. The first section describes various electrokinetic phenomena in soils and principles of electrokinetic soil processing. Also, the applications of this technology are discussed. In the second section, the principles and applications of various metal recovery technologies are briefly presented.

Electrokinetic Soil Processing

Electrokinetic phenomena in soils

Generally, clay particles have a negative surface charge in the clay-water-electrolyte system. This surface charge can be developed by different ways, including the presence of broken bonds and isomorphous substitution, and the cations are electrostatically attracted to the clay particle resulting in their accumulation close to the surface to satisfy electroneutrality. The accumulation of cations is opposed by the diffusional force which tries to equalize its concentration everywhere. Hence, the diffusional force acts in the direction opposite to that of the electrostatic force. The anions in the suspension will experience an electrostatic repulsion by the negatively charged clay surface and a diffusional force in the opposite direction tending to equalize the concentration. This results in the formation of a diffuse cloud of ions surrounding the clay particles. The charged surface and the distributed charge in the adjacent phase are together termed as the diffuse double layer. The concept of a diffuse double layer is schematically shown in Fig. 8.1.

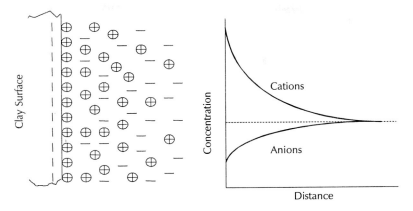

Fig. 8.1 Distribution of ions adjacent to clay surface and the diffuse double layer.

Coupling between electrical and hydraulic flows can be responsible for five electrokinetic phenomena in porous media such as fine-grained soils. Electrokinetics is defined as the physicochemical transport of charge, action of charged particles, and effects of applied electrical potential on fluid transport in porous media. In short, these electrokinetic phenomena can be defined with respect to the movement of one phase relative to the other. Each involves relative movements of electricity, charged surfaces, and liquid phases, as shown schematically in Fig. 8.2.

Electromigration or Ionic Migration

Ions and polar molecules in the pore fluid migrate under an electric field. This transport of species is called electromigration or ionic migration in Fig. 8.2(a). When an array of electrodes (cathodes and anodes) is inserted into the soil with a potential difference of a few hundred volts per meter approximately, cations (e.g., metal ions) move to the cathode whereas anions migrate to the anode.

Electro-osmosis

When an electrical potential is applied across a wet soil, there is a movement of water relative to the soil. This flow is termed as electro-osmosis in Fig. 8.2(b). As cations migrate towards the cathode and anions towards the anode, they carry their water of hydration and drag water molecules around them. Since the clay particles are negatively charged, there are more cations than anions. Hence, the net movement of water is toward the cathode. Electro-osmosis is only effective for low-permeability, fine-grained soil that has a hydraulic conductivity of less than 1×10^{-5} cm/sec. For a hydraulic conductivity greater than 1×10^{-5} cm/sec, the electro-osmotic effect is nullified by backflow from the cathode.

Streaming Potential

When water flows through a saturated clay soil under a hydraulic gradient, the water movement causes the displacement of double layer charges in the direction of

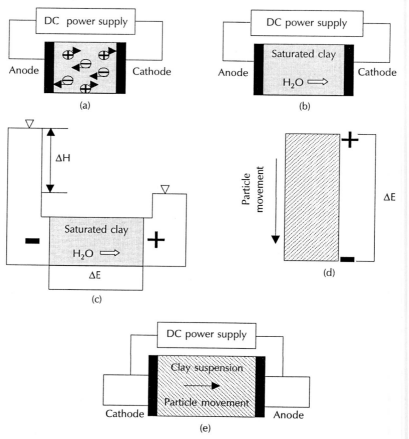

Fig. 8.2 Electrokinetic phenomena in soils (Mitchell, 1993) (a) Electromigration or Ionic migration. (b) Electro-osmosis. (c) Streaming potential. (d) Migration or Sedimentation potential. (e) Electrophoresis.

flow, resulting in an electrical potential gradient in the direction opposite to that of the flow of water. This water flow-induced electrical potential is denoted as the streaming potential in Fig. 8.2(c). Streaming potentials have been recorded in the order of millivolts in clays.

Migration or Sedimentation Potential

The movement of charged particles such as clay generates an electrical potential difference. An example is gravitational settling of sediments shown in Fig. 8.2(d). As the clay particle settles, the water drags and retards the movement of the diffuse layer cations relative to the particles. This results in the development of an electrical potential that is measured across the top and bottom of the specimen. This electrical potential generated by the movement of particles suspended in a liquid is called the migration or sedimentation potential.

Electrophoresis

When a direct current electric field is imposed on a suspension of charged clay particles or colloids, charged particles are electrostatically attracted to one of the electrode and repelled from the other. The negatively charged clay particles move towards the anode. This movement of charged solid particles or colloids is called electrophoresis in Fig. 8.2(e). In a compact system such as clay, electrophoresis is not significant, because colloids and solid particles are large in size and exhibit higher frictional drag force in motion.

Fundamental Principles of Electrokinetic Soil Processing

Electrokinetic soil processing uses a low level direct current in the order of mA/cm^2 of cross-sectional area between the electrodes to transport and remove contaminants from soils. The low level direct current results in physicochemical and hydrological changes in the soil mass, leading to species transports by coupled mechanisms. Electrolysis of water produces hydrogen ions in the anode compartment, which causes an acid front to migrate through the soil. This, in turn, causes contaminants to be desorbed from the surfaces of soil particles, and results in an initiation of electromigration. The applied electrical potential also leads to the process of electro-osmosis. Electromigration and electro-omosis are important mechanisms in electrokinetic soil processing which remove contaminants from soil. The principle of electrokinetic soil processing is schematically shown in Fig. 8.3.

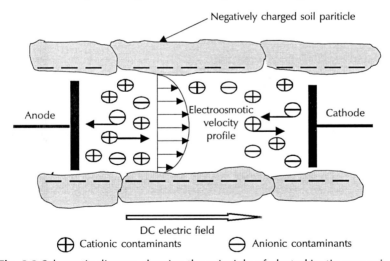

Fig. 8.3 Schematic diagram showing the principle of electrokinetic processing.

Electrolysis of Water

One of the important aspects of electrokinetic soil processing is the migration of an acid front from the anode to the cathode during the treatment. When electrolysis of water takes place in the surface of electrode, hydrogen ions are produced at the anode and hydroxyl ions at the cathode. At the electrodes, electrolysis of water takes place as follows:

At anode, $2H_2O - 4e^- \rightarrow 4H^+ + O_2(g)$ (1)

At cathode, $2H_2O + 2e^- \rightarrow 2OH^- + H_2(g)$ (2)

The production of H^+ ions at the anode decreases the pH by eq. (1) and the eq. (2) reaction increases the pH at the cathode by increasing OH^- ions. This electrolysis results in an acid front at the anode and an alkaline front at the cathode, which move to the cathode and the anode, respectively. The propagation of the acid and the base fronts through soil promotes the dissolution of metal ions near the anode and the precipitation of the metal ions near the cathode. These conditions significantly affect the pH and ionic strength of pore water, the mobility and solubility of metal contaminants, and charge conditions of soil particles. Since the ionic mobility of the hydrogen ions is twice as high as that of hydroxyl ion and the electromigration of hydrogen ions is enhanced by electro-osmosis toward the cathode, the acid front is dominant in soils and decreases soil pH. Increased hydrogen ions in soils may be the most useful in extraction of metal ions adsorbed on soil surface and dissolution of most metal precipitates in soils at low pH conditions. Having greater replacing power on soil surface than any other species, hydrogen ions in the acid front displace metal ions and organics adsorbed on soil surface, resulting in more mobile metal ions in pore water that can be transported by electromigration.

The variation of pH condition in soils by electrolysis of water in the electrode compartment has effects on ionic strength of pore water and soil surface properties such as cation exchange capacity, magnitude and sign of the electrokinetic zeta potential. Furthermore, speciation, mobility and solubility of contaminants are often varied with pH in soils during treatment, which may limit or enhance the treatment efficiencies.

Transport and removal of contaminants

As the acid front migrates through the soil bed, the species adsorbed on the soil surface are desorbed. These free chemical species present in the pore fluid transport towards the electrodes depending on their charge. Electromigration, electro-osmotic advection, and diffusion are the significant driving mechanisms contributing to the transport of species through the soil mass. However, both precipitation and sorption retard species movements at the high pH zones. Cations are collected at the cathode and anions at the anode as a result of the transport of chemical species in the soil pore fluid. In particular, metal and other cationic contaminants are removed from the soil with the cathode effluent solution and/or deposited at the cathode. Hence, the effluent should be treated for the complete removal and recovery of the metal contaminants.

Enhanced electrokinetic soil processing

In order to remove and avoid precipitation in the cathode compartment, various enhancement techniques have been proposed and used. Many researchers have suggested: (1) injection of enhancing agents such as acetic acid or use of hydroxyl ion selective membrane in the cathode reservoir to prevent precipitation or solubilize precipitates of cationic metal contaminants near the cathode (Acar and Alshawabkeh, 1993; Probstein and Hicks, 1993; Rødsand et al., 1995); (2) conditioning the anode and/or the cathode reservoirs to control pH and zeta potential, enhance desorption, and increase the electro-osmotic flow rate to increase mobility of contaminants

(Eykholt and Daniel, 1994; Pamukcu *et al.*, 1990; Probstein and Hicks, 1993; Shapiro and Probstein,1993); and (3) adding or mixing strongly complexing agents such as ammonia, citrate, and EDTA into soil, which compete with soil for metal contaminants to form soluble complexes (Eykholt, 1992; Pamukcu *et al.*, 1990; West and Stewart, 1995; Yeung *et al.*, 1996). Among these enhancement technologies, the scheme for prevention of metal precipitation has been mostly focused and experimentally evaluated.

Alshawabkeh (1994) summarized the characteristics of enhancement schemes: (1) the precipitate should be solubilized and/or precipitation should be avoided; (2) ionic conductivity across the soil specimen should not increase excessively in a short period of time both to avoid a premature decrease in electro-osmotic transport and to allow transference of species of interest; (3) the cathode reaction should possibly be depolarized to avoid generation of the hydroxide and its transport into the soil specimen; (4) such depolarization will also assist in decreasing the electrical potential difference across the electrodes leading to lower energy consumption; (5) if any chemical is used, the precipitate of the metal with this new chemical should be perfectly soluble within the pH ranges attained, and (6) any special chemicals introduced should not result in any increase in toxic residues in the soil mass.

Installation

Figure 8.4 presents a schematic diagram of installation of electrokinetic soil processing. The system mainly consists of two unit processes—electrokinetic removal and effluent treatment processes.

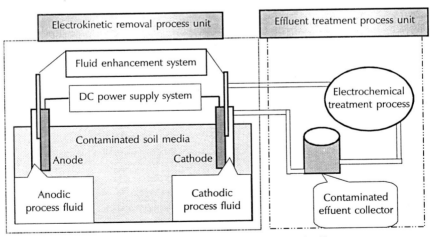

Fig. 8.4 Installation of electrokinetic soil processing and effluent treatment.

Practical applications

Removal of heavy metals from artificially spiked kaolinite and tailing soils

In order to explore the feasibility of electrokinetic soil processing on the removal of heavy metals from soils, a laboratory research was conducted using artificially spiked kaolinite. Also this technology was used for the removal of heavy metals from tailing soils taken from an abandoned mine area to investigate the applicability of this technology to naturally contaminated soils.

A schematic diagram of the experimental apparatus used in this study is shown in Fig. 8.5. The experimental apparatus consists of four principle parts; soil cell, electrode compartments, electrolyte solution reservoirs and power supply. The acryl soil cell measures 9 cm × 9 cm × 15 cm with a volume of 1215 cm³. Each end of the soil cell had 81 holes (diameter 0.5cm) to enhance uniform electro-osmotic flow. At both sides of the soil cell, two sheets of GF/B filter paper were inserted to prevent clay particles from flowing into the electrode compartments.

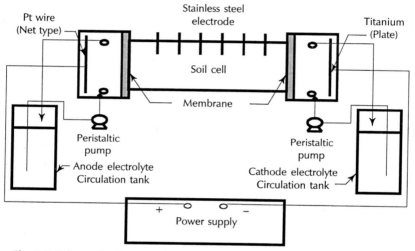

Fig. 8.5 Schematic diagram of experimental apparatus (Kim et al., 2000a,b).

Platinum wire that was plated like a net of 9 cm × 9 cm with the interval of 1 cm was used as the anode to prevent electrode-electrolysis reaction, and titanium plate, 11 cm × 11 cm, was used as the cathode. The electrode compartments contained 400 ml of electrolyte solution ensuring that sufficient volume was present to avoid sudden variations of electrolyte solution. The compartments also have a role to reject gas from the electrode and to provide water for electro-osmosis. Two mass cylinders (2l volume) were used as electrolyte solution reservoirs to measure the water volume transported. The electrolyte solutions were recirculated in both electrode compartments by peristaltic pumps and DC power supply (5-200 V, 0.01-2 A, 200 W) was used.

Soils used in this experiment were commercial kaolinite soils articficially contaminated with Pb and Cd and tailing-soils taken from the abandoned Gubong mining area which is located in the middle of South Korea. Tailing-soils were heavily contaminated with Pb, Cd, Zn and Cu. The kaolinite soils were passed through an ASTM No. 200 sieve of mesh size 75 μm. The physical properties of kaolinite soils used in this study are summarized in Table 8.1. In order to be valid for the analysis, tailing-soils were pretreated by No. 80 sieve of mesh size 180 μm.

Table 8.1. Physical properties of kaolinite soils used in this study

Physical property	Measured value
Group symbol according to USCS	CL
Liquid limit (%)	78
Plastic limit (%)	32
Specific gravity	2.64
pH of soil at 50% water content	4.93 − 5.20
Permeability (cm/sec)	1×10^{-7}
Initial water content (%)	50 - 54

Two soil samples were prepared for the removal experiments of heavy metals in soils. The first was the kaolinite soil which was artificially contaminated with $Pb(NO_3)_2$ and $Cd(NO_3)_2$ solutions. One litre of 1000 mg/l Pb(II) and Cd(II) solutions were prepared by dissolving 1.615 g of $Pb(NO_3)_2$ and 2.801 g of $Cd(NO_3)_2$ respectively in 1l of distilled water. Two kilogram samples of air-dried kaolinite soils were mixed with the prepared solution of 1000 mg/l Pb(II) and Cd(II) at 33 wt% water content. The other was the tailing soil taken from the abandoned mining area which was contaminated with Pb, Cd, Zn and Cu. Tailing-soil samples were analyzed by Inductively Coupled Plasma-Atomic Emission Spectrometry (ICP-AES) to determine the initial concentrations of Pb, Cd and other elements (Cu, Zn). Four kilogram of tailing-soils were mixed with 1l of distilled water to give 20 wt% water content. The slurries of kaolinite and tailing-soils were mixed mechanically with an electric stirrer for 1 hour, and these mixtures were allowed to settle down for more than 3 days to attain the uniform distribution of contaminants and to complete adsorption in the soil samples. Five grams of three samples was then taken from the prepared soils for the determination of the initial concentration of contaminants.

In order to compare the removal efficiencies of different contaminants and soil types under the same conditions, the equivalent anode purging and cathode electrolyte solutions were used. Anode purging solutions of 0.005 N H_2SO_4 solution (2 L) in kaolinite soil experiments, and 0.005 N H_2SO_4 solution (4 l) in tailing-soil experiments were used over a four-day period. Hydroxides were precipitated by hydroxyl ions generated by electrolysis of water in the cathode compartment. These precipitates prevented the removal of contaminants from the soil cell, and cathode electrolyte solutions may buffer the concentration of hydroxyl ions. Therefore, 1 l of 0.5 N H_2SO_4 solution was used as the cathode electrolyte solution in both kaolinite and tailing-soil experiments. In order to enhance the effectiveness of the process in removing heavy metals from soils, concentrations and volumes of anode purging and cathode electrolyte solutions were determined by preliminary experiments.

Three tests were conducted for the removal of heavy metals from soils by electrokinetic soil processing. Constant-current was used in all tests to maintain the net rates of the eletrolysis reactions constant and to minimize complicated current-boundary conditions during the experiment. All experiments were carried out under the equivalent conditions of applied current, area and length of soil cell, and duration of treatment. Anode purging and cathode electrolyte solutions were compared for Pb and Cd depending on the difference of the removal efficiencies. The parameters measured in each experiment are summarized in Table 8.2.

Table 8.2. Summary of testing program and measured parameters for the removal experiment of heavy metals

Parameters	Test 1	Test 2	Test 3
Soil specimen	Kaolinite	Kaolinite	Tailing
Contaminants	Lead ($Pb(NO_3)_2$)	Cadmium ($Cd(NO_3)_2$)	Pb, Cd
Initial concentration	391	367	1438(Pb)
of contaminants ($\mu g/g$)			22 (Cd)
Applied current (A)	0.1	0.1	0.1
Area of soil cell (cm^2)	81	81	81
Length of soil cell (cm)	15	15	15
Duration (hr)	96	96	96
	0.005N H_2SO_4	0.005N H_2SO_4	0.005N H_2SO_4
Anode purging solution	Solution 2l	Solution 2l	Solution 2l
Cathode electrolyte	0.5N H_2SO_4	0.5N H_2SO_4	0.5N H_2SO_4
Solution	Solution 1l	Solution 1l	Solution 1l

The prepared soil samples were packed into the soil cell and the acryl cover was set on the packed soil cell. The cover was then sealed tightly by silica bond in order to prevent the leakage of pore water. The anode and cathode electrolyte solutions were pumped into the electrode compartments for 30 minutes without electric currents to equalize the electrolyte solutions. During the experiments, the overall voltage drops of the soil cell and electrode compartments; pH and conductivity variations of anode purging and cathode electrolyte solutions; soil pH variation and transported pore water volume by electroosmotic flow were measured every 4 hours. Five samples were obtained from the soil bed using a stainless steel sampler (diameter 1.2 cm) at every 3 cm to analyze the concentrations of contaminants and soil pH. The duration of all tests was 96 hours. Heavy metals in the samples taken during the experiments were extracted with 0.1 N HCl ; wet soil samples were dried at 110 °C. Fifty ml of 0.1 N HCl solution was added into 5 g of each dry soil sample (dilution factor; 10) and agitated (100 rpm, 30 °C and 1 hour). The solutions were then analyzed by ICP-AES and Atomic Absorption Spectrometry (AAS).

Variations of pH in the soil cell are shown in Fig. 8.6. Electrolysis of water in the anode compartment generated hydrogen ions and those hydrogen ions migrated from anode to cathode. Over the period of the experiment, the overall soil pH decreased but the decrease was less marked toward the cathode compartment. It was this migration of the acid front that made the soil pH decrease. As shown in Fig. 8.6(a) and (b), corresponding to tests 1 and 2, one kind of anode purging solution and kaolinite soil was used, resulting in a similar pH pattern. However, test 3 was conducted in tailing-soil which had large pH buffering capacity and the overall soil pH in test 3 was higher than those in tests 1 and 2 (Fig. 8.6(c)). This large pH buffering capacity of tailing-soil decreased dissolution and desorption of adsorbed species on the soil surface in test 3. If the soil pH data are combined with the removal efficiency at different locations in the soil cell, the influence of the migration of the acidic front from anode to cathode should be taken into great account due to the effect of dissolution and desorption on heavy metal ions.

In electrokinetic soil processing, pore water is transported by electro-osmosis. The volume of transported water by electro-osmosis is shown in Fig. 8.7. Electro-osmosis is affected by soil pH, applied electric field intensity and soil permeability. Since the pH and conductivity of the anode purging solution determine the pH and

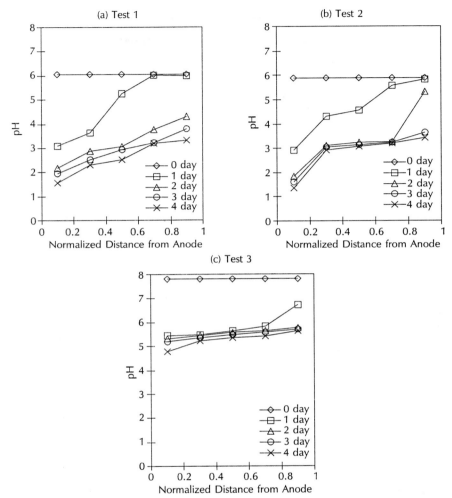

Fig. 8.6 Variation of pH at sampling points in the soil cell during the removal process.

conductivity of the soil bed, the anode purging solution may affect electro-osmosis in soils. Electro-osmotic velocity on a plane surface, U (m/sec), is expressed as:

$$U = - \varepsilon \zeta \, E_x / \mu \qquad (3)$$

where, ε is permittivity of the medium (C/V m), ζ is zeta potential (V), E_x is electric field (V/m) parallel to the direction of electro-osmotic flow and μ is viscosity of the medium (Nsec/m^2). This formula for the electo-osmotic velocity on a plane charged surface is known as the Helmholtz-Smoluchowski equation (Shapiro, 1990; Mitchell and Yeung, 1991; Pamukcu and Wittle, 1994; Probstein, 1994). According to the Helmholtz-Smoluchowski eq. (3), the electro-osmotic velocity (U) is proportional to zeta potential (ζ). If the cationic concentrations in pore water increase by increasing

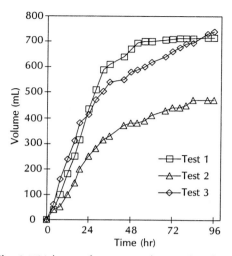

Fig. 8.7 Volume of transported water by electro-omosis.

the cationic concentrations in the anode purging solution, the cation concentrations adsorbed to the surface of the negative clay particles should increase. The increased concentration of adsorbed cations results in the reduction of the zeta potential. The electro-osmotic velocity then decreases by increasing of the cationic concentration in pore water. Consequenty, the decrease of the soil pH and the increase of the conductivity of pore water will make the electro-osmotic velocity decrease. As shown in Fig. 8.7, the volume of transported pore water decreased significantly from the middle stage of each test. Since the applied electric field intensity in test 1 was very high compared with other two tests, the amount of transported water was more than the other two tests. In test 3, soil pH was relatively high (>5) during the treatment, and it caused the increase of electro-osmotic flow. In test 2, the limited amount of transported water was due to the low soil pH and the low electric field intensity.

The variation in the amount of contaminants in the soil cell for each test is shown in Fig. 8.8. In the electric field, the acid front produced by electrolysis of water in the anode compartment migrated toward the cathode by electromigration and electroosmosis with desorbing contaminants adsorbed on the soil surface. The adsorption of hydrogen ions and the desorption of adsorbed species on the soil-surface were repeated in the soil bed during the treatment, and the contaminant species appear to be gradually transported toward the cathode in the electric field. The migration of desorbed species by hydrogen ions is shown in Fig. 8.8, and the overall amount of contaminants in the soil cell gradually decreased as time went on.

Table 8.3 presents the removal efficiency for each test. When the removal efficiency of the test using kaolinite soils was compared with that of the test using tailing-soils, the former showed higher efficiency than the latter. This is because tailing-soils had a larger void fraction than kaolinite soils. Hydrogen ions, therefore, have less chance of contact with the soil surface in tailing-soils than in kaolinite soils, and the chances of desorption of adsorbed species on the soil-surface decreased. In the tests on the removal of Pb(II), the lower amount of removed Pb(II)

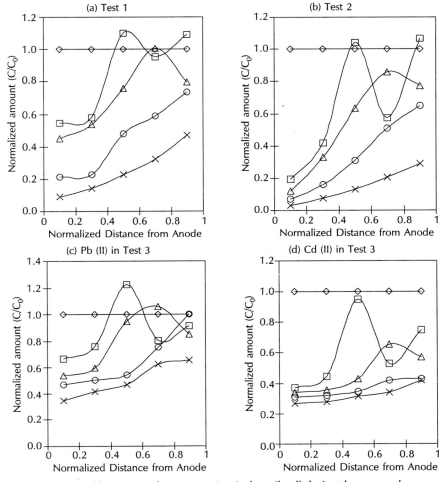

Fig. 8.8 Variaion of heavy metal concentration in the soil cell during the removal process.
($\Diamond$ = 0 day; $\square$ = 1 day; $\triangle$ = 2 day; $\bigcirc$ = 3 day; X = 4 day)

was caused by the greater adsorption capacity and the immobility of Pb(II) (Rose *et al.*, 1979; Alloway, 1995), as shown in Fig. 8.8(a) and (c). In the case of the Cd(II) removal test, the amount of Cd(II) removed was higher than that of Pb(II) owing to the lower adsorption capacity and higher mobility of Cd(II).

Table 8.3. Removal efficiency of heavy metals for each test to remove (unit %)

Time (day)	Test 1	Test 2	Test 3 (Pb)	Test 3 (Cd)
1	14.5	34.2	12.7	39.2
2	28.9	45.8	20.3	53.6
3	66.1	55.1	34.6	63.6
4	74.9	85.5	49.8	67.6

Table 8.4 presents the energy consumed during each test. In tests 1 and 2 for kaolinite soil, and adsorption mechanisms influenced the soil bed significantly, and more energy was consumed than in test 3 for tailing-soils. Because of the strong bonding of Pb(II) on the soil-surface, energy consumption in test 1 on the removal of Pb(II) was higher than that in test 2 for the removal of Cd(II). As shown in Table 8.4, energy consumption in the initial stage of the treatment was higher than that in the later stage, resulting from adsorption and desorption mechanisms in the soil cell. After the desorbed species began migrating, the rate of energy consumption decreased significantly.

Table 8.4. Energy consumption in each test (unit: kWh/ton)

Time (day)	Test 1	Test 2	Test 3
1	106.4	66.2	86.9
2	284.6	169.4	129.9
3	412.4	265.6	167.2
4	501.5	328.7	197.3

From the result of this study, it is suggested that heavy metal contaminants can be effectively removed from artificially contaminated kaolinite and tailing-soils taken from an abandoned mine by enhanced electrokinetic soil processing in which acidic solution was used as a solubilizing agent. The removal efficiencies of Cd (II) in kaolinite and tailing-soils were 86% and 68%, but those of Pb (II) were 75% and 50%, respectively. The relatively lower removal efficiency of Pb(II) compared to Cd(II) was explained by the greater affinity or adsorption onto soil surface and immobility of Pb(II) in soils. Energy consumption was mainly dependent on the conductivity of soil cell. As the adsorption and desorption mechanisms were predominant on removing heavy metals from soils, the species in soils could not migrate freely through the soil bed, and the conductivity of the soil cell decreased.

Removal of heavy metals from wastewater sludges

The sludge from wastewater treatment contains various toxic heavy metals, resulting in the environmental contamination from its landfill into waste disposal site and the unsuitability of its reuse as composts. In order to remove toxic heavy metals in the sludge, electrokinetic remediation technique was examined in a small pilot scale. The objectives of this study were to investigate the feasibility of electrokinetic technique on the removal of heavy metals from the sludge, to determine the speciation of heavy metals in the sludge and to examine the removal efficiency depending on chemical speciations of heavy metals.

The sludge used in this experiment was taken from a municipal wastewater treatment plant in Kwangju, Korea. For the electrokinetic removal experiment, 60 l of sludge was prepared to give 82 wt% water content. Triplicate samples (5 g of each) was then taken from the prepared sludges for the determination of initial concentration of heavy metals prior to the electrokinetic removal test. For the analysis of target metal contaminants such as Cd, Cr, Cu, and Pb in sludge before and after the electrokinetic remediation, metals were extracted from sludge using sequential (speciation analysis) and aqua regia (total concentration analysis) extraction methods.

Sequential extraction method used in this study (Table 8.5) was suggested by Tessier *et al.* (1979) and revised by Environmental Geochemistry Research Group at Imperial College, UK (Li *et al.*, 1995). The extracted solutions were analyzed by ICP-AES (Thermo Jarrel Ash, USA).

Table 8.5. The sequential extraction scheme used in this study

Fraction	Chemical extractant
Exchangeable	0.5 M $MgCl_2$ + NH_4OH/HOAc (pH=7)
Bound to carbonate or specially adsorbed	1 M NaOAc + + NH_4OH/HOAc (pH=5)
Bound to Fe and Mn oxides	0.04 M NH_2OH HCl in 25% HOAc
Bound to organics and sulfides	0.02 M HNO_3 + 30% H_2O_2 +
	3.2 M NH_2OAc in 20% HNO_3 (pH=2)
Residual	HF/$HClO_4$/HNO_3 (4:2:15)

The experimental apparatus used in this study consists of four major parts; sludge cell, electrode compartments, electrolyte solution reservoirs and power supply (Fig. 8.9). The acryl sludge cell measures 0.3 m × 0.4 m × 0.5 m with a volume of 60 L. Each end of the sludge cell had holes (diameter 0.5cm) to enhance uniform

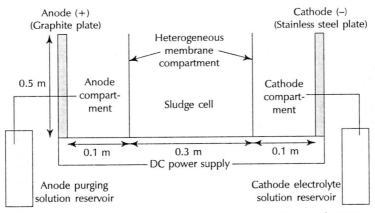

Fig. 8.9 Schematic diagram of experimental apparatus (Kim *et al.* 2000c).

electro-osmotic flow. At both sides of the sludge cell, two sheets of heterogeneous membrane were inserted to enhance the transport of ions toward the electrode compartments and to prevent sludge particles from flowing into the electrode compartments. The heterogeneous ion exchange membrane was semi-permeable.

Graphite plate (0.4 m × 0.5 m) was used as an anode to prevent electrode-electrolysis reaction, and stainless steel plate (0.4 m × 0.5 m) was used as a cathode. The electrode compartments contained 20 l of electrolyte solution ensuring that sufficient volume was present to avoid sudden variations of electrolyte solution. The compartments have a role to reject gas from the electrode, and to provide water for electroosmosis. Two mass cylinders (3 l volume) were used as electrolyte solution reservoirs to measure the water volume transported. Anode purging solution of 0.05 N H_2SO_4 (23 L) was used during the removal experiment. Hydroxides were precipitated by hydroxyl ions generated by electrolysis of water in the cathode compartment, and these precipitates prevented the removal of contaminants from the sludge cell. For this reason, 23 l of 0.5 N H_2SO_4 solution was used as the cathode electrolyte solution to buffer hydroxyl ions produced in the cathode compartment. In order to enhance the effectiveness of the process in removing heavy metals from the sludges, concentrations and volumes of anode purging and cathode electrolyte solutions were determined by preliminary experiments. Constant-current was used to keep the net rates of the electrolysis reactions constant and to minimize complicated current-boundary conditions during the experiment. The parameters used in this study are summarized in Table 8.6.

Table 8.6. Summary of experimental conditions for electrokinetic metal removal

Parameter	Condition
Contaminants	Cd, Cr, Cu, Pb
Applied current (A)	2
Area of sludge cell (m^2)	0.2
Length of sludge cell (m)	0.3
Duration (hr)	120
Anode purging solution	0.05 N H_2SO_4 solution 23 l
Cathode electrolyte solution	0.5 N H_2SO_4 solution 23 l

The prepared sludge sample was packed into a plexiglass cell covered by a lid. The anode and cathode electrolyte solutions were pumped into the electrode compartments for 30 minutes without electric currents to stabilize the condition. During the experiment, the overall voltage drops of the sludge cell and electrode compartments; pH variation of anode purging and cathode electrolyte solutions; sludge pH variation and transported pore water volume by electro-osmotic flow were measured. Five samples were obtained from the sludge bed using a stainless steel sampler (diameter; 2 cm) at every 6 cm to analyze the contaminants in the sludge. After 5 days of the experiment, the electro-osmosis flow became constant due to the decrease of the sludge pH in the anode compartment of the sludge cell. For this reason, the duration of the experiment was determined for 5 days. Heavy metals in the samples taken after the electrokinetic removal experiment were extracted with the sequential extraction method and the extracted solutions were analyzed by ICP-AES.

Initial concentrations and speciations of metal contaminants in sludges were determined using sequential and aqua regia extraction method (Table 8.7). Even though the sludges were taken from a municipal wastewater treatment plant, the total concentrations of target heavy metals were relatively high. The recovery of the sequential extraction procedure [(sum of all fractions/total metal concentration) × 100%] was in the range of 91-97%. Most of the heavy metals were enriched in the organic/sulfides and residual fractions, resulting from the anaerobic digestion of the sludges in the plant prior to coagulation and settling.

Table 8.7. Initial concentrations of metal contaminants analyzed by sequential and aqua regia extraction methods

Extraction method	Fraction	Concentration (mg/kg)			
		Cd	Cr	Cu	Pb
	Exchangeable	1.6	1.0	47.9	0.0
	Carbonates	0.3	1.0	21.4	5.2
Sequential extraction	Fe/Mn oxides	0.6	3.3	5.8	2.9
	Organic/Sulfides	0.0	46.0	120.7	33.5
	Residual	1.8	61.4	117.7	18.8
	Sum	4.0	112.7	313.5	59.9
Aqua regia extraction		4.3	120.0	343.5	62.8

During the electrokinetic removal experiment, the variations of pH in the anode purging and cathode electrolyte solutions were observed (Fig. 8.10(a)), and the variation of pH in the sludge cell was also measured at three different depths after the experiment (Fig. 8.10(b)). As shown in Fig. 8.10(b), there was no significant variation of the sludge pH with the depth. Electrolysis of water in the anode compartment generated hydrogen ions and those hydrogen ions migrated from anode to cathode. For this reason, the overall pH of the anode purging solution and sludge decreased over the period of the experiment. In spite of production of hydroxyl ions by the electrolysis in the cathode compartment, the cathode electrolyte solution was buffered by 0.5 N sulfuric acid, resulting in pH 2.5. During the experiment, the overall pH in the sludge cell did not decrease sufficiently due to relatively high initial sludge pH and large pH buffering capacity of the sludge. The trend of pH variation in the sludge cell is the most significant factor in predicting the effectiveness of the electrokinetic removal process of heavy metals.

Figure 8.11 shows the change of water volume in the cathode compartment due to elelctro-osmosis. The decrease of the sludge pH and the increase of the conductivity of pore water make the electro-osmotic velocity decrease. Consequently, the volume of transported pore water decreased significantly from the middle stage of the experiment.

Table 8.8 presents the variation of residual concentrations in the sludge cell determined by aqua regia analysis after electrokinetic treatment. In the electric field,

Table 8.8. Residual concentration after electrokinetic treatment

Aqua regia extraction analysis		Concentration (mg/kg)			
		Cd	Cr	Cu	Pb
Initial concentration		4.3	120.0	343.5	62.8
	Normalized distance from anode				
Residual concentration	0.2	0.0	61.9	202.4	43.4
	0.4	0.0	66.5	282.5	47.8
	0.6	0.9	83.6	298.8	52.6
	0.8	1.0	88.1	309.4	58.4
	1.0	2.7	101.7	330.5	58.9

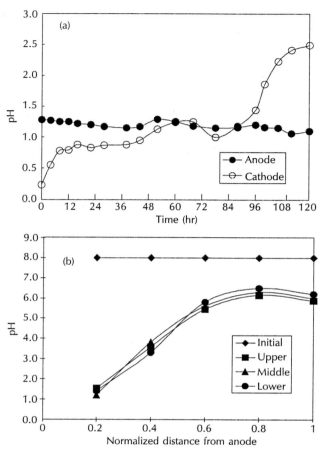

Fig. 8.10 (a) Anode purging and cathode electrolyte reservoir pH versus time; (b) sludge pH versus distance from anode after the experiment (upper; 13 cm from the top, lower; 13 cm from the bottom, middle; between upper and lower).

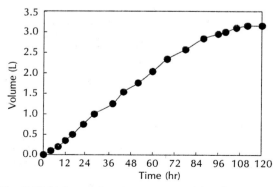

Fig. 8.11 Volume of water transported by electro-osmosis.

the acid front produced by electrolysis of water in the anode compartment migrated toward the cathode by electromigration and electroosmosis with dissolving and desorbing contaminants adsorbed and/or complexed in the sludge. The dissolution and desorption of species in the sludge cell occurred continuously along with the migration of the acid front during the treatment, and the contaminant species appeared to be gradually transported towards the cathode in the electric field.

The removal efficiencies of different contaminants were calculated by the ratio of the initial amount to the remained amount after the electrokinetic treatment depending on their speciations (Table 8.9). With the results of sequential extraction analysis, it was suggested that the more weakly bound fractions of metal contaminants were more easily removed by electrokinetic technique. From the result of high removal efficiencies of weakly bound fractions of contaminants, the electrokinetic technique can be applied to remove the fractions of highly mobile and bioavailable metal contaminants from sludges which may cause a significant environmental hazard due to landfilling. Although the total removal efficiencies calculated by the aqua regia analysis were relatively low, the removal efficiencies of the exchangeable fraction and fraction bound to carbonates were significantly high by electrokinetic technique.

Table 8.9. The removal efficiencies of metal contaminants after electrokinetic treatment

Extraction method	Fraction	Removal efficiency (%)			
		Cd	Cr	Cu	Pb
	Exchangeable	100.0	100.0	86.0	—
	Carbonates	91.7	100.0	75.4	68.8
Sequential	Fe/Mn oxides	74.3	70.0	66.7	68.9
extraction	Organic/Sulfides	—	32.9	25.6	33.6
	Residual	25.0	17.3	21.6	11.7
Aqua regia extraction		73.3	37.3	20.4	20.0

From the removal of heavy metals in sludges by electrokinetic technique, the following conclusions have been obtained;

(1) The sludges used in this study contained high concentration of heavy metals in various speciations. In order to estimate the environmental impacts of metal contaminants in sludges, the speciations and concentrations of metal contaminants were determined by sequential and aqua regia extraction methods. Except for Cd, most target metals (Cr, Cu and Pb) were mainly enriched in the organic/sulfide and residual fractions (72-95 %).

(2) The removal efficiencies of different metals were significantly dependent on their speciations in the sludge. The more strongly bound fractions such as organic and residual fractions, the less effectively removed by electrokinetic treatment. More than 65 % of exchangeable, carbonate and Fe/Mn oxide fractions in sludges were removed by electrokinetic technique with the comparison of less than 10-35 % of removal efficiency for organic/sulfide and residual fractions. In accordance with the result of sequential and aqua regia extraction analyses, the removal efficiency of Pb, one of the relatively immobile elements, was lower than those of other target metal contaminants.

(3) As the acid front generated by the electrolysis of water at the anode compartment migrated toward cathode gradually, the overall sludge pH decreased. The complexed and/or adsorbed metal species were dissolved and desorbed along with the migration of the acid front. The variation of sludge pH influences the removal efficiency in electrokinetic removal of heavy metals. The overall decrease of sludge pH was relatively small due to pH buffering capacity of the sludges. For this reason, the removal efficiency of metal contaminants was relatively low.

Removal of sodium from food wastes

Recently, the treatment of food waste has been one of the most serious problems. Major treatment method of food wastes is landfill (almost 95%) and only 2.1% of them are treated for being reused as animal food and/or recycled as fertilizer through composting in the Republic of Korea (Kim, 2000). The food waste causes many problems in the landfill sites, such as bad odour and slow stabilization, so that public agencies and governmental sectors are preparing of new regulations to prevent dumping of food waste in landfill sites. Table 8.10 shows the treatment methods of food wastes in Korea, and Table 8.11 presents the percentage in food waste in municipal solid waste of various countries (Kim, 2000).

Table 8.10. Treatment methods of food wastes in Korea (1995)

Total	Landfill	Incineration	Reuse/Recycling
15,075 ton/day (100%)	14,387 ton/day (95.3%)	372 ton/day (2.5%)	316 ton/day (2.1%)

Table 8.11. Percentage of food waste in municipal solid waste of various countries

Korea ('95)	USA ('90)	England ('90)	Germany ('89)
31.6 %	25 %	19 %	28 %

Recently, various technologies have been proposed for the treatment of food wastes. In particular, anaerobic digestion became one of the most promising technologies for the production of methane biogas as energy source as well as the reuse and recycling of food wastes. However, most of the food wastes are solid and contain large amount of sodium ion. The sodium ion is recognized as one of the inhibition compounds in the microbiological processes of food waste. The high sodium concentration of food wastes should be reduced prior to anaerobic process. For the pretreatment of this food waste, a laboratory experiment was conducted to remove the sodium ions using electrokinetic process.

The food waste specimen used in this study was collected at the restaurant of Kwangju Institute of Science and Technology (K-JIST), Republic of Korea. Table 8.12 shows the characteristics of food waste.

Table 8.12. Characteristics of food waste used in this study

Initial pH	5.7
COD (mg/l)	24,000
Alkalinity (mg/l)	2,000
Initial sodium concentration (mg/l)	32,000
Initial water content (%)	71
TKN (mg/l)	2,341
NO_2^- (mg/l)	2.1
NO_3^- (mg/l)	3.5

The experimental apparatus used in this study is presented in Fig. 8.5. The experiment was conducted for 56 hours. All the other experimental procedure and the operating conditions were equivalent as explained in the previous section 1.3.1, on the experiment for the removal of heavy metals from artificially spiked kaolinite and tailing soils. In order to measure the concentration of sodium ion in the food waste during and after the electrokinetic treatment, 10 g of food waste was sampled at 4 points of the electrokinetic cell after 22 and 56 hours. The samples obtained during the experiment were centrifuged and the extracted solutions were analyzed by ICP-AES. The variation of sodium concentration in the electrokinetic cell is shown in Fig. 8.12. It is suggested that the electrokinetic technique can be applied to remove the sodium ions from the food waste.

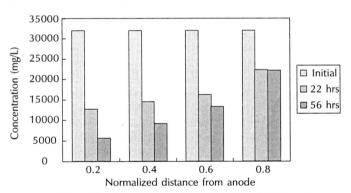

Fig. 8.12 Variation of sodium in the electrokinetic cell during the removal process.

Effluent Treatment Processing

The cathode effluent solutions are contaminated with metals from contaminated media, such as soils, sludges, and food wastes, as a result of the application of electrokinetic removal technology. Hence, these cathode effluents should be properly treated to recover metals contained in these solutions. Most of the recent technologies used for the treatment of effluent wastewaters involve the electrochemical processes. Obviously, a major advantage of electrochemical processes is that the metal can be recovered directly in its most valuable form, i.e. solid, pure metal. The competitive processes (e.g. precipitation of the metal ion as a hydroxide with sodium hydroxide, solvent extraction and ion exchange) essentially lead to a precipitated salt or another metal ion solution of increased concentration, while cementation produces very impure metals (Pletcher and Walsh, 1990). Herein, ion exchange and electrochemical technologies that have been most frequently used for the treatment of wastewaters containing metal contaminants are briefly introduced, focused on their basic and technical principles and applications.

Ion Exchange

Principle

Ion exchange is a powerful technology for the treatment of relatively dilute aqueous solutions. Although ion exchange has long been applied as a standard method of purifying wastewater, its potential and strength for the recovery of metals from wastewaters have only recently been envisioned. Ion exchange resins can be described simply as solid, insoluble acids or bases which are capable of entering into chemical reactions (Genders and Weinberg, 1992). Although many processes are applied for metal recovery by exchange of anionic complexes on anion exchangers, cation exchangers are significantly important in this text.

Cation exchange resins exchange metal cations such as copper, zinc, lead, and nickel for hydrogen ions. For example, passing a dilute metal sulfate solution through a cation exchange resins in the hydrogen form will generate a dilute sulfuric acid solution.

$$2RH + MeSO_4 \rightarrow R_2Me + H_2SO_4 \text{ (Me; metal cations)} \tag{4}$$

After treatment, the resins should be regenerated. Strong acids such as hydrochloric or sulfuric acids are usually contacted with the used resins.

$$R_2Me + H_2SO_4 \rightarrow 2RH + MeSO_4 \text{ (Me; metal cations)} \tag{5}$$

Recently, electrochemical ion exchange has been introduced for wastewater treatments. Electrochemical ion exchange (EIX) is an advanced ion exchange process, in which an exchange material has been incorporated into an electrode structure using a suitable binder. Ion exchange is controlled by application of an electrode potential between the EIX electrode and a counter electrode. The combination of the EIX electrode and the counter electrode, in various configurations, comprises the EIX cell. Ion exchange process has a number of advantages resulting from the application of electrical potential (Genders and Weinberg, 1992). These include the extra process control variable of applied potential which can lead to remote automatic control and increased safety; the minimization of additional waste by being able to conduct reactions without the need for chemical additions; mild operating conditions through the supply of process energy in a selective electrical form. Enhancement of process effectiveness through electrical control can also lead to smaller plants with a reduced energy consumption, in turn giving savings in both capital and running costs.

Applications

Numerous works have been done for the recovery of metals from wastewaters using ion exchange process. Adsorption properties of heavy metals on various ion exchange resins and the kinetics of mass transfer of ion exchange processes were investigated by researchers (Zarraa, 1992; Pesavento *et al.*, 1994; Ahuja *et al.*, 1995; Byers and Williams, 1996; Grebenyuk *et al.*, 1998; Rawat and Bhardwaj, 1999). There have been numerous researches on the removal and recovery of metals from wastewaters using ion exchange processes (Leinonen *et al.*, 1994; Chiarizia *et al.*, 1995; Kanzelmeyer and Adams, 1996; Jangbarwala, 1997; Abe, 1999; Berker *et al.*, 1999; Kabay *et al.*, 1999; Koivula *et al.*, 2000). Jones et al. (1992) and Zhou *et al.* (1996) conducted the researches on the simultaneous multi-metal recovery with electrochemical ion exchange process.

Electrolysis

Principle

In the electrolytic processes for the removal of metal contaminants from solutions, the main mechanism is the cathode reaction, i.e.

$$Me^{n+} + ne^- \rightarrow Me \text{ (Me: metal)} \tag{6}$$

Another mechanism is the precipitation of metal ions using cathodically generated hydroxide, although less common approach, e.g.

$$2Fe^{3+} + 6H_2O + 6e^- \rightarrow 2Fe(OH)_3 \downarrow + 3H_2 \tag{7}$$

The presence of complexing agents such as oxygen or other electroactive species in solution, as well as poor buffering, greatly complicates the task of recovery of heavy metals, particularly when the metal ion is present in low concentrations. Another difficulty is variability of the composition of the effluent solutions. According to Pletcher (1992), there are several factors influencing the success of electrolytic processes for effluent treatment. These factors can be summarized as follows:

a. Achievement of process objectives, i.e. does the electrolysis cell reduce the concentration of metal ions to below the limit for discharge?
b. Value of metal or chemicals recovered.
c. Energy consumption.
d. Rate of effluent treatment.
e. Cost and lifetimes of cells and components.
f. A strategy for the counter electrode chemical, e.g. evolve harmless gases, form useful product.
g. Reproducibility of process performance with all effluents likely to the handled (variation of concentration, $[O_2]$, pH, complexing agents, etc.).
h. Competing technology.
i. Patents and company experienced of electrolysis.

The merits of electrolytic process depend on a number of experimental variables. Hence, the overall performance of the electrolytic cell can be determined by a complex interplay of various parameters which may be used to optimize an electrolytic process. The main electrolysis parameters are as follows (Pletcher and Walsh, 1990):

1. Electrode potential

The electrode potential determines which electron transfer reactions can occur and also their absolute rates, i.e. current densities. The potential and/or current densities are major factors controlling the current efficiency and the product quality.

2. Electrode materials and structure

The ideal electrode material for most processes should be absolutely stable and inert in the electrolysis medium and can lead to the desired reaction with a high current efficiency at low potential. However, these inert electrodes, in practice, have a finite lifetime due to corrosion and physical wear. The recent trend is away from massive electrodes. The better materials are frequently expensive and it is therefore more common for the active material to be a coating on a cheaper and inert substrate. These coated electrodes, in addition to cheapness, have the advantage that they have high real-surface areas due to their microroughness and can be made with the different structures (various crystalline sizes) in order to improve activity.

The form of the electrode is also modified to meet the needs of particular processes. Accordingly, for example, electrodes are constructed commonly from meshes, expanded metal and related materials, leading to maximized surface area, reduce cost and weight and also enhance the release of gaseous product.

One of the newest types of porous, three-dimensional materials is reticulated (foam) metal (e.g. Ni, stainless steel, Cu) or carbon (reticulated vitreous carbon, RVC). The open-cell porous structure of these materials gives interesting properties:
   a. Reasonable isotropy of electrical conductivity, porosity and flow characteristics.
   b. Reasonable electrical conductivity.
   c. High porosity and, hence, low electrolyte pressure drop.
   d. Low effective density.
   e. Possibility of coating the material to provide a surface-modified electrode.

3. The concentration of electroactive species

The concentration of the electroactive species such as metal ions is the major parameter that determines the maximum feasible current density and the optimum space-time yield. Normally, this current is proportional to the concentration of electroactive species.

4. Electrolysis medium

The properties of the electrolysis medium can be determined by pH, the type of electrolytes, complexing agents, additives (i.e. species present in relatively low concentrations to modify the properties of the electrode-electrolyte interface), and reagents present to react with intermediates produced in the electrode reactions. The concentration of each constituent is also important.

5. Temperature and pressure

Generally, electrolysis is to be avoided at elevated or reduced pressure because of the complexity of cell design. In practice, the large-scale tasks of electrochemistry above atmospheric pressure are significantly limited to special water electrolysers and battery systems. On the other hand, an increased pressure may be desirable to minimize solvent loss, if volatile solvents are used. Temperature is an important parameter. Temperatures above ambient are frequently used because of their beneficial effects on the kinetics of all steps in an electrode process.

6. Cell design

Cell design is a significant parameter affecting all the merits for an electrolytic process. In designing an electrolysis cell, the principal factors influencing the performance are the presence or absence of a separator and its type (porous diaphragm or ion selective membrane), the arrangement and form of the electrodes, and the construction materials.

Applications

Electrolysis has been used for the recovery of heavy metals by numerous researchers. Almost all the researches have determined the effect of current density, estimated current efficiency, investigated the energy consumption, and observed the relationship between the concentration of contaminants and the processing performance (Huang *et al.*, 1993; Fuchs and Riedel, 1995; Zherebilov, 1996; Yang and Tsai, 1998; Juang and Lin, 2000). There have been several works focused on the cell configuration, the electrodes, and the complexing agents (Kuhn and Mason, 1996; Widner, 1997; Juang and Wang, 2000).

Huang *et al.* (1993) conducted a task for the recovery of heavy metals from scrap metal pickling wastewaters using electrolysis. The specific objectives of the study were to determine the effects of current density and influent surface loading of the copper ion in the cathode compartment on current efficiency, specific energy consumption, purity of copper and the electro-deposition rate of copper using the electrolysis reactor. According to the results, the electro-deposition rate of copper increased with the increasing current density when the current density ranging from 1.3 to 3.9 A/dm^2 was controlled, but declined when the current density was raised to 5.2 A/dm^2. The current efficiency of the electro-deposition rate of copper decreased with the increasing current density, while the specific energy consumption increased with the increasing current density. This study proposed the basic findings that should be considered in operating the electrolysis cell.

Widner *et al.* (1997) carried out the study on the electrolytic recovery of metals from industrial wastewater using porous cathode of reticulated vitreous carbon (RVC). The equipment was tested for Cu(II), Zn(II) and Pb(II) removal. By flowing effluent metal ions through porous cathodes, it was possible to obtain both high mass transfer rates and large surface area for the electrolytic reactions. Metal ions in such solutions were reduced at the inner surface of the cathodes as the electrolyte flows through. According to the researchers, RVC is chemically and electro-chemically inert over a wide range of potentials and chemicals. Also, it is inexpensive and easily shaped as required by cell design considerations. In addition, RVC has good mechanical resistance, high fluid permeability and specific surface area within the porous structure that is accessible to electrochemically active species. Development of the electrolytic cell for metal removal was carried out in two stages. Initially, a voltametric study of the Cu(II), Zn(II) and Pb(II) reduction reactions on a glassy carbon rotating disc electrode was performed to determine the range of potentials over which these reactions are controlled by mass transfer. Subsequently, a potential value within these intervals was selected and applied to a flow through electrolyte cell containing a reticulated vitreous carbon (RVC) cathode. The flow rate of the electrolyte and the cathode porosity were varied to optimize metal removal. After a 30 min. treatment, the concentration of Cu(II) decreased to 0.1 mg/l from 48 mg/l of initial concentration. The Pb(II) concentration decreased from 50 mg/l to less than 0.1 mg/l after 20 to 30 min. In case of Zn(II), the concentration reached 0.1 mg/l from 50 mg/l after 40 min. of electrolysis.

Juang and Wang (2000) conducted electrolytic recovery of binary Cu(II) and Pb(II) from solutions containing strong complexing agent EDTA (ethylenediaminetetraacetic acid) using a two-chamber cell separating with a cation exchange membrane. Strong complexing agents including EDTA, NTA (nitrilotriacetic acid), citrate, and tartrate complicate the treatment of industrial effluents and other polluted waters, which reduce the efficiency of metal removal by conventional chemical precipitation, ion exchange, and other processes. For this reason, the study was carried out using solutions containing strong complexing agent EDTA. The work was focused on examining the recovery (i.e. the ratio of actual amount of metals deposited onto the cathode to initial amount of metals in the catholyte) and current efficiency (i.e. the ratio of amount of metals deposited onto the cathode to the amount of metals deposited based on Faraday's law of electrolysis) of binary complexed solutions, Cu(II)- and Pb(II)-EDTA. Experiments were performed at equimolar solutions of EDTA and total metals, and as a function of current density, initial catholyte pH, total metal concentration and their concentration ratio. According to the results of this study, the recoveries of both Cu(II) and Pb(II) decreased with increasing initial chatholyte pH. Moreover, the higher the concentration ratio of Cu(II) to Pb(II), the higher the recovery was expected. Compared to single complexed solutions, the lower recovery and current efficiency were obtained in binary solutions regardless of what metal concentration ratios. Increasing current density leads to a monotonous decrease in current efficiency, while the recovery increased and the extent of increment for Cu(II) was larger than Pb(II). In other words, there was an optimal current density with respect to the recovery and current efficiency. Furthermore, to obtain an economically acceptable current efficiency and recovery, the feed complexed concentration should be high enough.

Electrodialysis

Principle

The principle of the electrodialysis process is shown in Fig. 8.13, which shows schematically a typical electrodialysis cell arrangement consisting of a series of

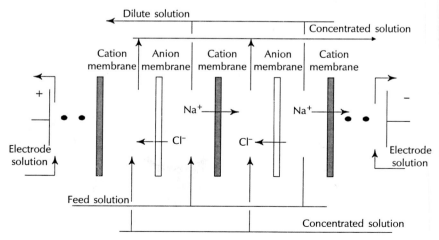

Fig. 8.13 Schematic diagram of the electrodialysis process.

anion- and cation-exchange membrane arranged in an alternating pattern between an anode and a cathode to form individual cells. Figure 8.13 shows only a few cation-exchange membranes and anion-exchange membranes. However, an actual electrodialysis stack may have a few hundred of such membranes.

A cell consists of a volume with two adjacent membranes. If an ionic solution is introduced through these cells and an electrical potential established between the anode and cathode, the positively charged cations migrate toward the cathode and the negatively charged anions towards the anode. The cations pass easily through the negatively charged cation-exchange membrane but are retained by the positively charged anion-exchange membrane. Likewise, the negatively charged anions pass through the anion-exchange membrane but are retained by the cation-exchange membrane. The overall result is an increase in the ion concentration in alternate compartments, while the other compartments simultaneously become depleted. The depleted solution is generally denoted as the diluate and the concentrated solution as the brine or the concentrate. The driving force of ion transport in the electrodialysis process is the applied electrical potential between the anode and cathode.

Like those in all electrochemical cells, the energy consumption of an electrodialysis cell is effectively determined by the cell voltage and the current efficiency. In electrodialysis process, these quantities are, however, estimated by quite different factors from electrolytic processes. The current efficiency is determined solely by the properties of the membrane and the faradaic processes occurring at the electrodes are not important. Clearly, the performance of an electrodialysis cell is significantly controlled by the properties of the membranes. Both the anion and cation membranes should have a low resistance and high chemical, mechanical and dimensional stability. In order to have the required mechanical and dimensional stability, the copolymerization reaction is initiated around a reinforcing mesh or within a porous polyethylene or thermoplastic sheet.

The maximum useful current density through the membrane is normally determined by a phenomenon known as 'polarization' (i.e. depletion of the transported ion at the membrane surface). This is a mass transport problem and it is necessary to avoid stagnant layers at the membrane-solution interface by operating the cell at a high Reynolds number or with turbulence promoters. This will clearly be most necessary as the dilute stream becomes depleted in ionic species. Electrodialysis cells are usually operated with a current density in the range 20-200 mA/cm^2 (Pletcher and Walsh, 1990).

Membranes suffer from a susceptibility to poisoning and fouling. Hence, certain ions have a high affinity for the active centers where they become irreversibly absorbed. This is poisoning and can only be controlled by removal of the offending ion before it contacts the membrane. Fouling is a similar but less serious problem caused by phenols, detergents and large organic acids at anion membranes. It can be minimized by controlling the pore structure of the membrane.

The energy consumption in electrodialysis plants is significantly dependent on local factors and the objectives of the process but for modern plant it normally lies in the range 1-2 kWh/dm^3 of solution treated (Pletcher and Walsh, 1990). The number of electrodialysis plants has increased. In particular, metal finishing processes offer numerous applications for electrodialysis in pollution control and material recovery. The rinse streams form such processes pose particularly troublesome pollution problems. They are usually too dilute for direct metal recovery and too concentrated for disposal.

Applications

Electrodialysis process has been applied for the recovery of heavy metals from various effluent solutions by numerous researchers (Audinos, 1986; Chapotot *et al.*, 1995; Takahashi *et al.*, 1995; Ramachandraiah *et al.*, 1995; Lumbroso, 1996; Rockstraw and Scamehorn, 1997; Klischenko *et al.*, 1999).

Audinos (1986) conducted a study for the improvement of metal recovery by electrodialysis. This task was one of the earliest studies on the application of electrodialysis to the treatment of effluent containing metals using electrodialysis. In order to improve the performance of electrolysis and electrodeposition, electrodialysis was used on effluents of the electrolysis bath and on solutions undergoing electrodeposition. A model in agreement with the experimental results gave the main parameters involved in the transmembrane transfer and the optimal condition. The researcher concluded that electrodialysis and related electromembrane techniques are to be fairly flexible tools to be used before, during or after electrolysis.

Takahashi *et al.* (1995) performed the study on multi-stage electrodialysis for the separation of two metal ion species. In the study, electrodialysis of a mixture of potassium chloride and sodium chloride with a cation exchange membrane was operated in a batchwise mode, the continuous mode and the multi-stage continuous mode. In the experiments with the multi-stage dialyzer, reflux flow from the top stage of the anode towards the bottom stage of the cathode was adopted and the effect of the reflux flow on the concentration ratios of potassium ion to sodium ion in the solution of top stage were examined as well as the effect of stage number. According to the results of the study, the permeation selectivity between potassium ion and sodium ion for the dialysis with the electric field was not different from that without the electric field in the system. For the separation with a multi-stage electrodialyzer, the ion concentration ratio in the top product increased with the reflux ratio and the stage number. When the reflux flow rate was zero, the effect of stage multiplying could not be obtained. These behaviors were same as in the distillation with a multistage column. The dialyzer used in the study corresponded to the enriching section of the distillation column. By adding the stripping section, it might be possible that one species of the two ions was taken from the top product and the other one from the bottom product.

A task was carried out for the recovery of acid, salt and heavy metal mixtures from aqueous solutions using electrodialysis by Rockstraw and Scamehorn (1997). The workers investigated the effect of solution composition on the electrodialysis design parameters of overall current efficiency and apparent stack resistance for mixtures of acids, monovalent salts, and divalent metals. According to the workers, overall current efficiencies of less than unity are related with a number of contributing factors: (1) the membrane may not have a permselectivity of unity and therefore allow passage (intrusion) of the co-ion; (2) parallel currents may exist across the membrane stack manifold (current leakage); (3) at high current densities or low solute concentrations, hydrogen and hydroxyl ions present in the aqueous media begin to participate in the current carrying process (transport by water dissociation ions); (4) the concentration gradient across a membrane separating the diluate and concentrate compartments drives a diffusive flux of electrolyte back into the diluate (counterion backdiffusion). With the results of the experiments, the workers concluded: (1) metals, salt, and mixtures of these components could be efficiently removed from aqueous solutions of moderate pH by electrodialysis. As the pH of solution was decreased, removal of the metal and salt cation became less efficient; (2) divalent cation had similar current efficiency and resistance under similar operating conditions, provided other effects (i.e. precipitation on or in a membrane) did not occur; (3) for the monovalent salts, divalent metals, and salt/metal mixtures, apparent stack resistance was essentially independent of the solute nature and was a function of the total electrolyte concentration in the diluate (by Ohm's law).

Electrodialysis process was applied for the recovery of metals from galvanic sewage (Klischenko *et al.*, 1999). The study was focused on the investigation of the effect of electric current density and the presence of a load (filler) in the desalination chamber on the deionization process of the rinsing water formed in the bath of standing rinse by galvanization. They examined the dependence of specific electromigrative transport of zinc and sodium ions across both the cation-exchange and anion-exchange membranes on the intensity of the polarization regime. According to the results, the dependence of Zn^{2+} electromigrative transport reached its maximum at 10-15 mA/cm^2 of current density and up to 25 % of zinc was transferred across the anion-exchange membrane as its anionic form. The workers concluded that the presence of the ion-exchange filler in the desalination chamber leads to an increase in zinc electromigrative transport and a decrease in its concentration in the dialyzate and the residual zinc concentration in the dialyzate decreased as the electric current dentsity increased.

Electrowinning

Principle

Electrowinning is one of the electrolytic processes that has been used in the manufacturing metals from their ores. In this text, however, this process will be described as a tool for the recovery of metals from aqueous solutions. Ion exchange is a powerful tool for the processing of relatively dilute aqueous solutions in applications such as boiler feed water makeup, while electrowinning has traditionally been used to recover gross quantities of metals from concentrated solutions. While various metals can be recovered from wastewater by electrowinning, copper, nickel, and zinc are interesting because they pose some rather typical challenges, which apply for other metals (Brown, 1992).

Of the base metals, copper is the easiest to electrowin because of its relatively high potential compared to hydrogen. If the copper and hydrogen ions are present in equal concentrations, then the plating out of copper metal (the least negative reaction) will take place almost exclusively. The reaction for electrowinning of copper and nickel from a sulfate media are as follows (Genders and Weinberg, 1992):
Anode reaction
$$2H_2O - 4e^- \rightarrow O_{2\,(gas)} + 4H^+ \tag{8}$$
Cathode reactions
$$Me^{2+} + 2e^- \rightarrow Me_{(metal)} \text{ (Me: Metals)} \tag{9}$$
$$2H^+ + 2e^- \rightarrow H_{2\,(gas)} \tag{10}$$

As shown in reaction (8), free acid is produced at the anode as the metal deposition proceeds at the cathode. The acid can sometimes be recycled to the metal recovery process.

Zinc electrowining is more difficult than copper because of zinc's position relative to hydrogen evolution on the electromotive series. At first glance, it would seem impossible to plate zinc because it requires 0.76 more volts than the evolution of hydrogen. The high overpotential for evolving hydrogen from a zinc cathode is the reason that it is possible to plate zinc. The potential required for evolving hydrogen on zinc metal is approximately 1 volt higher than on platinum, and at a zinc cathode the voltage for plating zinc is about the same as the voltage for evolving hydrogen.

Applications

By combining electrowinning with other processes, the disadvantages and limitations of each process can be overcome to a large degree and the range of applications, particularly in the field of metal recovery can be extended.

Brown (1992) conducted several studies on the recovery of metals from various effluent solutions using electrowinning combined with ion exchange process. He suggested the following categories of applications of the combined ion exchange and electrowinning:

a. Removal of contaminants from electrowinning electrolytes by ion exchange.
b. Pre-concentration of dilute metal streams by ion exchange, followed by electrowinning of metal from the ion exchange eluate.
c. Selective recovery of a particular metal from a solution containing other unwanted or non-toxic metals or chemicals, followed by electrowinning of the desired metal from the ion exchange eluate.
d. Conversion of chloride or nitrate metal salts to sulfates, followed by electrowinning of the metal from the sulfate electrolyte.

He recovered copper and zinc from brass pickling solution containing a mixture of sulfuric acid and hydrogen peroxide using ion exchange/electrowinning system. The copper values were previously recovered via crystallization of copper sulfate, but this was found to be a very inefficient and messy method of recovery and the market for recovered copper sulfate crystals was poor so that an alternative method of recovery was sought. The system was replaced with an ion exchange/ electrowinning system to recover sulfuric acid/peroxide pickle liquor as well as the copper values. According to the worker, advantages of the system, in addition to the value of the recovered copper, include a reduction of overall cleaning line maintenance time, an improvement in process bath life, a reduction in wastewater treatment operations and an overall stabilization of the cleaning line process.

Two electrochemical-based methods were evaluated to recover heavy metals from a tin/zinc electroplating rinse solution with subsequent recycle of the metals back into the original plating bath (Smith and Foreman, 1997). The first method used electrodialysis to move tin to the anolyte strip solution as an anionic citrato complex while zinc was distributed to both the anolyte and catholyte, showing it existed as both an anionic citrato complex and in free cationic form. Zinc can be recovered by scraping the loose deposit from the cathode and dissolving it in mineral acid. The second method was a combined electrowinning/ electrostripping technique. The process involved continuously flowing the plating rinse solution through a porous graphite cathode and removing the metal ions via electrodeposition. When a sufficient quantity of metal was deposited, the electrode was placed in a solution whose chemical composition was similar to that of the original plating bath and the metal ions were stripped from the electrode anodically. The resulting solution was then placed into the original plating bath. Table 8.13 shows results of a typical electrodialysis run.

Table 8.13. Results of electrodialysis of tin-zinc plating rinse solution

	Catholyte, 0.97 l		Feed, 1.97 l		Anolyte, 1.06 l	
Metal ions	Sn	Zn	Sn	Zn	Sn	Zn
Initial quantity, mg	0	0	240	140	0	0
After 9,400 Coulombs	0	90	100	40	140	10
pH	12.9		5.4		5.3	

According to the results of the operation of the combined processes, both methods successfully reduced the metal ion concentrations to less than 15 ppm from 100-300 ppm and approximately 70% of the tin and 100 % of the zinc were recovered by the methods. The workers concluded that both electrodialysis and electrowinning/electrostripping were viable methods for the recovery and recycle of metals from a tin-zinc plating bath rinse solution.

REFERENCES

1. Abe, H. (1999). Separation of free acids and metal salts by ion exchange resin, *J.Surf. Finish. Soc. Japan*, **50**(8): 687-691.

2. Acar, Y.B. and Alshawabkeh, A.N. (1993). Principles of electrokinetic remediation, *J. Env. Sci. Technol.*, **27**(13): 2638-2647.

3. Acar, Y.B., Gale, R.J., and Alshawabkeh, A.N.(1995). Electrokinetic remediation:Basics and technology status, *J. Hazard. Mat.*, **40**: 117-137.

4. Adriano, D.C. (1986). *Trace Elements in the Terrestrial Environment*, Springer-Verlag, 7-14.

5. Ahuja, M., Gupta, S., and Mathur, P.N. (1995). Selective adsorption of metal ions on a new chelating ion exchange resin chemically derived from Guaran, *J. Polym. Mater.*, **12**(4): 257-262.

6. Alloway, B.J. (1995). Heavy metals in soils (2nd Ed.), Blackie Academic and Professional, U.K., pp.11-35.

7. Alshawabkeh, A.N. (1994). Theoretical and experimental modeling of removing contaminants from soils by an electric field, Ph.D. dissertation, The Louisiana State Univ., Baton Rouge, LA, U.S.A.

8. Alshawabkeh, A.N., Yeung, A.T., and Bricka, M.R. (1999). Practical aspects of in-situ electrokinetic extraction, *J. Env. Engrg.*, **125**: 27-35.

9. Audinos, R. (1986). Improvement of metal recovery by electrodialysis, *J. Membr. Sci.*, **27**(2): p.143.

10. Berker, U.G., Guener, F.S., Dizman, M., and Erciyes, A.T. (1999). Heavy metal removal by ion exchanger based on hydroxyethyl cellouse, *J. Appl. Polym. Sci.*, **74**(14): 3501-3506.

11. Brown, C.J. (1992). Electrochemical ion exchange, in: Electrochemistry for a cleaner environment, Genders, J.D. and Weinberg, N.L. (Eds.), The Electrosynthesis Company Inc., New York, U.S.A., pp.183-205.

12. Byers, C.H. and Williams, D.F. (1996). Continuous ion exchange-selective sorption of metal ions, *Spec. publ.-R . Soc .Chem.*, **182**:460.

13. Chambers, D. C., Willis, J., Giti-Pour, S., Zieleniewski, J. L., Rickabaugh, J. F., Mecca, M. I., Pasin, B., Sims, R. C., Sorensen, D. L., Sims, J. L., McLean, J. E., Mahmood, R., Dupont, R.R., and Wangner, K. (1991). *In Situ Treatment of Hazardous Waste Contaminated Soils* (2nd Ed.), Noyes Data Corp. (NDC), New Jersey, U.S.A., pp.98.

14. Chapotot, A., Lopez, V., Lindheimer, A., and Aouad, N. (1995). Electrodialysis of acid solutions with metallic divalent salts: cation-exchange membrane with improved permeability to protons, *Desalination*, **101**(2):141.

15. Chiarizia, R., Ferraro, J.R., D'Arcy, K.A., and Horwitz, E.P. (1995). Uptake of metal ions by a new chelating ion-exchange resin. VII. Alkaline earth cations, *Solv. Extr. Ion Exch.*, **13**(6): 1063.

16. Eykholt, G.R. (1992). Driving and complicating features of electrokinetic treatment of soils, Ph.D. dissertation, Univ. of Texas at Austin, Texas, U.S.A.

17. Eykholt, G.R. and Daniel, D.E. (1994). Impact of system chemistry on electroosmosis in contaminated soils, *J. Geotech. Engrg., ASCE*, **120**(5): 797-815.

18. Fuchs, M. and Riedel, F.H.(1995). Electrolysis for the treatment of wastewaters containing metals, *Gas-und Wasserfach. Wasser, Abwasser :GWF*, **136**(6): 296.

19. Genders, J.D. and Weinberg, N.L. (1992). *Electrochemistry for a cleaner environment*, The Electrosynthesis Company Inc., New York, U.S.A., pp.184-185.

20. Grevenyuk, V.D., Verbich, S.V., and Linkov, N.A. (1998). Adsorption of heavy metal ions by aminocarboxyl ion exchanger ANKB-35, *Desalination*, **115**(3): 239.

21. Hsu, C.N. (1997). Electrokinetic remediation of heavy metal contaminated soils, Ph.D. dissertation, Texas A&M Univ., College Station, Texas, U.S.A.

22. Huang, J.S., Lee, I.C., and Lin, B.J. (1993). Recovery of heavy metal from scrap metal pickling wastewater by electrolysis, *Wat. Sci. Technol.*, **28**(7): 223-229.

23. Jangbarwala, J. (1997). Ion exchange resins for metal finishing wastes, *Met. Finish.*, **95**(11): 33.

24. Jones, C.P., Neville, M.D., and Turner, A.D. (1992). Electrochemical ion exchange, in: Electrochemistry for a cleaner environment, Genders, J.D. and Weinberg, N.L. (Eds.), The Electrosynthesis Company Inc., New York, U.S.A., pp. 207-219.

25. Juang, R.S. and Lin, L.C. (2000). Efficiencies of electrolytic treatment of complexed metal solutions in a stirred cell having a membrane separator, *J. Membr. Sci.*, **171**(1): 19-29.

26. Juang, R.S. and Wang, S.W. (2000). Electrolytic recovery of binary metals and EDTA from strong complexed solutions, *Water Res.*, **34**(12): 3179-3185.

27. Kabay, N., Demircioglu, M., Yayli, S., Yuksel, M., Saglam, M., and Levison, P.R. (1999). Removal of metal ions from aqueous solution by cellulose ion exchangers, *Sep. Sci. Technol.*, **34**(1): 41.

28. Kanzelmeyer, T.J. and Adams, C.D. (1996). Removal of copper from a metal-complex dye by oxidative pretreatment and ion exchange, *Water Environ. Res.*, **68**(2): 222.

29. Kim, D.H. (2000). Anaerobic solid digestion under thermophilic condition, M.S. thesis, Kwangju Institute of Science and Technology (K-JIST), Kwangju, Republic of Korea.

30. Kim, S.O., Moon, S.H., and Kim, K.W. (2000a). Enhanced electrokinetic soil remediation for removal of organic contaminants, *Env. Technol.*, **21**: 417-426.

31. Kim, S.O., Moon, S.H., and Kim, K.W. (2000b). Removal of heavy metals from soils using enhanced electrokinetic soil processing, *Water, Air, and Soil Pollut.***125**: 259-272.

32. Kim, S.O., Moon, S.H., and Kim, K.W. (2000c). Removal of heavy metals from wastewater sludge using an electrokinetic technique, *Proc. of the 5th Inter. Symp. on Env. Geotech. and Global Sustain. Develop.*, COD 284, ID235.

33. Klischenko, R., Kornilovich, B., Chebotaryova, R., and Linkov, V. (1999). Purification of galvanic sewage from metals by electrodialysis, *Desalination*, **126**: 159-162.

34. Koivula, R., Lehto, J., Pajo, L., Gale, T., and Leinonen, H. (2000). Purification of metal plating rinse waters with chelating ion exchangers, *Sep. Sci. Technol.*, **34**(1): 41.

35. Kuhn, A. and Mason, R. (1996). Using a spreadsheet to characterize the performance of electrolytic metal recovery cells, *Met. Finish.*, **94**(4): 57.

36. Lageman, R. (1993). Electro-Reclamation: Applications in the Netherlands, *J. Env. Sci. Technol.*, **27**(13): 2648-2650.

37. Leinonen, H., Lehto, J., and Makela, A. (1994). Purification of nickel and zinc from wastewaters of metal-plating plants by ion exchange, *Reactive polymers*, **23**(2):221.

38. Li, X., Coles, B.J, Ramsey, M.H., and Thornton, I. (1995). Sequential extraction of soils for multielement analysis by ICP-AES, *Chem. Geol.*, **124** :109-123.

39. Lumbroso, R. (1996). A new method of producing metal hydroxides by electrodialysis, *Spec. publ.-Rus. Soc. Chem.*, **182**: 421.

40. Mitchell, J. K. and Yeung, T. C. (1991). Electrokinetic flow barriers in compacted clay. *Tranpotation Research Records, No. 1289,* National Research Council, Washington DC. 1-9.

41. Mitchell, J.K. (1993). *Fundamentals of soil behavior*, John Wiley & Sons, Inc., pp.228-292.

42. Pamukcu, S., Khan, L.I., and Fang, H-Y. (1990). Zinc detoxification of soils by electroosmosis, *Geotech. Engrg 1990. Transportation Research Record 1288*, Transportation Research Board, National Research Council, Washington, D.C., U.S.A., 41-46.

43. Pamukcu, S. and Wittle, J. K. (1994). Electrokinetically enhanced in situ soil decontami-Nation, in: *Remediation of Hazardous Waste Contaminated Soils*, D.L. Wise and D.J. Trantolo (Eds.), Marcel Dekker, Inc., New York, U.S.A., 245-298.

44. Pesavento, M., Biesuz, R., and Cortina, J.L. (1994). Sorption of metal ions on a weak acid cation-exchange resin containing carboxylic groups, *Analytica Chimic .acta*, **298**(2): 225.

45. Pletcher, D., and Walsh, F.C. (1990). *Industrial electrochemistry* (2nd Ed.), Chapman and Hall, New York.

46. Pletcher, D. (1992). Electrochemical technology for a cleaner environment-fundamental considerations, in: Electrochemistry for a cleaner environment, Genders, J.D. and Weinberg, N.L. (Eds.), The Electrosynthesis Company Inc., New York, U.S.A., pp. 207-219.

47. Probstein, R. F. and Renaud, P. C. (1986). *Proc. Workshop on Electrokinetic Treatment and its Application in Environmental Geotechnical Engineering for Hazardous Waste Site Remediation.* Univ. of Washington, Seattle, Aug. 4-5. Hazardous Waste Engineering Research Laboratory, EPA, Cincinnati, U.S.A.

48. Probstein, R.F. and Hicks, R.E. (1993). Removal of contaminants from soils by electric fields, *Science*, **260**(5107): 498-503.

49. Probstein, R. F. (1994). *Physicochemical Hydrodynamics*: An Introduction (2nd Ed.) John Wiley & Sons, Inc., New York, U.S.A., 203- 207.

50. Ramachandraiah, G., Thampy, S.K., and Narayanan, P.K. (1995). Separation and concentraion of metals present in industrial effluent and sludge samples by using electrodialysis, coulometry and photocatalysis, *Water Treat.*, **10**(3): 235.

51. Rawat, J. P. and Bhardwaj, M. (1999). Sorption equilibria of some transitional metal ions on a chelating ion-exchange resin, Duolite ES 467, *Adsorp. Sci. Technol.*, **17**(9): 741-760.

52. Rødsand, T., Acar, Y.B., and Breedveld, G. (1995). Electrokinetic extraction of lead from spiked Norwegian marine clay: Characterization, containment, remediation, and performance in environmental geotechnics, *Geotechnical Special Publication, ASCE,* New York, **2**(46): 1518-1534.

53. Rockstraw, D.A. and Scamehorn, J.F. (1997). Use of electrodialysis to remove acid, salt, and heavy metal mixtures from aqueous solutions, *Sep. Sci. Technol.*, **32**(11): 1861.

54. Rose, A.W., Hawkes, H.E., and Webb, J.S. (1979). *Geochemistry in Mineral Exploration* (2nd Ed.), Academic Press, pp.195-206.

55. Shapiro, A. P. (1990). Electroosmotic Purging of Contaminants from Saturated Soils. Ph.D. dissertation, Massachusetts Institute of Technology, Cambridge, U.S.A.

56. Shapiro, A.P. and Probstein, R. (1993). Removal of contaminants from saturated clay by electroosmosis, *J. Env. Sci. Technol.*, **27**(2): 283-291.

57. Smith, W.H. and Foreman, T. (1997). Electrowinning/electrostripping and electrodialysis processes for the recovery and recycling of metals from plating rinse solutions, *Sep. Sci. Technol.*, **32**(1-4): 669-679.

58. Takahashi, K., Sakurai, H., Nii, S., and Sugiura, K. (1995). Multi stage electrodialysis for separation of two metal ion species, *J. Chem. Eng. Japan*, **28**(2): 154.

59. Tessier, A., Campbell, P.G.C., and Bisson, M. (1979). Sequential extraction procedure for the speciation of particulate trace metals, *Anal. Chem.*, **51**: 844-851.

60. West, L.J. and Stewart, D.I. (1995). Effect of zeta potential on soil electrokinesis: Characterization, containment, remediation, and performance in environmental geotechnics, *Geotechnical. Special Publication, ASCE*, New York, **2**(46): 1535-1549.

61. Widner, R.C., Lanza, M.R.V., Sousa, M.F.B., and Bertazzoli, R. (1997). Electrolytic removal of metals from industrial wastewater, *Plating Surf. Finish.*, **84**(10): 59.

62. Yang, G.C.C. and Tsai, C.M. (1998). A study on heavy metal extractability and subsequent recovery for a municipal incinerator fly ash, *J. Hazard Mater.*, **58**(1): 103.

63. Yeung, A.T., Hsu, C., and Menon, R.M. (1996). EDTA-enhanced electrokinetic extraction of lead, *J. Geotech. Engrg., ASCE*, **122**(8): 666-673.

64. Zarraa, M.A. (1992). Mass transfer during the removal of dissolved heavy metals from wastewater flows in fluidized beds of ion exchange resins, *Chem. Eng. Technol.*, **15**(1): 21.

65. Zherebilov, A.F., Bek, R.Y., and Zamyatin, A.P. (1996). Electrolytic recovery of noble metals from dilute cyanide solutions, *Russ. J. Appl. Chem.*, **69**(7): 960.

66. Zhou, C.D., Stortz, E.C., Taylor, E.J., and Renz, R.P. (1996). Simultaneous multi-metal recovery with an electrochemical ion exchange process, *Proc. AESF Annu. Technol. Conf.*, **83**: 689.

Section III

Case Studies

9

The TEAM (TERI's Enhanced Acidification and Methanation) Process for Biomethanation of Organic Municipal Solid Waste

BINDIYA GOEL, DINESH PANT, KUSUM JAIN AND V.V.N. KISHORE

*Tata Energy Research Institute, Darbari Seth Block, Habitat Place,
Lodhi Road, New Delhi-110003, India*

Introduction

Municipal solid wastes (MSW) are heterogeneous mixtures of paper, plastic, cloth, metal, glass, organic matter, etc. generated from households, commercial establishments, and markets.

MSW management is gaining more and more attention these days because of its direct implication on human health and environment. Though various technologies (both thermal and biological) are available for converting solid waste into useful forms, (fuels pellets, fuel gases like biogas and producer gas, compost etc.), all are not applicable to all waste types with same efficiency. Biomethanation, a biological process for generating gaseous fuel called biogas (mainly methane gas) is most desirable for the organic fraction of municipal solid waste because of its high temperature content, good nutritive value and high biodegradability.

Realizing the potential of energy generation from solid waste in India and the fact that there are no indigenous standardized, efficient and low-cost reactor designs for treating such fibrous and semi-solid wastes, TERI began work on the development of a high-rate digester for biomethanation of organic solid waste in 1996. TEAM (TERI Enhanced Acidification and Methanation) process, for which a patent has been filed, is the culmination of these efforts.

This chapter deals with waste-to-resources technology options for MSW processing, TERI's TEAM process for biomethanation of MSW with brief introduction and problems associated with other European technologies existing in India. The chapter also emphasizes on policy and research needs in the field of solid waste management.

Technology Options for MSW Processing

Thermal Method

Various thermal methods like pyrolysis, pelletization or RDF technology; gasification, incineration have been developed for solid waste treatment. However these are not very successful with Indian MSW because of its high moisture content and low calorific value which increase the cost of initial drying. Pelletization is generally used for low moisture content (<15%) wastes. Incineration is more suited for high calorific value wastes like papers, plastics and hazardous wastes like hospital wastes, etc. Excessive moisture and inert content of MSW decreases the net energy gain from the process. Nevertheless, in the present century, the viability of incineration is decreasing both economically as well as environmentally. Emission of particulate, SOx and NOx, chlorinated compounds ranging from HCL to dioxins (a very powerful air pollutant) and heavy metals are a cause of concern for which very sophisticated testing and pollution control equipment are required which increase the cost of operation. Accumulation of toxic metals in ash residue is another problem which has subjected the method to more scrutiny.

Biological Method

Composting, vermicomposting and biomethanation by anaerobic digestion are the most popular biological options of the present times.

Composting

Scientific composting of various organic components of the MSW can suppress the GHG emissions and leachate formation from the waste. Many countries, for the landfilling of MSW are considering this option. The product contains nutrients and micro-nutrients essential for plant growth. However, the composting plants have to be maintained properly as they may affect the surrounding and the neighbouring areas. They are potential sources of odour, breeding places for flies and rodents and can produce immature manure if proper aerobic conditions are not maintained.

Composting pile bulk density (porosity) nutrients, balance (C:N ratio), PH, oxygen and moisture content, temperature and retention time are the important process parameters for composting process. Excessive moisture in composting piles often leads to anaerobic conditions resulting in formation of volatile organic acids, which cause odour related problems. Being heavier than air, they tend to stay close to the ground and can travel for miles. Insufficient moisture is unable to hold ammonia (generated by degradation of protein in MSW) in solution resulting in slow microbial activity, loss of nitrogen and smell problem due to evaporation of ammonia into the air. In a forced aeration system, where blowers are used to provide air, excessive airflow can be cause of moisture loss. The problem is aggravated by the channeling of air within the piles at high airflow rates. The overall result is non-uniform compost that contains immature material.

Bulk density of the composed pile is another criteria in successful composting. If the material is too dense (bulk density greater than 1000 Ibs/cy) frequent turning and agitation would be needed to maintain air in the pile. For densities over 1200 Ibs/cy, the material would be too dense to adequately aerate even with turning. A poorly mixed pile that has hot and cold spots would retain pathogens and weed seeds.

Materials that contain oils, such as some types of fish, can cause problems in the composting process. They can be composted in amounts less than 10-20% of the mix weight. Additional curing time may be necessary to obtain mature compost from oily feedstock as they break down more slowly. Addition of fast decomposing material like horse manure may be required to accelerate the process. There is also the persistence of an oily odor that lasts for weeks into the composting process, even after the material is no longer visible. The oil imparts a "wet look" that makes an otherwise dry pile appear moist which can be misleading to the operator.

The compost is a soil conditioner only when its properly prepared and readily biodegradable constitutes have been converted into stable form. Normally it takes six weeks of active composting and additional curing time to get mature compost. An immature product continues to heat even inside the bags. It may contain high levels of VOAs, ammonia and readily available carbon, the first two being harmful to plants and germinating crops. There should be strict and proper procedures to check that the product quality meets the requisite standards before it leaves the plant site as manure.

Biomethanation

In the past few years, biomethanation by anaerobic digestion has become another important tool in the management of MSW. Its a key method for both solid waste reduction as well as recovery of renewable fuel and other valuable product. In comparison to composting it's a net energy producing process and the land requirement is less though the initial cost is comparatively high. Apart from being a source of renewable energy, there are other benefits like pathogen reduction, complete organic stabilization after post-composting, and production of high quality compost. The considerable reduction in the volume of MSW is an added incentive.

DRANCO process (Belgium), VALORGA process (France), BIMA process (Austria) two phase digestion (IGT, UAS), the slurry based process (Italy/Spain) etc. are some of the processes developed by European countries for biomethanation of solid wastes. However, the application of this technology in the field of solid waste management is still at developmental stages in India and other developing countries.

Preliminary Work Undertaken at TERI

Realising the high biomethanation potential of organic fractions of MSW, TERI began work on development of a suitable process in 1996. The initial research undertaken in the batch reactors at constant temperature indicated a high possibility of extraction of the organic content by periodic recirculation of water through a fixed bed of organic solid waste. The continuous digestion studies were carried out at different flow rates and waste to liquid ratio, in order to optimize the system for maximum extraction of organics. The liquid extracted during the waste bed digestion was tested in a UASB reactor for methanation. The result was production of a high purity biogas. All these studies led to development of a high rate process for digestion of fibrous and semi solid organic waste, named as TEAM (TERI's Enhanced Acidification and Methanation) process. Presently a 50-kg/day capacity TEAM Process plant is successfully operational at TERI's Gual Pahari campus, Gurgaon , Haryana. The plant is generating high calorific value gaseous fuel (biogas) and manure with high mineral value from vegetable market waste from the past two years without any process or operational problem.

The TEAM Process

TEAM is a two-phase approach to the biomethanation of organic solid waste. The process flow chart is attached as Fig. 9.1. In the first phase, known as the acidification phase, hydrolysis and fermentation of the organic matter takes place under anaerobic conditions. The shredded waste (2-3 cm) is fed into the acidification reactor. Raw water is sprinkled over the bed to wash off the organic products (mainly lower fatty acids) formed as a result of bed digestion. This also helps in uniform distribution of nutrients and micro-organisms over the entire bed. The filtrate is periodically (daily thrice for 20 minutes each time) recirculated through the bed at a constant rate for 6 days. The final product is a low pH and high organic strength (in terms of COD) liquor called leachate. The solid residue left behind is good quality compost after drying. Since the acidification phase has a retention time of 6 days, 6 reactors are provided to have a continuous operation.

Battery of Acidification Reactors

The second phase, i.e. Methanation takes place in the UASB reactor at a pH of 6.5-7.5

The lechate is neutralized by addition of NaOH. The neutralized leachate is fed into the UASB reactor. The microbial consortia present in the UASB sludge destroy 90% of the COD forming biogas. The gas was analyzed for 70% - 75% methane (a high calorific value fuel), carbondioxide and nitrogen. Once the process is stabilized, a part of UASB overflow is used for neutralizing the leachate. Table 9.1 gives the operational parameters of the present TEAM process plant.

Salient Features

- Low retention time (6 days) decreases the total space and time requirement for the TEAM process as compared to conventional single phase processes (30-40 days) or composting (3 months) thereby making it economically viable.
- There is complete elimination of engineering problems like scum formation, floating of feed material leading to incomplete digestion, blockage of inlet and out let pipes, difficulty in feed flow, etc. as faced in small-scale commercial plants.

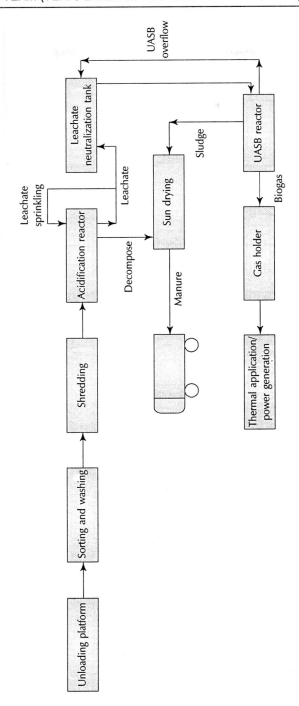

Fig. 9.1 Process Flow chart for the TEAM process.

Table 9.1. Oprational parameters for TEAM process

Acidification Reactor

No. of modules	6
Retention time	6 days
Maximum COD	20.07 g/l
Total VFA	20.14 g/l
TS extraction	30%
Manure generation(dry wt.)	3-5 kg/day

UASB

No. of reactors	1
Retention time	16 h
COD reduction	90 %
Biogas production	0.45 ml/kg COD reduced (1.1 ml/day max.)
Methane content	70-75 %
Average granule size	3 mm
Sludge volume index (SVI)	5 ml/g

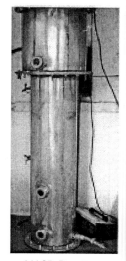

UASB Reactor

- The UASB reactor overflow is reused for leachate neutralization and in the acidification reactor, which decreases the water requirement for the process.
- Low pH in the first stage kills the pathogens and weed seeds.
- The decrease in total volume of the feed stock after decomposition is more than 50%.
- High calorific value biogas is generated, which can be used for power generation or thermal applications.

Sludge granules

- There is decrease in the cost of gas purification, gas holder size, booster blower capacity, gas conveying pipeline and burner size because of partial removal of Carbon dioxide (having no fuel value) in the first phase resulting in the high methane content in the biogas.
- High quality compost is generated. Table 9.2 gives a comparative study of the manure with other fertilizers.
- Technology suitable for adoption by small entrepreneurs.

Table 9.2. Comparison of acid reactor residue with other manures.

Nutrients	Acidification reactor residue	Farm yard manure	Farm litter compost	Compost (urban)	Compost (rural)	Cattle dung slurry	Green manure
Nitrogen (%)	2.1	0.5-1.5	0.5	1.0-2.0	0.4-0.8	1.6-1.8	0.5-0.7
Phosphorus (P_2O_5%)	1.6	0.4-0.8	0.15	1.0	0.3-0.6	1.1-2.0	0.1-0.2
Potassium (K_2O%)	2.4	0.5-1.9	0.5	1.5	0.7-1.0	0.8-1.2	0.6-0.8

Problems, Policy and Research Needs

In order to facilitate and ensure the successful implementation of any waste to resource technology, it is necessary that source segregation of the waste be practised. The Indian MSW has only 42% organic matter 18-20% recyclable and rest all is inert. There is a need to have decentralized plants for the treatment of organic fraction of MSW before it gets mixed with other components forming a heterogeneous mixer which is difficult to handle.

DRANCO process, BIMA digester, and Wabio process are the three European technologies available in India for anaerobic digestion of organic solid waste. The problem associated with these imported technologies is that their performance has not yet been established since there are no operating plants in India and the composition of Indian MSW is different from US and European MSW. These technologies are not suitable for small-scale decentralized application. Moreover the cost of technical know how is very high. In order to promote development of indigenous technology, R&D activities and full-scale development of these technologies till they establish themselves as commercial successes, need institutional support. Training and financial support should be provided for small entrepreneurs to help commercialisation and dissemination of the indigenous technology.

10

Industrial Biomethanation Practices for Decentralized Energy from Waste: Options for Sri Lanka

Ajith de Alwis

*Department of Chemical and Process Engineering, University of Moratuwa,
Moratuwa, Sri Lanka*

Introduction

This paper is presented with a view towards stimulating interest and awareness. With the impending energy crisis (due to resource limitations) and with the continuous necessity to fulfill immediate requirements as well as satisfy future growth Sri Lanka will have to look at options which so far have been looked down upon! Waste offers a unique opportunity of extracting at least part of our energy requirement. The parallel advantage from exploiting this option is the feasibility of effectively taking care of the rapid environmental degradation that is taking place right around us (de Alwis, 2000). Central Bank, 1999 states pollution as one problem facing the Sri Lankan industry. Industrial biomethanation thus offers twin benefits of energy and environmental management.

Resource Limitation—A Fact to Acknowledge

Energy is the primary requirement for all our activities. The country's per capita energy consumption indicates its position with respect to development, standard of living, type of activities carried out, etc. Sri Lankan per capita electricity consumption, which is 4 GJ, is only ahead of Nepal and Bangladesh and this is indicative of our position. We do not have access to conventional resources such as oil and coal that powered industrial nations to their present positions. All these countries have realized the limitations of their fossil fuel resources and have initiated various programs to develop renewable energy technologies, demand side management (DSM) practices and to exploit ways of obtaining energy in a decentralized manner with much better and inherent efficiencies. It was stated by a 1992 Japanese study that the resources of oil, natural gas and coal last, using the 1992 consumption rate and economics of extraction, to 42, 65 and 250 years respectively. It is seen that out of traditional fuel resources only coal gives us any breathing space to develop and utilize other options. Sri Lanka should understand this serious limitation that is going to face us and should plan accordingly. Our transportation sector is totally dependant on petroleum fuels and the future electricity demand is expected to be met mainly through thermal means (via fossil fuels again) as the exploitable hydro resources are more or less at an end.

Sri Lankan Energy Situation

Today the total electricity generation capacity is 1691 MW (Hydro 1143, Thermal 545 and Wind 3). In 1999 total power generation inclusive of self-generation was 6184 GWh. The Ceylon Electricity Board (CEB) is responsible primarily for electricity generation and distribution. The losses in this system have increased to 20.7% from 18.8% in 1998 due to an increase in distribution losses. This situation itself is unacceptable. Ceylon Petroleum Corporation is primarily responsible for the petroleum fuel sector in the country except for LPG, which is handled by Shell Gas Lanka Ltd. 85% of the country's electricity consumption is in urban areas. Electrification in the Western province where the bulk of the industries are found is comparatively higher and is at 73%.

The SL Governments policy on electricity is biased towards centralized main grid supply. It lays prominence on large-scale power generation and with little emphasis laid on micro-level power supply. Alternative energy options have not been pursued by governments, mainly because they do not provide energy on large scales, to satisfy economic growth rates. Within the CEB the division entrusted with the pursuing of these objectives has been termed the pre-electrification unit. However, it is not certain whether the CEB considers significant industrialization or the industrial demand as an important element in their generation planning process. The main emphasis is on home electrification, which is also a political goal.

In energy generation, today there are options available to simultaneously take both energy and environment into consideration. The world is slowly but surely moving into a renewable energy driven economy even though it would still take some time for a total switching to take place. However, what one should note is that energy demand drivers of 70's have been more or less replaced by environmental drivers. It is also important to realize that energy lies at the core of this environmental dilemma. Hence instead of solving problems piecemeal and separately it may be wiser and profitable for Sri Lanka to look at tackling both problems simultaneously.

Inclusion of power generation via garbage or in engineering parlance, waste-to-energy plants in the generating plan is a possibility. This would mean solving two problems. Solid and liquid waste management and simultaneous energy generation. In some instances improved efficiencies could be had via co-generation systems. Neither supporting infrastructure, nor policy support is yet available in implementing this in Sri Lanka.

Waste as a Resource

Waste as an energy generating raw material concept is not new. Developed countries have practised this method for many years. There are many options available to the user when looking at waste-to-energy and many more options are emerging. Direct energy generation as well as indirect energy conservation is possible with utilizing waste.

Methods of Indirect energy conservation via waste

For example if the glass industry utilizes cullet the industry stands to save vast amounts of energy while lengthening the natural resource availability. Similarly if plastic waste is recycled the crude oil resource is saved, as well as saving energy in producing pellets. These are examples of environmentally sound waste management strategies that conserve energy simultaneously. It is important to realize that energy and environment are intricately linked and are not two separate entities.

For Sri Lanka it is important to realize that conservation is an ideal option at the short- and medium-term energy planning. The inherent wasteful resource utilization should be examined and rectified.

Methods of direct energy generation via waste

The following are the main means of producing energy via waste:
- Briquetting of agricultural residues, residues from timber operations (Refuse Derived Fuel - RDF)
- Biomass gasification
- Anaerobic digestion of wet biomass
- Fermentation of wet biomass substrates.

Except for the last option Sri Lanka has experience in projects and systems that looked into the first three activities. However, most of them had never reached a sustained commercial application stage. In the first case of biomass briquetting it has been the economics and cost of electricity to carry out briquetting that has affected the viability of the process. There have been some commercial ventures into this area, though the status is not well known. The only available evidence is that in certain parts we have significant problems due to agricultural residue (paddy husk) and timber operations (saw dust). If incentives are given for solid fuel fired systems to be used in industrial processes the economic outlook for this may change. The waste mounts of coir dust is now being consumed for briquetting to be used in horticulture and also for activated carbon production and thus no longer is a problem. Figure 10.1 shows disposal of sawdust by industrialists directly on the sea front causing significant water pollution. This is serious waste of waste!

Fig. 10.1 Sawdust disposal by timber processors along the coastal belt.

Anaerobic digestion is a mixed culture operation. Biogas, a mixture of methane and carbon dioxide, results from the first operation. This is a potentially useful gas which has many applications. Various waste products can act as substrates to this process—animal wastes, human waste, agricultural refuse and municipal solid waste. Industries should also look into waste management practices based on anaerobic digestion. All this potential is decentralized energy generation potential. The options available to the Sri Lankan industry are analyzed here.

Using biogas as an option to energy delivery systems was identified a long time ago as feasible and desirable for Sri Lanka. In fact a program of the United Nations adopted the resolutions of the *Colombo Declaration* way back in April 1974, which stated that one of the urgent priorities in the region is energy. Subsequently, a project for the development of anaerobic digestion throughout Asia was approved in November 1974 (Amaratunga, 1977; Barnett *et al*, 1978). However, much of Asia had really not benefited from developments in this area and energy shortage problems are quite common within the region. However, significant developments are now taking place in India with the utlization of GEF (Global Environmental Facility) funding.

With biogas the energy extraction and use automatically suggests decentralized option. If the energy use is purely thermal then the gas generation and use has to be placed quite close together for the process to be effective. Biogas extraction and use has become a vibrant industrial technology that is now finding applications due to variety of reasons. Some of these are;

Conventional energy sources are depleting
The world has realized that there is only a finite time for conventional fuel resources and that the time available is also short before this resource base is depleted. Thus the need to develop renewable energy resources have intensified and it is important to have proven, reliable and commercially viable energy systems available when the crunch occurs;

Uncontrolled methane emissions—as emitted from solid and liquid wastes of organic composition under anaerobic condition – are not acceptable today as methane is a powerful greenhouse gas (GHG). Thus control of these emissions are essential from a climate change point of view. The solution that has been forwarded is the capture of this gas and utilizing it via combustion thus converting it to CO_2 and obtaining energy as well.

As a tool for sustainable development - Today the emphasis is on sustainable development and both the ways energy is used and the environment is managed have been given equal consideration. In satisfying these twin needs the biogas option provides a neat and clean development mechanism.

Figure 10.2 shows an industry in Sri Lanka (Pelwatte Sugar Company) where anaerobic digestion has been utilized in a limited manner to treat wastewater but a huge potential still exists.

The Industrial Sector: Biogas Potential from Industrial Wastes

Industrial wastewaters and solid waste streams today present a serious pollution threat to the environment. This is a of concern in Sri Lanka. As a developing country Sri Lanka is looking at industrialization for economic development and growth, and if we do not take appropriate safeguards from the outset the environment is bound to suffer adversely. Though appropriate waste management

Fig. 10.2 Pelwatte Sugar Company and Distillery.

techniques are suggested very little is done due to factors such as financial difficulties, non-availability of technology at reasonable costs and simply due to lack of understanding of the necessity and to some extent lack of commitment.

In developed countries the technology for utilizing biomethanation on an industrial scale is well developed and understood. Any planned industrialization in Sri Lanka can consider this mode of energy generation as a supplementary source and the other direct benefit would be sound environmental management.

Industrial wastewaters can be treated anaerobically and this would generate biogas, which could potentially be used *insitu*. Not all industries can make use of the anaerobic digestion process due to the requirement of the presence of high strength organic waste. The normal accepted value is that if a process generates a waste with COD greater than 2000 mg/l, there is potential for using the anaerobic process for waste treatment. Figure 10.3 indicates the scheme recommended for industries.

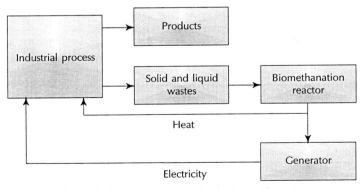

Fig. 10.3 Objective of Industrial Biomenthanation.

There is a belief that anaerobic technology is not appropriate for treatment of most industrial wastewaters due to 'toxic' compounds frequently present in such wastes. This assumption has greatly hindered widespread application of these processes to industrial wastewater. Today however, with introduction of the concepts of bioaugmentation, and with more understanding of the methanogenic bacteria and their metabolic needs, anaerobic digestion is increasingly applied to industrial wastes. Various methods of anaerobic degradation are being investigated and applied in reducing or eliminating the many toxic compounds found in chemical and industrial effluents.

Aerobic processes which use oxygen from air, are the most popular treatment processes. The following advantages of anaerobic treatment processes in comparison to aerobic treatment processes are: No energy necessary for aeration, considerably less biomass production, (thereby lowering biomass disposal costs and reducing mechanical maintenance costs) and methane gas production. It is natural that in waste treatment net energy generating processes are favoured over other type of processes, any expenditure made purely to achieve compliance from an economic position is not directly productive.

The rate of generation of biogas is important when planning such an anaerobic system for an industry. Long residence times associated with low-rate systems means that significant storage will be needed to handle waste generated at this rate. Thus the option available is to utilize high-rate systems. The adoption of high-rate systems also means that combined treatment of solids and liquids will have to be adopted.

The industrial sector data for Sri Lanka has been obtained from various project study reports (CEA, 1992 etc.) and annual surveys of industries. However, obtaining up to date data is difficult and the surveys by the Department of Census and Statistics, Sri Lanka are not frequent. This is also a deficiency which planners had to address for proper industrial development.

An Assessment of Biogas Energy Potential

This section presents an assessment of biogas generation potential for some selected industrial sectors of Sri Lanka. The intention is not to be overly accurate but to highlight the possibility. The particular locations and areas considered in this assessment are indicated in Fig. 10.4.

Industrial wastewater streams with COD levels exceeding 2000 mg/l are amenable and are more suited for anaerobic digestion. Sri Lanka has many industrial wastewater streams having this characteristic, as the following summary indicates.

Industry Type	Typical COD	No. of Factories
Tannery (leather processing)	5000	15
Food - General	3000	200
Distilleries	25,000 – 90,000	09
Breweries	1500-100,000	04
Rubber processing		
Crepe rubber	6000	
Conc. latex	25,000	
		229
Desiccated coconut	6000	65
Sugar	—	04

If these industrial establishments are made to adapt/adopt anaerobic treatment methodologies sound environmental management as well as energy generation aspects may be realized. The *insitu* generation or CHP type plants may be implemented, boosting the overall efficiency of the industry. *Most publications dealing with waste treatment technologies for Sri Lanka recommend aerobic plants for the treatment of these wastes. This is not the logical choice.* It is also seen that most plants installed are also aerobic treatment plants, which should not have been the proper choice.

Some calculations are presented below for some industrial sectors to indicate the biogas potential:

The useful part of the energy of biogas is the calorific value of its methane (CH_4) content. The other components mainly absorb energy from the combustion of CH_4. For 100% CH_4 CV – 36000 kJ/Nm^3; each 10% of CH_4 content in biogas additional CV change is 3600 kJ/Nm^3. The actual calorific value of the biogas is a function of the methane percentatge, the absolute pressure and temperature. The actual value is the most important parameter for the performance of an engine, a burner or any other application using biogas as a fuel. It is assumed in the subsequent calculations that the CV of the biogas produced to be 25.2 MJ/m^3 (70% methane content). An efficiency of 25% in mechanical to electrical energy conversion is also assumed.

One area of activity necessary in Sri Lankan industry is waste characterization. Much of the work done had been related to achieve license or for project studies and a systematic study of waste characterization reflecting the work practices, level of technology adopted had hardly taken place. Much of the generated data are not available for a serious scientific study and stagnate in close documentation. This also is an impediment to proper development.

Brewery and Distillery Industry

Breweries

Sri Lanka has several large breweries producing beer. Some factories came up to exploit the unprecedented consumption of the brew after the demand for beer going up following a reduction in price (1995 Budget of GOSL). These industries do generate a significant quantity of solid (spent grain, brewers' yeast and spent hops) and liquid wastes, which are mainly biodegradable. With careful planning this waste can be used to generate biogas and subsequently electricity for the facility.

Published literature on breweries indicate the following of brewery wastewater (as a percentage of the total volume):

Services	68.8%
Packaging	10.5%
Beer and yeast recovery	0.6%
Brew house	8.8%
Fermentation	7.6%
Chilling and conditioning	3.7%

Average characteristics of wastewater from breweries indicated are:

COD	1500-3500 mg/l	2105
SS	600-1295	840
Temp	19-27°C	22
pH	7-12	

It is important to note that these production methods and housekeeping play an important role in determining wastewater strength. Beer spillage can greatly alter the BOD_5 as beer typically has a BOD_5 of 60,000 to 90,000 mg/l.

Two options are possible in brewery waste utilization in anaerobic digestion:

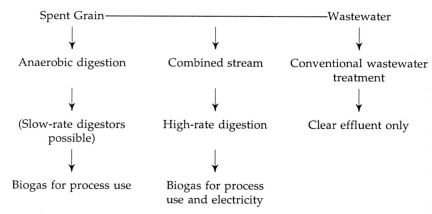

The solid waste stream of yeast should potentially be utilized in a process to produce products such as Marmite (which is an yeast extract).

This only illustrates, using a single situation, the possibilities available for an industry to handle the generated waste.

For example a factory is known to produce about 20 tonnes of spent grain per day during production time. In other countries this is sold, as animal feed though the practice is not so widespread here due to the absence of a well-developed animal feed industry. It is therefore profitable to look at the option of generating biogas.

A simple calculation for brewery waste in biogas production (using solid waste only):

Basis : 1 day

Waste solids (spent grain) : 20,000 kg

Volatile solids - 65% (It is considered that 70% of volatile solids gets reduced)

Available VS kg - 9100

For gas generation it is necessary to use practical data available or where such data are absent some generally accepted values.

1 kg of VS will yield 500 litres of biogas (70% CH_4)

(the composition of the waste, i.e. protein, carbohydrate and fats, affects the methane percentage)

Biogas produced during the process is 4550 m^3/kg VS/day

Total heat energy - 4550 × 25.2 MJ/day – 114660 MJ per day (i. e.; 1.33 MW or 31.9 MWh/day)

Electrical energy possible from this waste per day – 0.333 MWe or 8.0 MWh

For the four breweries currently operational the possible electrical energy (assuming equal capacities) – 1.33 MWe.

In many countries the conventional option adopted in brewery waste-solid handling is selling the waste as animal feed. This practice had developed over the years as breweries normally have long histories of operations. Generally yeast is pasteurised and sold as chicken feed and the spent grain is fed to cattle as a protein supplement. However, in Sri Lanka these two options cannot be practised profitably as there is neither organized large-scale livestock nor an animal feed industry. Sajeewarekha (1999) has practically and systematically investigated this hypothesis with positive results to demonstrate the theoretical calculation provided above.

Distilleries

Distilleries could either operate on molasses or toddy. The main pollutants that are emitted from a molasses based distillery are the stillage or slop and the washing up waters of the distillery vats. These could be collectively termed as spent wash. The other distilleries operate on toddy and the waste are similar.

In a distillery the anaerobic digestion of spent wash could generate energy. Toddy is the main input to the arrack distillation process in Sri Lankan distilleries. A pilot project with support from UNIDO had targetted this industry (UNIDO, 1995). The objective of this project had been the anaerobic digestion of arrack distillery and bottling plant effluents to reduce their pollution loads and to produce methane (biogas) to be used as a boiler fuel. Though many detailed studies have been done and the feasibility of the project established, its final implementation is yet to materialize though a number of years have passed.

The distilleries are in several locations (1 major facility at Seeduwa in the Gampaha district and a further 8 in the Kalutara district) and each industry has to manage the digestion process. An example for one facility of this type is presented below.
Spent wash analysis of fermented toddy:

COD (Settled)   – 25,000 mg/l
BOD_5 – 15,000 mg/l
Suspended solids   – 9,900 mg/l
Dissolved solids   – 22,200 mg/l

Measured nutrient balance (COD:N:P) - 500:0.69:1.8 (nutrient supplement is necessary to carry out the digestion process)
Effluent flow rates:
Production varies from month to month as it is tied up to the toddy supply and the effluent flow rates will vary as a result.

Average effluent production level - 91 m³/day (The wastewater flow has been noted to vary up to 151 m³ per day in a distillery)
COD load - 2275 kg /day
COD reduction in effluent - 85%
Biogas production (70% CH_4, 30% CO_2) - 0.7 m³ biogas/kg COD removed) – 1353.6 m³
The Energy Content - (@ 25.2 MJ per m³ of Biogas) – 34111.35 MJ per day
The Energy Potential (thermal) - 0.395 MW(th)
The Energy Potential (electrical) - 0.1 MWe; for 9 distilleries approx 0.9 MWe

The potential from molasses based distilleries would be higher as the strength is very much higher. The gas can be utilized in the boiler instead of using it for generating electricity.

Sugar Industry

Sugar factories

Cultivation of sugarcane in Sri Lanka began in the 1960s in Higurana and Kantalai and later at Pelwatta and Sevenagala (in the Uda-Walawe region). Higurana and Kantalai processing factories (in 1957) resulted from the Government's policy of achieving self-sufficiency in basic food items like sugar and rice though this is still to materialize. At present only the plantations at Pelwatte and Sevenagala are functioning satisfactorily with Hingurana and Kantalai factories lying idle. The total extent under sugarcane is only around 10,000 Ha with an annual production of about 65,220 MT (1999). This only constitutes 15% of the national sugar requirement. There are distilleries based on molasses functioning at both Sevenagala and Pelwatte factory sites in association with the sugar factories.

For biogas generation the pressmud stream is considered. Pressmud, which is the filtercake, is a byproduct of sugar factories. Another potential byproduct is sugarcane tops. Filter mud results during the vacuum filtration of hot sugarcane juice and contains 5-15% sugar, fibre 15-20% and wax and fat 5-14% and is a good substrate. The overall composition is 75% organic matter and 29% total solid content of which 65% is volatile solids. This mud has not much value and experimentation in India has shown the potential of this as a raw material for biogas generation. Potential for biogas production from pressmud has been demonstrated and it is reported that its biogas yield is two to three times that of cow dung. The biogas obtained can be used as a fuel for boilers in sugar factories. The outlet slurry has proved to be a better soil conditioner than pressmud. At present, the pressmud obtained from Sri Lankan factories is spread in paddy fields and sugarcane plantations as fertiliser. It does not have much value as a 'fertiliser'.

The potential for power generation using pressmud biogas plants:

No. of sugar factories:	02 (Sevenagala, Pelwatta)
Total sugarcane crushed (1999 data):	745,000 MT
(For individual factories separate harvest data could be used)	
Availability of pressmud (4% of the sugarcane crushed):	29800 MT
Annual biogas production (100 m^3/tonne) with feed having a ratio of pressmud:water::1:5	2,980,000 m^3
Daily gas production (assuming 350 working days)	8514 m^3
The gas quality is assumed to be 70% methane with 25.2 MJ/m^3 CV	
Thermal energy generated per day	2.48 MWth
Electrical energy generation potential	0.62 MWe

The Indian experiments have been conducted with the traditional KVIC systems (4 nos.) of 85m^3 capacity. The gas generation and the required mixing data are taken from Indian applications and applied to the local condition. High rate reactor systems would have better gas production rates.

The conventional approach is to send the filter cake (as pressmud is also known) back to the sugarcane fields. The Sugarcane Research Institute of Sri Lanka has commenced some work on biogas generation from sugarcane tops.

Food Industry (Food processing and vegetable oil)

All types of wastes from the food industry represent a potentially large-scale resource for biogas production. Different processes have been used for biogas production from different types of food processing wastes, such as sugar refining, dairying, cheese production, brewing, distilling and fruit and vegetable process wastes. Some positive results have been achieved from biscuit and chocolate production (wastes generated from mixing and packaging units) which are rich in biodegradable carbohydrates and fats. According to the statistics of the Department of Census and Statistics there about 2300 firms in this sector, of which 75% are small firms employing less than 25.

This sector can have significantly more wastewater emissions. The industry itself is widely scattered and a proper quantification is difficult from available data. This value of COD will be a considerable underestimation. Data are available from 47 establishments in Colombo, Kalutara districts which have been studied for their wastewater load (ERM, 1992). It has been identified that this contributes a load of 4,114 m^3 per day of total wastewater, with COD load of 12,333 kg/day.

In one industrial food processing facility where the possibility of anaerobic digestion has been proved for waste-to-energy in many facilities worldwide, the overseas principals have identified aerobic systems for wastewater treatment for their Sri Lankan enterprise. Whether the decision was intentional, as sometimes the decisions for Sri Lanka are made on the basis that systems of this nature are expected to operate by default rather than by design, is not clear. However, finally the plant has been made to treat the wastewater and the plant uses additional energy to satisfy the environmental aspects.

Desiccated coconut (DC)sector

There are 65 establishments producing DC in Sri Lanka. This is an important industry for Sri Lanka and ranks second in the league table of DC producers behind the Philippines and ahead of Indonesia.

In a typical DC mill there are three major sources of wastewater—washing of white kernel pieces, coconut water and the sterilization of ground kernel particles.

Combined wastewater characteristics are pH – 4-5.5, BOD 1000-5000, COD 4000-8000 and oil 4000 (mg/l). For this industry there are some established emission factors:

It has been estimated that 6800 nuts yield 1 tonne of desiccated coconut and 1 nut releases one litre of wastewater.

COD load estimated: 7200 kg/day (average BOD 3500 mg/l and COD 6000 mg/l)

Most of these factories are located in rural areas between the cities of Colombo, Kurunagala and Puttalam (what is known as the coconut triangle of Sri Lanka).

Leather Industry: The Tanning Sector

There are 15 establishments in this sector. All these factcries are in Colombo and Gampaha districts. However, there is a move to relocate all these factories to a single location – i.e to an industrial estate in Hambantota (at Bata-atta). This would certainly facilitate this proposal of carrying out anaerobic digestion as all waste streams would be available at a single location. At present the existing tanneries have been restricted from expanding their operations due to the environmental reasons and relocation had risen as the solution to the problem.

Emission factors:

    1 tonne of hides - 1429 sq. ft hides (cow, buffalo, goat)

    1 tonne of hides - 52 m^3 of wastewater

    Estimated Pollution load: COD load 8070 kg/day

Rubber Processing

There are about 229 medium- and large-scale establishments involved in rubber processing. However, there are several thousand of small holders involved in the industry producing dry rubber products in small scale using latex. Rubber industry is a traditional industry in Sri Lanka and it is estimated that this contributes the highest wastewater load of any industry in Sri Lanka. Being a traditional industry, practices vary significantly.

    Emission factor: 1 tonne of rubber = 29 m^3 of wastewater

    COD load: 29,040 kg/day (average BOD 2000 mg/l and COD 6000 mg/l)

Industries are concentrated in the south-western parts of the country.

The Rubber Research Institute, Sri Lanka has designed and constructed an anaerobic filter system for the treatment of rubber wastewaters and development and optimization studies are in progress. Anaerobic systems have been built in about 20–30 factories though none had been developed with the aim of utilizing energy via optimization of the quality of biogas generated.

Recent Indian research has shown that rubber processing effluent when mixed with 10% sawdust (100 g sawdust in 1 liter of effluent) is capable of producing biogas. It is stated that the sawdust left after cultivation of mushrooms or decayed sawdust is better for production of biogas. If some waste exchange mechanism is set in place, it would be possible to access the vast quantity of sawdust that is available in the Moratuwa region. There are about 1100 saw mills and furniture manufactures in this locality and their common method of disposal of sawdust is dumping into lagoons, lakes and rivers as shown in Fig. 10.1. Experimental data is not available to quantify the potential yield of biogas from this method.

The total energy potential from treating the waste from the identified sectors (about 353 industrial establishments) is as follows:

Total COD load per day:	– 56,643 kg per day
As total COD load will not degrade, assuming a 85% load reduction COD load for gas generation	– 48146.6 kg per day
1 kg COD yields 0.35 m^3 of methane with a calorific value of 36 MJ/m^3	
The total potential *combustible* gas yield:	– 16851.3 m^3
Assuming 70% of methane yield the Biogas generation	– 24073.3 m^3
The potential energy yield (@25.2 MJ/m^3):	– 606647.2 MJ/d or 7.02 MWth
The potential electricity yield from these four industrial sectors	– 1.76 MWe

Table 10.1. Summary of the Calculations Presented Earlier and the total Estimated Potential

Industry	
Sugar	0.62 MWe
Brewery	1.33 MWe
Distillery	0.9 MWe
Other (Tannery, food and drink, rubber, dessiccated coconut)	1.76 MWe
Total	4.61 MW

The calculations are of theoretical interest, although they pave the way to many questions and some answers. They demonstrate the applicability of biogas generation, the need for its decentralized exploitation and emphasises the independence or partial independence achieved by the user, potential savings in bought energy and the bringing about of environmental compliance with positive gain as some of the obvious advantages of following the anaerobic waste treatment strategy at an industrial level.

Some possible industries where some segregation at source or pre-treatment is necessary, such as textile, paper and pulp and pharmaceutical industries, are not included in this calculation. These industries can also utilize the anaerobic digestion technologies. It should be noted that about 75 % of the industries in Sri Lanka have no treatment systems of any type, and discharge their wastes into common drains, nearby streams, rivers, canals, etc. Most available systems utilize aerobic treatment facilities. The sludge generated is disposed again in an ad-hoc manner without much control. These primary and secondary sludges can be profitably utilized when harnessing anaerobic systems.

The total generation potential (electrical) for the industrial sector, is shown to be around 4.61 MWe. The values given above can be modified in the course of further studies. It is important to note that a similar study for UK (DTI, 1993) has come out with a total theoretical energy resource from treating the UK's industrial wastewaters by anaerobic digestion to be in the region of 500 to 700 MW (electrical). This is an indication also of the state of industrialization between the two countries. Thus with more industrialization and especially with developments in the food and agro-processing sector the anaerobic treatment options and potential would increase.

The country is looking forward for development and growth in this sector and these methods would have a major role to play in improving the sectoral performance and competitiveness as a whole.

A strategy for implementation of Anaerobic digestion as decentralised energy delivery systems in Industry

A biogas system's value as an environmental management technique is a key factor in utilizing the technique across sectors such as the urban environment, industrial waste treatment and livestock management facilities. The utilization of a biogas system within these sectors will remove these places at least partially from the main energy supply chain. It is for this partial displacement that we should be planning.

Location of Breweries

Nuwara–Eliya	[1]
Meegoda	[2]
Biyagama	[3]
Mawathagama	[4]

Distilleries

Seeduwa	[5]
Kalutara district	[6]
[all toddy distilleries]	
Molasses based distilleries	
Pelwatte	[7]
Sevenagala	[8]

Sugar

Hingurana	[9]
Kantalei	[10]
Pelwatte	[7]
Sevenagala	[8]

Desiccated Coconut

DC triangle – Kurunagala, Puttalam, Colombo	[11]

Tanneries

Colombo distrist

Bata–atta	[12]

Food Processing

Colombo, Gampaha Kalutara

Districts (only considered)	[6]

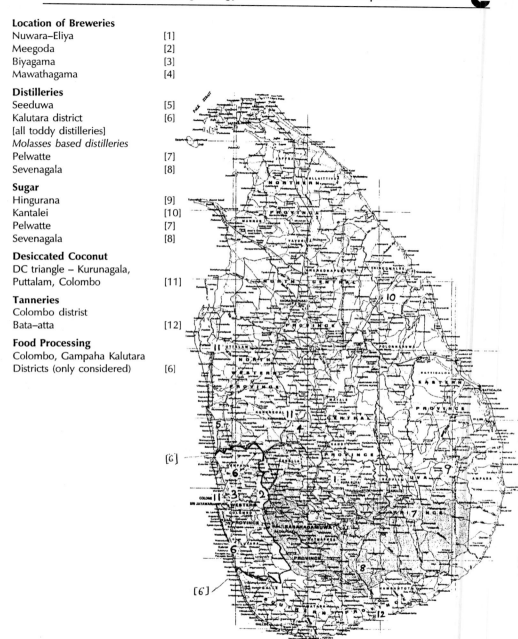

Fig. 10.4 Sri Lanka – Identification of Places considered in this paper
[no. identify the location in the map]

However, to realise the biogas potential, several important additions are necessary to what is currently available. These should form essential steps in an overall strategy.

1. Awareness of the problem (in terms of both energy and environment) and the consequent need for innovative solutions. As the industry is becoming more and more interested in certification schemes such as ISO14000 it is important that they realize the benefits of schemes such as integrated waste and energy management.

2. Industry should engaged in applied R&D and in pilot projects in this area. Once the potential is understood the successful implementation needs further development and this aspect should also be understood. The government should have necessary supporting taxation systems for industries engaging in such efforts.

3. Emergence of private entrepreneurs with the ability to install and support biogas systems. For industrial systems it is possible to look at making the technology currently available externally, such as BIOENERGY (from Biomechanics), CLEAR (Clear), ANAMET (AC-Biotechnics), BIOMASS (Biomass Int.), BIOFAR (Degremont), HYAN (Gore & Storrie), under license to the local market. The market and monitoring mechanisms practiced in Nepal are of interest for small systems in Sri Lanka. It is important to understand that high-rate systems are necessary in the industry sector if biomethanation is to make an any impact (Nyes and Thomas, 1998).

4. Since food and agro industries have the potential to dominate the Sri Lankan production and manufacturing sector, one can always envisage that if successful use of biogas systems is demonstrated, our emerging industries will stand to benefit from the knowledge base that will be readily available within the country. Successes at the industrial sector level may in turn encourage the small-scale use of low-rate systems at societal level.

REFERENCES

1. Amaratunga M (1977), Biogas in an Integrated Farming System, *Agricultural Engineering*, Vol. 1, No. 1, 1977 University of Peradeniya.

2. Barnett A., Pyle L. and Subramanian SK (1978), *Biogas Technology in the Third World: A Multidisciplinary Review*, IDRC Publication.

3. CEA (1992), Industrial Pollution control guidelines No 1-5 (Natural rubber, concentrated latex, desiccated coconut, leather and dairy).

4. Central bank (1999), *Central Bank of Sri Lanka Annual Report*.

5. DTI (1993), *Anaerobic Digestion of Industrial Wastewater: A Survey of Potential Applications in the United Kingdom Industry*; ETSU B 1294.

6. De Alwis AAP (1999), *Solid Waste Management in Sri Lanka: A Country Perspective, Solid Waste Management* (VI Grover, BK Guha, W. Hogland, S.G. McRae, Eds), pp. 203-233, Oxford & IBH Publishing Co. Pvt Ltd., India.

7. Nynes and Thomas S. (1998), *Biogas from Waste and Wastewater Treatment*, CD from James & James, UK.

8. Sajeewarekha S (1999), *Biomethanation as a process waste recycling technique with special reference to brewery spent grain*, MPhil Thesis, The Open University of Sri Lanka.

9. UNIDO (1997), *United Nations Industrial Development Organisation Project DG/SLR/91/019*.

11

Energy from Landfill Gas Utilization in Canada

EDWARD A. MCBEAN[1], RICK MOSHER[1] AND ALAIN DAVID[2]

[1]Conestoga-Rovers & Associates, 651 Colby Drive Waterloo, Ontario NZVICZ Canada
[2]Environment Canada, 351 St. Joseph Hull, Quebec, KIA OH3, Canada

Background

Although Canada as a country is large, approximately 65 % of the population lives in the largest 25 cities. Hence, there is a significant concentration of population in relatively few urban centres and along with this concentration, much of the municipal refuse generated by society. A substantial majority of the municipal solid waste (MSW) has been landfilled in the environs of the largest urban centres.

The proximity of the MSW landfills to urban centres in combination with Canada's stringent environmental regulations for protection of humans and environment, lead to a rapid evolution in the technologies utilized in landfill disposal practices. This included implementation of substantial control to prevent excursions of leachate releases to groundwater and gas emissions through the soil media or into the ambient air, thereby avoiding odour impacts for the neighbouring residents. In the past landfill gas has been collected for preventing offsite migration of landfill gas as a result of the explosion hazards and the ensuing health risks to neighbours.

As a consequence of these issues, combined with awareness of the energy in the gases arising from decomposition of organics within landfills, there has been a lengthy history of landfill gas collection for purposes of utilization. Canada's first utilization of landfill gas occurred at the Ottawa Street Landfill in Kitchener, Ontario in the mid-1970s when there was concern of offsite migration of landfill gas at hazardous levels to a nearby school and residential development. In response, there was collection and delivery of the landfill gas (LFG) to a nearby cement kiln, for both the destruction of the landfill gas odours and hazard potential and for utilizing the energy content of the LFG to offset the otherwise incurred fuel for use in the cement kiln.

Building from this initial undertaking, there have been a number of landfill gas collection projects undertaken in Canada. This included the construction and operation of a greenhouse in St. Thomas, Ontario (Crutcher *et al.*, 1981a and b) where the landfill gas was utilized to heat the greenhouse gas facility for several years for demonstrating the growing of tomatoes.

It is important to emphasize that interest in landfill gas recovery in the 1970s arose as a result of concern for protecting adjacent properties (humans and the environment) and the escalating energy prices being encountered at the time. Interest in energy generation from waste continued for these same reasons over the next two decades at a modest level. However, with the advent of climate change concerns over the last five years and escalating cost of energy, the intensity of interest in landfill gas collection has started to change dramatically. Interest in energy from landfill gas utilization landfills is poised to dramatically change the number of projects involving recovery and utilization of landfill gas in Canada as described below.

Landfill Gas as an Energy Source

The generation of landfill gas (LFG) is a result of the biological decomposition of solid wastes placed in a landfill. Biological decomposition takes place in stages and all the decomposition phases will occur simultaneously in different zones of any active landfill. However, within a few months to years following site closure, the anaerobic stage typically becomes and remains dominant while organic matter for decomposition is available. As the landfill refuse ages, gas production rates decrease over time as a result of changes in the refuse. The changes which occur in the refuse arise because of the refuse itself (the diminishing quantities of easily biodegradable organics from which the LFG is derived) and secondly, the impacts of design features of the landfill. The design features may inhibit, for example, the influx of moisture, a critical element for the decomposition processes. Hence, the total amount of LFG generated depends upon the organic content of the refuse and to varying extent, on other parameters that control LFG generation rates.

The rate at which landfill gas is generated depends on many factors. Continuing decomposition and gas production can be expected for up to 30 to 100 years. The peak rate of decomposition and associated gas production typically occurs within 2 years following site closure and then begins a gradual decline thereafter. LFG yield and LFG generation rates are the fundamental parameters that form the basis for estimating LFG production. The LFG generation rate over time has been widely accepted as characterized by the Scholl Canyon model, of the form

$$Y = k\, L_0\, e^{-kt} \tag{1}$$

where Y = LFG yield (l/kg.yr)
         L_0 = LFG production potential (l/kg)
         k = rate coefficient (1/yr)

The input parameters to the mathematical model in [1] are dependent upon the composition of the solid waste, the age, and the temperature.

Because of the uncertainty of the parameter definitions for utilization in [1], it is common practice to develop a range of projected LFG production rates, such as indicated in Fig. 11.1. The results illustrated in Fig. 11.1 typify the situation in Canada wherein a landfill is operated for a period of more than 25 years which, because of the time over which generation of the degraded gases occurs, results in the opportunity for utilization, encompassing a very lengthy period of time, potentially in the order of 40 to 50 years.

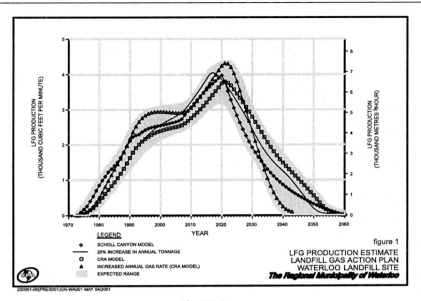

Fig. 11.1

As a result of the ongoing biological processes within the landfill refuse, and the relative absence of oxygen under conditions normally encountered in landfills, the predominant gaseous byproducts are methane (CH_4) and carbon dioxide (CO_2), and much lower quantities of hydrogen sulfide, nitrogen, and hydrogen. The net result is LFG which is comprised of approximately 50 % CO_2 and 50 % CH_4. As a consequence of the presence of methane, raw LFG typically has a caloric or heating value per unit volume approximately one-half that of natural gas. Although the caloric value of LFG is lower than natural gas, it is still very substantial. The fact that caloric value of LFG is lower than that of natural gas is due to the lower CH_4 content. The caloric value of LFG (50 % CO_2 and 50 % CH_4) is 16.9 MJ/m^3 (450 Btu/ft^3) as compared to 37.3 MJ/m^3 (1000 Btu/ft^3) for natural gas. This high caloric value of LFG is the basis why LFG is of interest as a potential energy source and also one of the reasons why LFG is considered a hazard to nearby residents due to its potential for collection to explosive concentrations.

Increased Interest in Landfill Gas Recovery as a Result of Global Warming

Interest in landfill gas recovery underwent a significant increment during the mid 1990s. During this period it became widely recognized that global increases in atmospheric concentrations of greenhouse gases (GHGs) such as CO_2 and CH_4 have adverse influences on climate through global warming. Opportunities for reduction of LFG emissions to the atmosphere, with its composition of global warming gases such as CO_2 and CH_4, have become one of the many initiatives for reducing anthropogenic influences on global climate change.

The global warming potential (GWP) provides a measure of the relative effects of emissions of GHGs to the atmosphere. The benefit of LFG utilization to the environment is that each mass unit of CH_4 that is captured and utilized reduces the GWP by approximately 21 times the equivalent potential of each mass unit of CO_2.

This occurs because of the higher GWP associated with CH_4 in comparison with CO_2. Hence, LFG combustion represents a significant opportunity for reducing global warming concerns in an economically responsible manner using proven technology. LFG is a relatively "clean burning" fuel when compared to most other fuels.

LFG's contribution to global warming can the reduced by any combustion technology. Hence, active LFG capture through an LFG collection system and flaring systems are highly effective for mitigation of on-site and off-site LFG impacts and also for reduction of GHG emissions. Utilization of LFG as an energy source reduces emissions of CO_2 equivalents just as much as flaring the gas. Further indirect benefits are realized if LFG is utilized to replace some other fossil fuels.

It is noted that landfill gas generated through the anaerobic decomposition of organic wastes in landfills is one of the most significant sources of anthropogenic (man-made) methane in Canada (approximately 25 % of the total). Hence, in response to the Kyoto Protocol, there is a great deal of interest in LFG recovery and utilization. As a result, LFG recovery and combustion has the potential to become an important element of Canadian initiatives in the federal climate change strategy for action plans to reduce emissions of GHGs in each major sector that is contributing to the generation of GHGs in Canada. This initiative contributes toward Canada's commitment made in Kyoto in 1997 to reduce GHG emissions to 6 % below 1990 levels during the period 2008 to 2012. The municipal solid waste sector has been identified as having significant potential to achieve substantial, early reductions in GHG emissions.

Of interest, however, in terms of identifying future changes in Canada's capture of energy from waste, it is important to consider the potential economic implications of GHG cutbacks being considered for the Kyoto Protocol. Climate change and the Kyoto Protocol are changing business conditions by potentially making GHG a commodity that carries a cost to those who generate GHGs and a value to those who can reduce it. In effect, the Kyoto Protocol is expected to eventually be responsible for creating a new tangible, tradable currency: GHG credits. Since industries may be required to reduce their GHG emissions, the potential exists to trade credits where the industries, governments and municipalities purchase the credits associated with the landfills since control of emissions from landfills represents a more technically viable and less costly alternative than undertaking their own reductions.

The methane component of LFG is a potential energy resource and a powerful GHG that contributes to global climate change. The rate of increase of atmospheric methane is among the highest of all GHGs. Increases in methane emissions are responsible for more than 20 % of the overall increase in GHGs in Canada during the period 1990 to 1995.

The capture and use of LFG as an energy resource therefore offers important environmental benefits, including the reduction of GHG emissions about which there may be economic credits. The economic returns from utilizing the recovered LFG could thus defray some of the costs associated with building, operating and maintaining a landfill.

To summarize, the capture and utilization of LFG as an energy source possesses numerous possiblities of revenue in Canada such as:

Value of Energy

The benefits of LFG recovery and utilization include the foregone costs of fuel which would otherwise be needed.

GHG Credits

The benefits of GHG emission reductions and the associated value of credits are becoming clearer. While the "rules of the game" that will govern the implementation of the Kyoto Protocol in Canada have yet to be determined, it is already evident that LFG based power will have more value to buyers.

Green Power

Related to the issue of GHG credits is the notion of "Green Power" which is defined as power generated from sustainable (small hydro, biomass, wind, solar, etc.) sources. It is the inherent value of green power itself that has led to the sale of this power at premium prices. The Green Power factor is becoming a reality in the US that has already gone through energy market deregulation. The designation of energy procured from LFG projects as "Green Power" would open up a new area of valuation and coupled with possible exemption to wheeling charges, direct-to-buyer sales and GHG credits, would make many projects cost competitive.

GHG Gas Emission Reductions from Landfills in Canada

The LFG generated must be captured from a landfill using an LFG-extraction system. For those landfills which have not needed to construct an LFG-extraction system, as a result of regulations and/or health and environmental quality issues, the systems will need to be designed and constructed.

It has been estimated that in 1997, Canada was capturing approximately 25 % of the LFG generated at Canadian landfill sites. A significant portion of this LFG collection was achieved at a small number of large sites, largely due to the need to control localized impacts and as a result of the advantages offered by economies of scale for LFG utilization projects.

At those facilities at which there is extraction of the LFG, the economic values associated with the value of the energy, GHG credits and green power, each of which are contributing to the revenue stream that can offset some or all of the costs of LFG collection and operation. Production of electrical power or use of LFG as a heating fuel (natural gas replacement or supplement) are two LFG utilization approaches that have been widely applied and proven to be technically sound. Therefore, there is confidence in LFG as a reliable fuel resource. Further, there is recognition of the potential for increasing the collection of LFG from more effective landfilling (CRA, 2000).

Examples of Projects Involving Landfill Gas Energy Generation in Canada

In Canada, there were 42 landfills with active gas collection systems as of December 1999. These landfills captured a total of 280 kt per year of landfill methane. Of these, 26 landfills flared the captured gas, while the remaining 16 facilities utilized the gas to generate electricity or for direct combustion. The majority (68 % or 192 kt of CH_4) of the captured LFG was utilized and the remaining 32 % (90 kt/year of CH_4) was flared.

Of the 16 installations which utilized the gas, eight facilities generated a total of 85.3 MW of electricity from 180 kt/year of CH_4. The remaining eight facilities utilized the remaining 12 kt/year of CH_4 for heating applications; these included the heating of buildings, fuel for a gypsum manufacturing plant, a steel refinery, a greenhouse and a recycling plant.

Examples of Utilization
Utilization for Purposes of Electricity Generation

Landfill	Province	Electricity generation
La Compagnie Meloche Inc.	Quebec	1.6 MW
Saint Michel	Quebec	25 MW
Brock West	Ontario	10 MW
Keele Valley	Ontario	33 MW
Beare Road	Ontario	3.5 MW
Waterloo	Ontario	3.7 MW
Lachenaie	Quebec	4 MW
Clover Bar	Alberta	6 MW

Utilization for Combustion

Landfill	Province	Type of utilization
Magog	Quebec	For heating of a maintenance garage
Port Mann	B.C.	For use in wallboard plant
Jackman Landfill	B.C.	For greenhouse operation
Coquitlam Landfill	B.C.	For a recycling facility
Cambridge Landfill	Ontario	For use in a steel mill
Vancouver Landfill	B.C.	For office and shop heating

Flaring of LFG

Landfill	Province	Flaring (m³/day)
Cook Landfill	Quebec	3,300
Cache Creek	B.C.	30,000
Cottonwood Landfill	B.C.	2,300
Beverly Landfill	Alberta	6,000
Sherbrooke Landfill	Quebec	18,350
Ste-Genevieve de Berthier	Quebec	122,300
Hartland Landfill	B.C.	45,000
Upper Sackville	Nova Scotia	10,000
Cornwall Landfill	Ontario	12,230

(Contd.)

(Contd.)

Landfill	Province	Flaring (m³/day)
Storrington Landfill	Ontario	20,000
Tom Howe Landfill	Ontario	16,300
Trail Road & Nepean	Ontario	57,000
North Sheridan	Ontario	71,780
Glenridge Quarry	Ontario	7,300
East Quarry	Ontario	16,300
Aurora Landfill	Ontario	20,000
Britannia Landfill	Ontario	36,700
Mirabel	Quebec	1,835
Ste Cecile de Milton	Quebec	6,672
Dunham	Quebec	18,000
Cedar Road	B.C.	4,500
Ste. Sophie	Quebec	61,165
Saint Nicephone	Quebec	50,970

These figures indicate that the collection and utilization/flaring of LFG is spread through a number of different provinces throughout Canada and the current capture and destruction rates are substantial.

The remaining potential for GHG reduction at the more numerous small to medium sized landfills is significant. Numerous opportunities are being considered for these additional projects including heating fuel for industrial boilers, dryers, kilns or gas furnaces, paper-making, sewage treatment plants, brick manufacture and thermal power plants. The ideal end user of LFG would have a consistent and adequate fuel demand and is located in close proximity (less than 10 km) to the site.

Constraints to Future Development of Energy from Landfill Projects

Currently the primary barriers to LFG capture and utilization are almost exclusively economic considerations. Specifically, a number of projects are severely impacted by limitations with respect to electrical power sale opportunities. The Canadian market for electrical power is largely subject to monopolistic control that limits access by many projects that could use LFG as the fuel source. Similarly, the franchise system for natural gas is such that it has effectively removed the use of LFG as a distributable fuel supply for industrial developments as a potential option for most sites.

The electricity market across North America is undergoing significant change. Recent action in the United States is likely to influence and transform the Canadian market and in fact, Alberta and Ontario are following some of the trends already seen in the US. In particular, there is movement from crown-owned or private monopolies to deregulated markets. The deregulated structures adopted or proposed to be adopted by each province vary in rules, controls and opportunities. Several examples of these changes are:

(i) The Province of Alberta works through a pool system where all electricity is bought and sold through the pool, but is not proposing to deregulate the transmission side of the market;

(ii) The Province of Ontario is proposing to set up a hybrid system with the spot market operating through the pool, with the option for bilateral contracts. Ontario is also not proposing to deregulate the transmission side of the market;

(iii) The Province of British Columbia has already opened up its wholesale market, but is not planning to deregulate generation or transmission in the near future;

(iv) The Province of Quebec allows wheeling to export electricity, but maintains its control over Quebec residents.

The level of uncertainty surrounding electricity markets in Canada is profound. There are many more questions than answers:

(i) How will timing, evolving rules and frameworks and competitiveness factors affect new electricity generation opportunities within each province?

(ii) How will commitments to the Kyoto Protocol affect the whole scene? Will carbon-based generation be penalized? By how much?

(iii) Will electricity prices increase?

(iv) Will Green Power establish itself in the market?

(v) How can accurate 20-year projections of the electricity market be developed to determine the feasibility of landfill projects?

There is significant uncertainty and risk in predicting how electricity markets will react over any 21 period. As evolving electricity markets and changes in valuation of carbon-based generation sources as commitments to the Kyoto Protocol and GHG reductions play out, it is nearly impossible to predict how the markets and frameworks will react and evolve in the next two decades. Assumptions and speculations can be made how things will evolve, but in fact, it is just that: speculation.

Nevertheless, uncertainty should not prevent efforts to influence rules and regulations to allow new generation, particularly cost-effective cleaner sources of energy, from entering the marketplace. To prepare for the carbon-constrained future, Canadians are looking at how the market operates today.

Conclusions

Energy from landfill gas utilization in Canada, is well proven. The major constraint influencing the development of additional sites exists because of economics. The availability of greenhouse gas credits, greater access to the electricity market from any generated electrical power, and designation as Green Power, are dimensions which would greatly assist the economic viability of additional projects. It is likely that the next few years will see significant increases in the numbers of landfill gas to energy projects coming on-line in Canada.

ACKNOWLEDGEMENTS

Support from the Erskine Fellowship Award at the University of Canterbury, New Zealand, for providing the opportunity to prepare this paper is gratefully acknowledged.

REFERENCES

1. Crutcher, A.J., McBean, E.A. and Rovers, R.A., (1981a), "The Impact of Gas Extraction on Landfill-Generated Methane Gas Levels", *Journal of Water, Air and Soil Pollution*, **113/1**, pp. 55-66.
2. Crutcher, A.J., Rovers, F.A. and McBean, E.A., (1981b), "Methane Utilization from a Landfill Site", *ASCE - Journal of Energy Division*, Vol. 107, No. EY1, May, pp. 95-102.
3. Conestoga-Rovers & Associates, "Design and Operation of Non-Hazardous Solid Waste Landfills to Optimize the Generation and Recovery of Landfill Gas and Energy", *Report prepared for Environment Canada*, January 2000.
4. Environment Canada, "Guidance Document For Landfill Gas Management", *Report EPS 2/UP/5E*, March 1996.

12

Landfill Biogas Management: Case of Chilean Sanitary Landfills

JOSE ARELLANO

Av. Larrain 9975, La Reina, Santiago, Chile

The Chilean Experience

Since 1977 the use of sanitary landfills has increased in Chile, for the disposal of municipal solid waste. Currently, about 80% of urban solid waste is disposed in similar facilities. It is supposed that soon this figure will be 100%.

During the stabilization of waste, a generation of biogas is observed. This biogas is made up of methane and carbon dioxide. The first one is a combustible gas. The calorific power is about 4,800 Kcal/m^3, under normal conditions of pressure and temperature.

The combustible and explosive characteristic of the biogas mix is considered for the design and operation of the Chilean landfill sites. These facilities have evacuation systems (or control release system) which enable the utilization of energy released by combusting this resource. These systems also protect the environment from uncontrolled release of gas from the facilities.

Since 1980 biogas has been utilized in the two main cities of Chile, Santiago and Valparaiso, as an energy source for domestic use (cooking and heating), as well as industrial use, mainly to dry raw material in the food industry. At the same time small projects were developed in Rancagua and Concepción. In the first one, it was only possible to try industrial applications, because the public gas pipeline distribution was unavailable; use was mainly focused on drying vegetables as a source of energy for industrial ovens. However, only the cases of Santiago and Valparaiso succeeded.

In Santiago, from 1982 to 1984, the use of biogas was tested as a substitute for combustible, used by the municipal waste collection system. However this idea did not work well, as it was necessary for each truck to have a big and heavy combustible tank, in order to have an appropriate hauling radio. Also it was an expensive system, because it was necessary to maintain high pressure in order to store enough gas in the tank system.

In Santiago the main project was developed in the sanitary landfill called *Lo Errazuriz*, which received the municipal waste of about 2,000,000 people. The rate of extraction was about 120,000 m^3/day, this biogas was used in the public gas pipeline, and in fact it became the most successful experience until now. At the same time the other sanitary landfill, *Cerros de Renca,* which received a similar amount of waste, had a rate of just 30,000 m^3/day; however this biogas was not appropriate for commercial use because of the low calorific power and a high concentration of pollutant, specially H_2S and water.

During 1980 a ratio of 100 m^3 of biogas was calculated for one ton of waste disposed, during a period of 20 years. It is important to mention the composition of this waste was about 50 % (weight base) organic matter, mainly food waste and green waste. Also the humidity reached 50 %.

During 1990 the use of biogas decreased, with development of related projects. The main reason being the construction of an international gas pipeline from Argentina, which made natural gas accessible. However due to increased price of oil, the use of biogas may become popular in the near future.

The current situation is that all big landfills have facilities and equipment for appropriate management of biogas, mainly focussed on its transformation into water and CO_2.

Central Burn Facility of Biogas Coraslos Colorados Landfill (Santiago, Chile)

Technology

In order to design and operate an efficient extraction system it is necessary to consider several factors,

- The generation of biogas is not uniform in the entire area of the landfill. Because of the characteristics of the waste disposed, the landfill operation system and the age of the waste itself.

- The generation rate of biogas is more intensive during the beginning of the decomposition and gradually decreases. The main factor is the amount of organic matter easily degradable.

In Chile, the typical extraction system could be summarized in the following cases:

(a) *Collecting System by Liquid Drainage* (Fig.12.1): It is a system, which uses the leachate collection system. It is known that the releasing points of leachate are also releasing points of biogas. For that reason it is possible to collect at these points. However this system of collection has proven to be inconvenient since the biogass mixes with air, an the ensuing risk of explosion.

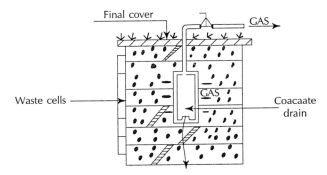

Fig. 12.1. System for leachate collection.

(b) *Collecting by Drills* (Fig. 12.2): This system consists of several drills of 2 inches diameter and 10 m length. The drills are connected to gas pipelines. These drills are built after the landfill is finished.

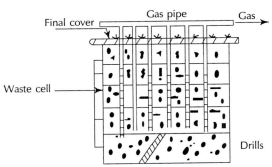

Fig. 12.2. Gas Extration Through.

(c) *Drilled Tubular Well* (Fig. 12.3): This system consists of burying some iron or plastic pipes in a finished landfill. The top of the pipe should be closed to avoid entrance of air. The problem this system faces is that the drills tend to get clogged with the accumulation of leachate. The typical pipe diameter ranges from 6 to 12 inches.

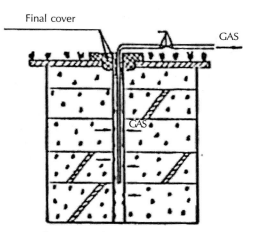

Fig.12.3 Perforated Pipes.

(d) *Collecting system by chimneys* (Fig. 12.4): This system is used in landfills still open, and is built as the landfill is disposed. The construction includes the use of cement rings or wood structures. Both cases are a kind of modular component. The inside of this elements should be filled with stone and gravel, in order to assure the structural quality of these units. The leachate also presents a problem, however here it is less accute than in the previous case. The wooden system is the main system used in Chile.

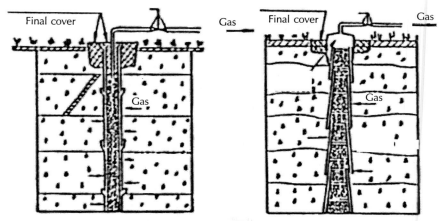

(a) Gas Collection Through Cement Pipes.   (b) Gas Collection through Wood Structure.

Fig. 12.4

(e) *Collecting system by mobile molding* (Fig. 12.5): This system is built using a mobile cylindrical molding. The molding is located near the bottom and it

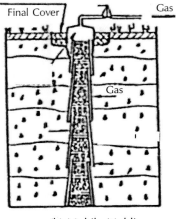

(b) Mobile Molding

Fig. 12.5

moves to the top, leaving the structure (cinformed) just by stone and gravel and without any other structural extra support. Diameter is about 1.2 m. It is supposed that the structure (resits) because of the internal pressure of the waste against the stone and gravel.

(f) *Collection at the borderline of the landfill* (Fig. 12.6): This consists in the use of periferical drainage system of the landfill site. In order to use this system the top should be sealed.

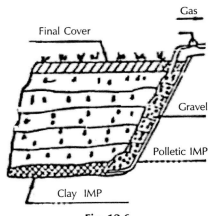

Fig. 12.6

Case Study: Biogas at the Lo Errazuriz Sanitry Landfill (1984-1996)

The Compañia de Gas –GASCO– (Gas Company) of Santiago made a full reutilization of this non-traditional, renewable energy source, mixing it with the gas produced from the cracking of the gasoline according to the following percentages: 30% biogas, 70% gas from cracking gasoline.

The main characteristic of the extraction system were the following:
- Estimated percentage of reutilization: 70% of gas produced
- Pipes' material: high density of polyethylene
- Action range of the wells: 25 m approximately
- Average suction of the wells: 25 cm of water column
- Machinery: 4 root compressors of 1,500 m^3/hr
    - 2 heat exchangers
    - 4 root measuring devices of 1,500 m^3 N/hr
- Main pipeline for transfer to factory: 355 mm in diameter and 4,000 m in length approximately
- Beginning of stable production: 1 year after first deposit of waste

About the Gas:
- Temperature when reaching compressors: 18 °C
- Temperature when being compressed: 60 °C
- Temperature after cooling: 20 °C
- Calorific power under normal conditions: 4,600 Kcal/m^3
- Composition:
    CH_4 : 50-60%
    CO_2 : 35-45%
    N_2 : 3-8%
    O_2 : 0-1%
- Main pollutants:
    H_2S : 5.6 gr/100 m^3
    NH_3 : 1.065 gr/100 m^3
    CS_2 : 0.317 gr/100 m^3

The landfill closed in 1996, however still generating significant amounts of biogas.

13

Energy from Waste: Case Study from Brazil

T. Cássia de Brito Galvão[1] and Willer H. Pos,[2]

[1]School of Engineering—Federal University of Minas Gerais, Av. Contorno 842/104,
30110-060, Belo Horizonte, Brazil
[2]Instituto Mineiro de Gestão das Agusa-Head Rua Santa Catarina,
1354, 4° andar, 3070-081, Belo Horizonte, Brazil

Introduction

A growing concern of environmental issues is a very important characteristic of the last 40 years in the whole world. This concern is well justified by the immediate problems: population growth, depletion of productivity of agricultural lands, watershed management, high-carbon energy system, climate changes, and waste management practices. In the past few years some conferences proposed by the United Nations had discussed some of the immediate problems left to be solved in the next millennium, and described above: Earth Summit-92 in Rio, Population growth in Cairo (1994), the future of cities in Istanbul in 1995. Also, in 1995 about 500 diplomats from 130 countries gathered in Berlin to negotiate an agreement for climate protection. In 1996, climate researchers had pointed out that the floods in China, and the third greatest drought in America, in this century, are the trigger mechanisms for the next climatic geohazard: deicing of the poles and the heightening of the sea level, as consequences of the great amount of polluted gas in our atmosphere- (CO_2, CFC –clorofluorcarbon, methane and nitrous oxide).

In Brazil, the clean technologies have been developed mainly due to economical needs. An example is the Alcohol Program, developed due to Petroleum Crisis, which increased the price of a barrel in four times, in 1973. However, a research done in 1998 by the Brazilian Institutions "Confederação Nacional da Indústria (CNI), the Banco Nacional de Desenvolvimento Econômico e Social (BNDES) and Sebrae", named "Research of Environmental Management in Brazilian Industry", showed that about 75% of the companies intend to invest on environmental research; 90% of the biggest companies and more than one third of the microcompanies invested in the environmental sector in 1997, and more than 39% of those companies used Brazilian consultants. This research shows the growing concern towards environmentally clean technologies, that means, lower consumption of materials and energy with lesser environmental impacts. At present, the modern industrial societies of the big cities need great amounts of petroleum for mass transportation, but future societies won't be able to count on this form of energy and, probably, would rely on solar, nuclear or thermonuclear energy, or develop methods to use energy from waste in a more efficient and economical way.

In the particular case of Brazil, preservation of our environment should be a national priority due to the fact that a third of all tropical forests and biodiversity of this planet are located in Brazil. Developing technologies based less on fossil-fuel, which produce less greenhouse gas emissions and developing more technologies based on the principle of the "clean-burning economical technologies" to preserve our environment, is an imperative need for the country.

This chapter encloses a state-of-the-art from the Brazilian experiences and developments on clean technologies and on the potential to develop clean technologies regarding energy production, which contemplates much more the environmental needs than the economical needs. An emphasis is given on the Brazilian Experiences on Biomass and Landfilling Process and on the Use of Landfill Gas. Although those technologies benefit the environment, and represent a necessary condition towards safe life conditions in this planet, they hardly will shake the billionaire electric power industry – one of the world's largest—in the next few years.

For the next 7 years, the Brazilian Government plans to invest about US$ 1.7 billion dollars in the construction of transmission lines for electricity, about 0.7 billion dollars in the construction of gas conduits, about 10 billion dollars in thermo electrical power plants and about 15 million dollars in the construction of hydro electrical power plants. No money for alternative sources of energy has been planned, so far.

Traditional Sources of Energy in Brazil

Brazil is a country rich in water resources (more than 12% of the available water resources of this planet), mineral resources, plants resources and renewable forest resources. Besides petroleum and its derived products we have: firewood; vegetal and mineral coal; sugarcane which gives us alcohol; "babaçu" plant, turf, vegetal residues, biogas, uranium, etc.

Traditionally, the energy for the growing Brazilian market was provided by harnessing its water resources potential, by means of constructing dams. Hydroelectric power plants cater to 92% of all energy requirements and produce about 65, 000 MWatts/hour. The majority of the Brazilian population and industries are located in the south and southeastern regions. The hydraulic potential of the region is well utilized, to the tune of > 80% via power harnessed from the rivers Paraná and São Francisco. What remains to be used for the next generation is less than 17%. The only region in the whole country, which remains to be explored, is the Amazonas. However, the negative environmental impacts of exploiting water resources of the region outweigh the benefits and the endeavour to do so. In order to fill a dam water reservoir it is necessary to flood extensive areas of forest killing most of the existing biodiversity. It requires also, flooding agricultural lands, villages and small cities, as well as displacing entire populations to other areas. The cultural and social costs involved in such activities are hardly greater than the advantages that a hydraulic power plant might offer. As an example, we have the hydraulic power plant of Balbina in the Amazonas region, which emits to the atmosphere a great amount of methane due to the biodegradation of the flooded biomass of the forest in an effect similar to a thermo-electric power plant of similar capacity. Also, it is known that in São Paulo, the thermo-electric power plant of Piratininga alone produces more sulfur dioxide than all the public transportation.

The second source of energy is petroleum, and Brazil is presently a world leader in extracting petroleum from great depths (more than 1000 m) and produces a total amount of 1,108,000 barrels a day.

The third source is natural gas and it represents 2.6% of the total energy matrix. It comes from national beds and from Bolivia, through conduits called "Gasoduto Bolivia-Brasil". It is estimated that Brazil will receive about 16 million m^3/day from Bolivia and could reach a maximum of 30 million m^3/day. In 1996, the natural gas reserves in Brazil were of the order of 157.7 billion m^3. With the gas importation from Bolivia, the goal is to go from 2.6% to 12%, of the total Brazilian market, by the year 2010. In Brazil's energy planning, for 2001, it is predicted a 10% rise in the natural gas consumption.

Although gas is basically transported by conduits in its gaseous state, it could be transported in liquid state, where its volume is reduced about 600 times, by diminishing its temperature to –160 °C. Natural gas is basically a compound of light hydrocarbons and in the standard conditions of temperature and pressure its natural state is in the form of gas. Although the chemical composition of natural gas varies, it is a compound of methane, ethane and propane. Usually, it is found in underground reservoirs associated or not with petroleum. If it is mixed with petroleum, we have the natural associated gas. The non-associated gas is characterized by the absence of oil or in very small amounts. In small amounts, one can find in its composition nitrogen, carbondioxide, and sulphur. Natural gas is odourless, colorless, inflammable, and suffocating. Then, as a matter of safety, the gas is impregnated with sulphur, to identify its presence by the odour, and has no corrosive properties. The components of associated and non-associated gas are presented in Table 13.1.

Table 13.1. Gas composition

Components	Associated gas (%)	Non-associated gas (%)
· Methane	81.57	85.48
Ethane	9.17	8.26
Propane	5.13	3.06
Isobutene	0.94	0.47
Butane	1.45	0.85
Isopentane	0.26	0.20
Pentane	0.30	0.24
Hexane	0.15	0.21
Heptane and above	0.12	0.06
Nitrogen	0.52	0.53
Carbon dioxide	0.39	0.64
Density	0.71	0.69

Source: Petrobras

Alternative Sources of Energy

The petroleum crisis of 1973 promoted a global awakening towards the fact that this is an exhaustible source of energy. In 1976, in Brazil, began researches on the development of alternative sources of energy, the most successful is the Alcohol Program that will be discussed later.

The other alternative sources of energy comprise solar energy, wind energy, tide energy, and biomass energy. Although a high amount of solar radiation is received in the country, specially the north and northeastern States, this potential is hardly exploited for heating purposes. COELCE—Brazilian Company of Electricity of Ceará has inaugurated a great wind plant in Fortaleza, which intends to generate 3.8 million kW/h annually. In 1987, CHESF- Company Hydroelectric of São Francisco commissioned a research to gauge the solar potential in the northeastern regions of Brazil, and this research showed that the entire region promises a high solar potential of about 11,400 MW/year in an area of 1000 km². The optimal utilization of such a natural resource needs to be addressed.

From 1976 until 1980, the Brazilian Electrical Company, Eletrobrás, and the Navy Ministry have conducted studies in Maranhão and Pará shores for installation of a tide-energy powered plant at the mouth of Bacanga River, in São Luís do Maranhão Island. Studies have shown 42 potential places to be at very advantageous positions. Among the cited advantages of such tide powered plants, were the development of renewable environmentally clean sources of energy, total independence upon the precipitation conditions of that region, the cost of the energy staying constant in the same life time of the equipments, and finally, the basin being natural, does not require any flooding of additional areas. The studies have also shown the potential of installation of a 36,000 MW tide-power plant. In the studied regions the tides are 5-8 m high, and could rise to 13 m.

Biomass energy is produced directly through the burning of wood, pulps, peels, residues and cakes or through the production of gases in biogestors. The gas produced by the burning of biomass causes lesser environmental impact than the pollution of hydroelectric power plants. In Brazil, the *babaçu* plant is considered to be an excellent fuel and can substitute coke in the steel industry and combustible oil in cement industries. The states of Maranhão and Piaui have great Brazilian potential in *babaçu* plants.

It is important to point out that there are some research areas that need to be investigated before the development of a wider biomass energy program, these are:
- a global rising of the potential of electricity generated by biomass and urban solid wastes in different parts of the country;
- government subsidies on biomass and urban solid wastes power plants also on transportation and distribution to the other regions of the country as well discounts on taxes and equipments; and
- minimal degree of efficiency for generation and co-generation of energy from biomass, encouraging technologies into the market.

Alcohol Program

Brazil has successfully developed the greatest worldwide program of utilizing in large scale, biomass in the form of sugarcane transformed in alcohol, which can be used as car combustible. It is important to point out the existence of technology for fabricating automobiles moved by hydrated alcohol. From the Brazilian point of view this is a very important program due to the following reasons: (i) it helps to save fossil combustible, (ii) it gives employment to people with lower levels of education; (iii) it reduces air-pollution. At the Fourth Meeting of Renewable Sources of Energy, held at Recife (October, 1998) they came to the following recommendations, which would significantly reduce greenhouse gas emissions and be beneficial for the economy,

- The gradative introduction of alcohol mixtures in Brazilian diesel
- The immediate decision of adding 2% anidro-etilic alcohol carburant into diesel oil, following recommendations of several international producers of diesel motors. This mixture is considered anti-freezing.
- The establishment of goals and procedures based on research observations, to set levels (above 2%) of optimized mixtures, by use of anidro or hydrated alcohol, by means of stabilizer, which could be co-solvent or emulsion.

The above recommendations are based on the following considerations, among others:

- This country spends, 900 million dollars, annually, importing 15% of diesel oil, which it needs
- Diesel motors contribute to the emission of a black smoke and other pollutants, which pollute major Brazilian cities
- The Brazilian society has invested great amounts of money in a national program of substitution of fuel by alcohol. Today, this program is responsible for more than 1 million jobs;
- That the addition of ethanol in diesel could be a way of stabilizing the seasonal fluctuations in diesel supply and demand;
- That it is possible to admit as a fact that the optimized use of mixtures alcohol-diesel can help in reducing the particulates in about 35%, the carbon monoxide and hydrocarbons in about in 15%; sulphur oxides in 10% and nitrogen oxides in 5%.

Biogas Production from Landfills

Landfills are a great source of methane emission, which is one of the principal greenhouse gases. After EPA (1996) methane is responsible for roughly 18% of the total contribution. Biogas is generated by the biodegradation of organic matter from vegetal or animal origin by anaerobic microorganisms. As an end product of this biological process one can have a gas composed of methane (CH_4) and carbon dioxide (CO_2) known as biogas. Anaerobic processes could either occur naturally or in a controlled environment such as a biogas plant. Depending on the type of waste and the system design, 55 to 75% of pure methane can be obtained from biogas. The process of anaerobic digestion consists of three steps: (i) the first step is the decomposition (hydrolysis) of organic matter, where higher molecular mass compounds (e.g. lipids, polysaccharides, proteins and nucleic acids) are transformed into intermediate mass compounds, which are much more suitable for the microorganisms as source of energy and cell carbon; (ii) the second one is the conversion of decomposed matter to organic acids—in this phase, the existing microorganisms convert the intermediate-molecular mass compounds into lower-molecular mass compounds such as acetic acid, smaller concentrations of fulvic acids, and other complex organic acids; and the (iii) third is the acetic acids getting converted into methane gas—in this phase anaerobic microorganisms transform the acetic acid into mainly CH_4 and CO_2 gases.

Gas amount and composition is thus supposed to vary according to many factors. However, parameters such as waste composition, operational procedures, temperature, humidity, pH, E_h, are those that mainly impact production of gases and composition. Biogas production is also a function of the presence of microorganisms capable of degrading the organic matter in an anaerobic way. If the following factors such as low temperatures, presence of air, excess of water, are controlled, then one can expect a stable production of biogas. In order that the solid waste receives a small impact from the environment, it is advisable for the cell to have a minimum height of 10 m and a daily solid waste disposition of more than 100 t.

In general, it is estimated that a landfill can produce biogas in an amount that varies from 200 to 300 l/kg. However, Brazil is estimated to produce about 50 to 80 m^3/ton biogas from solid waste, in 10 years, approximately.

In Brazil, the gas produced by anaerobic digestion of organic vegetal residues represents a technology meant for public consumption and has been applied specially in the northeastern Brazil, where more than 300 unities were distributed in the State of Maranhão, Ceará, Pernambuco, Paraíba and Bahia. The greatest poential to be exploited is the agricultural activities, and besides, tapping energy from the gas, fertilizers could also be obtained.

Solid Waste Facilities in Brazil

Brazil produces about 0.2 million tons of solid waste a day. 76% is placed in open dumps, such as erosional areas, 13% is placed in controlled areas, 10% is placed in sanitary landfills, 0.9% presents a composting plant (or anaerobic digestor), and 0.1% has an incineration power plant (IPT, 1995). "Landfilling" is not the main method for disposal of municipal and household solid wastes or refuse in Brazil, as shown by the statistics.

The country has 4,974 municipal districts (IBAM, 1993), from them 3,611 have less than 20,000 inhabitants, and the 21 biggest municipal districts (population bigger than 0.6 million people) have a total amount of 34 million inhabitants.

Only 37 municipal districts in the total of 4,974 have facilities for anaerobic digestion (composting plant). In 1990, 17 of those 37 facilities were closed or disabled, five were under annually, about construction and 15 were operational. The reasons (IPT, 1995) to close or disable those facilities were, among others:

- Bad planning of the installations, through credit given by the National Bank of Development—BNDES, which brought dispute among the constructors, who did not take into much considerations the real needs of the municipal districts;
- Total absence of management training in order to conduct the activities better;
- Inadequate placement of the landfills;
- Absence of financial integration and operation between the landfill and the Public Service of Urban Cleaning.

Assuming that 60% of the total generated solid waste in Brazil is organic - 0.14 million ton/day, this organic waste would generate about 2,160 MW of energy.

Main Composition of Solid Waste

As an example, Table 13.2 shows the solid waste composition of Belo Horizonte, the fourth biggest city in Brazil, especially because its composition has not varied along last years. One can observe that the waste composition in Belo Horizonte's landfill has a very small percent of paper, due to an intensive program, which involves a great amount of people, in separating paper before reaching the landfill facilities. The bulk density of this waste is 254 kg/m³. The landfills of the rest of the country have a great composition of paper of about 25%, organic matter about 52%, glass about 1%, glass about 1%, plastics 12% and 6% of others.

Table 13.2. Solid Waste Composition from Belo Horizonte Landfill (Source: SLU, 2000)

Material	Amount
Organic matter (%)	54.06
Paper (%)	12.50
Cardboard (%)	6.00
Plastics (%)	5.90
Cans (%)	2.70
Glass (%)	3.15
Textile (%)	6.20
Gum (%)	1.07
Metals (%)	0.60
Wood (%)	1.30
Leather (%)	1.09
Tiles and stones (%)	1.60
Solid (%)	2.50
Others (%)	1.33
pH	5.3
Volatile solids (% dry weight)	72.00
Carbon (% dry weight)	38.00
Nitrogen (% dry weight)	1.03
P_2O_5 (% dry weight)	1.30
CaO (% dry weight)	2.30
Mg (% dry weight)	1.02

Biogas Collection Systems

The biogas generated by landfills is either extracted through a collection system drilled into the waste, after which it is treated and stored, or it undergoes combustion by flare.

The gas collection system consists of a series of wells drilled into the landfill and connected by a plastic piping system. The collection systems for biogas mainly used are:

- collection in the drainage system for liquids
- extraction by means of probes
- extraction by perforated wells
- collection on the edges of the cell
- collection by using horizontal drains

The collection used mostly in Brazil is made by means of drains at the inoculation points, to check through the gas composition the stage of biodegradation of the landfill.

Table 13.3 shows a typical composition of the biogas (the low-Btu) from Brazilian landfills.

Table 13.3. Gas Composition of Brazilian landfills.

Methane	15-80 %
Carbon dioxide	35-45 %
Nitrogen	3-8 %
Oxygen	0-16 %
Sulphuric acid (H_2S)	$5.6g/m^3$
Organic Sulphur	$0.317/100\ m^3$

To purify the biogas, some usual procedures are used:
- For separating methane from carbon dioxide, in order to improve the combustion power of methane, biogas is saturated with water, and that water must be removed prior to further processing. The amount of water used is about 460 liter for $1\ m^3$ biogas (assuming that the amount of CO_2 is 35% and at standards conditions of pressure and temperature, which are 1 atm and 20 °C). When the pressure is increased and the temperature is lowered, the solubility conditions improve, but the residual water become very acidic, and cause corrosion. The following treatment could also be used: $Ca(OH)_2$, organic solvents such as monoethanoamine, diethanoamine and tri-ethanoamine, etc.
- The second step is to remove sulphur dioxide from the gas since it results in corrosion within the combustion equipment. The most economical way of removing the sulphur dioxide is through filings of iron, which is regenerated after air exposure for 3-4 days.
- The almost pure methane (without CO, H_2S, O_2, N_2 and hydrocarbons) can be obtained through the use of molecular panels.

The Experience of Belo Horizonte

The city of Belo Horizonte is the fourth largest city in Brazil with an average population of 2.3 million. About 4464 t solid wastes is generated daily and collected by the Municipal Services. Only about 2000 t/day of solid waste is placed in landfills, the rest in recycled or re-used, upon arrival in the landfill area. The solid waste facility occupies an area of 144 ha, and has 5 cells. The final average height of each cell is about 50 m.

The cell 5 corresponds to an old open dump, which people have been using since 1973 to dispose their garbage, and has no impermeabilization or gas layer.

The others cells received an impermeabilization layer and are much more controlled and monitored.

The waste treatment and disposal are done through a more controlled unit (cell 2); its leachate is collected and used for the bioremediation process of the cell 5. This process was proposed by a local enterprise (LM, 1998) and is intended to accelerate the production of methane (CH_4) and carbon dioxide (CO_2). It is based on the treatment, inoculation and recycling of leachate enriched with specially developed bacteria by LM (1998), capable of acquiring the final stabilization of the wastes within three years, followed by the mining of reusable materials and liberating the area for the treatment of new wastes. Figure 13.1 shows the Belo Horizonte Landfill area, and photos 2 to 3 show the cell 2 in 1992 e 1998, respectively. This cell has dimensions of $200 \times 200\ m^2$. Figure 13.1 presents a view of the landfill, with its five cells, and the location of the leachate and composting plants.

The biogas was collected by the gas company Gasmig, treated and stored in cylinders, until 1995. Nowadays, this service was suspended and the gas is simply combusted in a flare system showed in Photo 3.

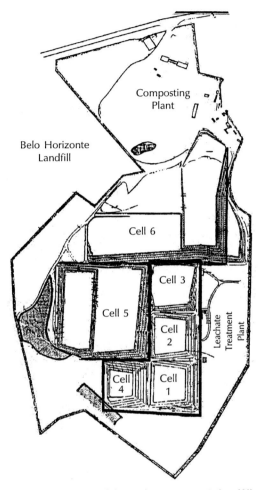

Fig. 13.1 View of the Belo Horizonte's landfill.

Carbon dioxide and methane analysis was performed by a gas-chromatographic system containing a thermo-conductivity detector (TCD). The chromatograph consisted of a PE-Autosystem unit held at a constant temperature of 50 °C. Helium was used as a carrier gas at performed at the Laboratory of Sanitary Engineering from Engineering School—Universidade Federal de Minas Gerais and the average results are shown in Table 13.4 for cell 2.

Photo 13.1 Cell 2, in 1992

Photo 13.2 Cell 2 in 1998

Table 13.4. Results of Methane and Carbon Dioxide from Cell 2

Methane	35-55%
Carbon dioxide	20-45%

Photo 13.3 Global view of cell 2, in 1998

Photo 13.4 Flare of cell 2

REFERENCES

1. Balderrama, L. (1993). Estudo de impacto ambiental causado por aterro sanitário via migração de gases. Master Program Dissertation, UNICAMP, 79 pp + annexes.
2. Engelman, R. (1994). Stabilizing the atmosphere: population, consumption and greenhouse gases' - Population and Environmental Program, Population Action International, 47 pp.
3. IBAM (1993). Instituto Brasileroi de Municipios. Relatório Interno.
4. IPT (1995). Instituto de Pesquisas Tecnológicas de São Paulo. Lixo Municipal. Manual de Gerenciamento Integrado, 1st Edition, IPT: CEMPRE, S.P.
5. LM Tratamento de Resíduos (1998). Implantação do Aterro Celular da Central de Tratamento de Resíduos da BR-040—Projeto Básico. Belo Horizonte-MG.
6. Monitoramento Ambiental do Aterro Sanitário da BR-040 - 1°, 2°, 3°, 4°, 5° Relatórios—Convênio SLU/FCO (DESA-ETG)—Ago/98-Dez/99—Belo Horizonte—MG.

7. Parker, A. (1983): Chapter 7. Behavior of Wastes in Landfill Leachate. Chapter 8. Behavior of Wastes in Landfill - Methane Generation. In: Practical Waste Management, J. R. Holmes (ed.), John Wiley & Sons, Chichester, England.

8. Palmisano, A. and Barlaz, M. (eds.) (1996). Microbiology of solid waste. In: CRC Press, Boca Raton (USA), 224 p.

9. Pohland, F.G. (1987). Critical Review and Summary of Leachate and Gas Production from Landfills. EPA/600/S2-86/073, U.S. EPA Hazardous Waste Engineering Research Laboratory, Cincinnati, OH.

14

Current Situation of Waste Incineration and Energy Recovery in Germany

BERNT JOHNKE

Umweltbundesamt Berlin, Seecktstr. 6-10, D-13581, Berlin

Waste Incineration

The role of waste incineration differs in the countries of the world. While in industrialized countries such as Europe as well as in Japan, the USA and Canada the proportion of waste burned in waste incineration plants can be very high (up to 100%), in most developing countries landfilling is the more common waste management practice.

Status of Waste Incineration in Various EU Member States

The role of municipal waste incineration in European countries varies (see Table 14.1). The compilation presented below shows the amounts of municipal waste incinerated in waste incineration plants of countries in western Europe. It has been taken from an EU report on waste incineration which has been prepared for the European Commission by the Netherlands-based TNO, with 1993 as the reference year. The figures for Germany, Portugal, Luxemburg, Austria and some other countries have been updated to reflect the status in 1998.

The compilation presented below shows the calorific values of mixed municipal solid waste in other countries differs very much and ranges from 3,500 to 15,000 kJ/kg. (see Table 14.2)

MSW Incineration in Germany

The thermal treatment of solid municipal waste mostly takes place in plants equipped with grate firing systems; in individual cases, in pyrolysis, gasification or fluidized bed plants or in plants using a combination of these process stages. Residual municipal waste (domestic refuse, commercial waste similar to domestic refuse, bulky waste, road sweepings, market waste, etc.) is delivered to grate furnaces as a heterogeneous mixture of wastes. Combustible components account for about 40-60 wt.%. Since the municipal waste incinerated is a heterogeneous mixture of wastes, in terms of sources of CO_2, a distinction is drawn between carbon of biogenic and carbon of fossil origin.

Table 14.1. MSW—Incineration plants in Europe

Country	Incineration capacity per country $Mg \times 10^6/y$	Share of incineration	No. of MSW incinerators
Austria	0.513	~20%	3
Belgium	2.5	~35%	18
Denmark	3.82	~75%	34
Finland	0.07	~4%	1
France	13.3	~45%	258
Greece	0	—	0
Germany	14	~32%	61
Ireland	0	—	0
Italy	1.71	~7%	32
Luxemburg	0.15	~95%	1
Netherlands	5.7	~27%	11
Norway	0.5	n.d.	18
Portugal	0.9	n.d.	3
Spain	1.13	~5%	9
Sweden	1.9	~40%	22
Switzerland	2.84	~100%	30
UK	2.71	~2%	17
West-Europe total	45.748	—	497
EU total	42.408	—	449

Table 14.2. Calorific values of municipal solid waste in other countries

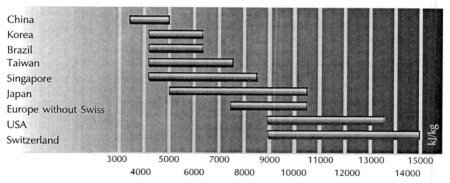

Source: Martin GmbH, München company brochure "Thermische Behandlung und energetische Verwertung von Abfall", page 5, 1997

The calorific value of mixed waste ranges from 7,500 to 11,000 kJ/kg. The waste's carbon content is generally in the range of 28-40 wt% (averages, related to dry matter).

Treatment in incineration plants is an output-controlled process (geared, as a rule, to steam output). The combustion temperature of the gases in the combustion chamber as measured for at least two seconds after the last injection of combustion air is usually at least 850 °C. The oxygen necessary for incineration is supplied via ambient air, as primary, secondary and/or tertiary air. The volume of air supplied to the incinerator is between 3,000 and 4,500 m³ (dry) per Mg of waste. This gives a waste gas volume of 3,500-5,500 m³ (dry) per Mg of waste.

At almost all municipal waste incineration plants, the heat produced during incineration is utilized for steam generation. Upon reaching the end of the steam generator the temperature of the waste gas has been reduced to 200 °C. The steam produced in municipal waste incinerators exhibits pressures between 14 and 120 bar and temperatures between 196 and 525 °C. Common steam parameters are 40 bar and 400 °C. A high heat utilization efficiency can only be achieved if incineration is controlled so that the produced amounts of steam can be made available continuously for direct supply of heat and electricity to an industrial plant or for use in a heating station or cogeneration plant. Table 14.3 shows the development of the throughput/capacity of German MSW incineration plants.

Table 14.3. Development of throughput/capacity of German municipal solid waste incineration plants

Year	Amount of plants	Yearly throughput of all plants × 1000 Mg/y	Average throughput per plant × 1000 Mg/y
1965	7	718	103
1970	24	2.829	118
1975	33	4.582	139
1980	42	6.343	151
1985	46	7.877	171
1990	48	9.200	191
1992	50	9.500	190
1993	49	9.420	192
1995	52	10.870	202
1998	53	11.900	225
2000	61	13.999	230
2005*	75	17.600	234

Hazardous Waste Incineration in Germany

Hazardous waste is treated almost exclusively by incineration. Incineration must be understood here as an element of comprehensive logistics for the treatment of those wastes which due to their harmful nature have to be managed separately from municipal waste.

Hazardous waste is waste requiring particular supervision, which by its nature, condition or amount poses a particular hazard to health, air and/or water or is particularly explosive, or may contain or bring forth pathogens of communicable diseases.

Since hazardous waste is generated for the most part in industrial production, notably the chemical industry, it is also referred to as industrial waste or industrial residue.

Hazardous wastes occur, for example, as residues from petrochemical distillation processes, as undesirable by-products of syntheses processes of the basic organic chemical industry and the pharmaceutical industry as well as in the recovery and disposal of contaminated or post-expiration-date products such as solvents, paints or waste oil. In addition, environmental protection measures such as regulations prohibiting PCBs, CFCs or halons may generate streams of hazardous waste.

The waste going to incineration is usually a mixture of waste types which may differ in composition and be present in solid, semi-liquid or liquid form. Its chemical description differs from that of municipal waste.

As hazardous wastes are of varying consistency, the rotary kiln is widely used as a universally applicable incineration process. Only in exceptional cases are hazardous wastes incinerated in a conventional combustion chamber, a muffle-type furnace or other types of incineration systems.

The rotary kiln operates according to the parallel-flow principle, in which the material being incinerated and the combustion gas are transported in the same direction, from the cold to the hot side. With combustion temperatures between 800 and 1200 °C, the residence time of solids in the rotary kiln is up to 1 hour while for the combustion gases it is only a few seconds. The waste gas generated during the combustion process is fed to an afterburning chamber, in which the minimum temperature of between 850 and 1200 °C is maintained for a residence time of at least 2 seconds. The waste gas volume from this process is generally assumed to be about 7,000 m³ (dry) per Mg of waste.

At nearly all hazardous-waste or residues incineration plants, the heat produced during incineration is utilized for steam generation downstream from the afterburner. Upon reaching the end of the steam generator the temperature of the waste gas has been reduced to 200-300 °C. The steam from hazardous-waste incineration exhibits pressures between 17 and 30 bar and temperatures between 250 and 300 °C.

Mono-incineration of Sewage Sludge in Germany

The system mainly used for the incineration of sewage sludge is fluidized-bed combustion. Most plants are stationary fluidized-bed furnaces, but there are also multiple-hearth furnaces and multiple-hearth fluidized-bed furnaces in use. Fluidized-bed furnaces for the incineration of sewage sludge are usually operated at combustion temperatures in the range of 850 °C and 900 °C. The waste gas volume from this process is generally assumed to be about 8,000 m³ (dry) per Mg. of sewage sludge (dry matter). Modern plants are equipped with a steam generator downstream from incineration, producing wet steam with a pressure of 10 bar and a temperature of 180 °C. Most plants use the produced steam to meet in-plant requirements (e.g. for sludge drying).

The sewage sludge delivered to the incineration plants in dewatered and/or partially dried condition usually has a water content of 50-70%. The calorific value of dewatered sludge averages 3,500 kJ/kg in the case of raw sludge (25% dry matter) and 2,500 kJ/kg in the case of digested sludge (25% dry matter). The content of mineral and inorganic components in sludge can be as high as 30%. The carbon content of sludge is generally about 30%.

Co-incineration in Germany

In the future, the use of waste in plants other than waste incineration plants will be gaining in importance as a waste management option. The object of co-incinerating high-calorific waste as substitute fuel (so-called waste for energy recovery) in production (e.g. cement works, brick manufacture, blast furnace), power plants (e.g. use of sewage sludge in coal-fired power plants) and industrial boilers is the substitution of regular fuel (coal, fuel oil, etc.) and to reduce energy costs.

The climate-relevant emissions of a waste incineration plant are made up of a proportion to be allocated to the waste's contribution to the thermal output and that of the remaining regular fuel. Therefore, a proportions calculation has to be carried out to determine the proportion of those climate-relevant emissions which result from the co-incineration of the waste.

Other Kinds of Waste Incineration

In most European countries the use of incineration plants for medical waste or as crematoria is for the cumbustion capacity and the climate-relevant emission of flue gas stream not so relevant. For that reason this kind of incineration will not be considered in this paper.

Energy Utilization Efficiency of (Energy Supplied by) Different MSW Incineration Plants

Status of Waste Incineration and Energy Use in Germany

In general, about 300 to 600 kWh of electricity can be produced in a MSW incineration plant from 1 Mg of municipal waste, depending on plant size, steam parameters and steam utilization efficiency. In the case of the co-generation of electricity and heat, about 1,250 kWh of heat per Mg of waste can be produced in addition and supplied to external users, depending on the incineration plant's site-dependent heat supply opportunities as well as the geographical location of the country and the (long-distance) heat utilization periods usual for that country (e.g. in Germany, 1,300-1,500 h/year out of a possible 8,760 h/year). Given ideal site conditions with favourable opportunities for utilization and supply in the form of steam, electricity and hot water or exclusively steam, the transformation/recovery efficiency of an incineration plant operating at base load can be increased to a maximum of 75-83% of the energy input (calorific value). In this energetically favourable case, about 2 MWh, as energy mix (electricity and heat), per Mg of waste can be produced and supplied to external users.

Actual energy transformation efficiencies are shown in the Table 14.4 below, ranging from a site with minimum supply of energy (electricity only) to sites with normal or optimized power/heat co-generation or exclusively supplying heat. This broad range of variation among existing plants illustrates that the energy transformation efficiency as well as the proportion of energy actually supplied by waste incinerators to substitute for fossil energy sources, as estimated on the basis of it, and the resultant emissions, are of major importance to the calculation of climate-relevant emissions.

In Germany approx. 14×10^6 Mg/a of residual waste is subjected to thermal treatment (status: 1999). For waste incineration plants in Germany a theoretical gross energy content in municipal waste, in kJ/a, can be specified by multiplying an assumed average calorific value of about 9×10^6 kJ per Mg of waste, roughly

Table 14.4. Energy recovery efficiencies of MSW incinerators in Germany

Energy transformation/recovery efficiencies (W) of the gross energy input of MSW incinerators (without deduction of energy consumed by itself)

	Minimum energy recovery	Normal energy recovery	Optimized energy recovery		
			(electricity/ heat)	(only heat)	(only electricity)
W thermal (%)	<1	11	15 - 55	70-76	0
W electrical (%)	13	14	20	0	24.5
W total (%)	13	25	35 - 75	70-76	24.5

reflecting as a rule the calorific value of low-quality lignite, by the amount of waste incinerated annually. From this, the energy made available by waste incineration can be calculated as a function of the assumed energy transformation and utilization efficiency (the average from all plants in Germany is in the range of 20% to 35%) to derive conclusions as to the relevance of waste-derived energy for the substitution of climate-relevant emissions from fossil energy sources. The waste incineration plant's own energy (electricity, etc.) requirements (e.g. for waste gas treatment) are considered to be external energy which in the ideal case would be met through in-plant electricity production.

Taking Germany as an example, the credit for energy from waste incineration is calculated as follows:

Amount of residual waste subjected to thermal treatment: approx. 14×10^6 Mg/a, multiplied by an average calorific value of approx. 9×10^6 kJ/Mg waste

gives 126×10^{12} kJ, or 126×10^6 GJ/a;

divided by 3.6 GJ/MWh

gives 35×10^6 MWh/a.

Of this amount, an average of 35×10^6 MWh/a $\times$ 0.25 total energy transformation efficiency is utilized, which brings the average allowable substituted net energy potential to 8.75×10^6 MWh/a (normal energy recovery from table 14.4).

Based on 1992 operating data taken from the ISWA's data compilation "Energy from Waste Plants 1994", the energy supplied to external users in the form of heat and electricity by waste incineration plants in Germany (incineration capacity 7.3×10^6 Mg/a) can be calculated at approx. 4.8×10^6 MWh/a.

If the total emission of the non-biogenic CO_2 is estimate with 0.414 Mg CO_2 per Mg waste and the average total energy transformation efficiency of all waste incineration plants with energy recovery were optimized to reach a value equal to or greater than about 0.25, the allowable substituted net energy potential would increase to 8.75×10^6 MWh/a (normal energy recovery), leading to neutrality in climate-relevant emissions from waste incineration due to the emission credits from the power plant mix.

15

Thermal Treatment of Waste in Vienna: An Ecological Solution?

SENATSRAT DIPL.-ING. HELMUT LÖFFLER

*Head of the Environmental Protection Department of the Municipality of Vienna,
Florianigasse 47, A1080 Wien, Austria*

It is a basic principle of an ecologically-oriented waste management policy, that waste which can neither be avoided nor recycled nor reused has to be subjected to a special treatment before its ultimate disposal to prevent any future environmental burden; i.e. it must neither end up in abandoned polluted areas nor must a long-term supervision of the disposal site be required.

The most flexible and commonly used process to meet these ecological requirements is the thermal treatment of unavoidable and unrecoverable residues in modern waste incineration plants equipped with special flue gas cleaning processes.

In these plants, the pollutants of the flue gases are eliminated to such an extent that they do not cause any environmental burden. At the same time, the volume of the residues is reduced to one tenth, the energy content of waste is reused and fossil fuels are replaced.

However, to prevent any future environmental burdens resulting from contemporary waste disposal, we have to treat all residues, thermally treated as well as untreated, in such a way that their eluates do not exceed the limiting values for potable water, which means that they will definitely not have a negative impact on groundwater.

To fulfill the requirements of the Viennese waste management policy, Vienna has set new standards for waste incineration plants:

1. The MWI Spittelau is the first waste incineration plant worldwide that has been equipped with selective catalytic No_x reduction (SCR). The No_x emissions of the MWI Spittelau are in the range between 30 and 40 mg No_x/m^3 flue gas.
2. The MWI Spittelau is the first incineration plant worldwide with a dioxin emission limit of 0,1 ng TE per m^3 flue gas and with a catalytic destruction of dioxins.
3. The flue gas cleaning systems of the MWI Spittelau have such an excellent performance that the real emission figures for SO_2 and dust lie at approximately one tenth of the required emission limits.
4. To convince the public of the excellent performance of the MWI, the actual emission figures are displayed on large screens in public places so that every Viennese can inform himself about the emissions of the MWIs.

A further proof of the environmentally sound working of the MWI Spittelau is the fact that the famous Austrian artist Friedensreich Hundertwasser, an environmentally oriented person, was convinced after a year-long discussion with the head of the Environmental Protection Department of Vienna, that waste incineration, as performed in Vienna, is an important contribution to a good environment and he agreed to undertake the new architectural design of the plant's facade without accepting remuneration.

The MWI Spittelau therefore is not only a new architectural landmark in Vienna but also a legacy of the artist demonstrating that the city of Vienna has undertaken all possible measures to make waste treatment and incineration, in particular, as environmentally sound as possible.

Improvement of Air Quality in Vienna during the Last Decades

As can be seen from Fig. 15.1, the air quality in Vienna improved significantly during the last decades. The annual sulphur dioxide mean values were reduced to less than one tenth of the values of the seventies.

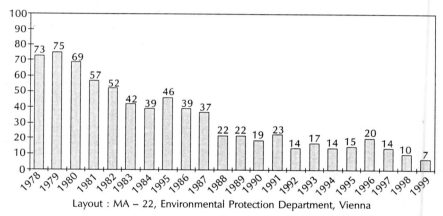

Layout : MA – 22, Environmental Protection Department, Vienna

Fig. 15.1 Sulphur dioxide—Annual Mean Values, Vienna 1978–1999

This sulphur dioxide reduction has been attained by:
• equipping power plants with desulfurization equipment
• reducing sulphur content in oil
• eliminating household fires and substituing with district heating
• using mainly waste heat from power plants for district heating
• running waste incineration plants with best available flue gas cleaning systems.

Ecological Requirements of Waste Treatment

There is a great number of waste management models (Leitbild für die Schweizerische Abfallwirtschaft, Juni 1986 /1/ and Fortschreibung des Wiener Abfallwirtschaftskonzeptes, 1988 /2/ or Leitlinien zur Abfallwirtschaft issued by the Austrian

Federal Ministry for Environment, Youth and Family) which define the fundamental requirements an ecologically oriented waste management policy is supposed to meet.

All these ideas and claims articulated in the eighties are well summarized in §1 of the waste management law of June 1990 and the amendment of August 1996 given below.

BUNDESGESETZBLATT
FÜR DIE REPUBLIK ÖSTERREICH

Jahrgang 1990   ausgegeben am 26. Juni 1990   140. Stück

325. Bundesgesetz: Abfallwirtschaftsgesetz - AWG (und *fett* Novelle vom August 1996)

Artikel I I. ABSCHNITT **Allgemeine Bestimmungen** **Ziele und Grundsätze der Abfallwirtschaft**	Chapter 1 **General Provisions** **Objectives and Principles of Waste Management**

§1 (1) Die Abfallwirtschaft ist danach auszurichten, daß

1. schädliche, nachteilige oder sonst das allgemeine menschliche Wohlbefinden beeinträchtigende Einwirkungen auf Menschen sowie auf Tiere, Pflanzen, deren Lebensgrundlagen und deren natürliche Umwelt so gering wie möglich gehalten werden,
2. Rohstoff- und Energiereserven geschont werden,
3. der Verbrauch von Deponievolumen so gering wie möglich gehalten wird,
4. nur solche Stoffe als Abfälle zurückbleiben, deren Ablagerung kein Gefährdungspotential für nachfolgende Generationen darstellt (*Vorsorgeprinzip*).

(2) Für die Abfallwirtschaft gelten folgende Grundsätze:

1. Die Abfallmengen und deren Schadstoffgehalt sind so gering wie möglich zu halten (*Abfallvermeidung*);
2. Abfälle sind stofflich **oder thermisch** zu verwerten, soweit dies ökologisch vorteilhaft und technisch möglich

§1 (1) Waste mangament shall be based on the understanding that

1. hazardous, negative or other effects that impair the general well-being of man, animals, vegetation, their basis of existence and their natural environment, shall be kept as low as possible,
2. raw-material and energy reserves shall be conserved,
3. the use of landfill volumes shall be kept as low as possible,
4. only such materials shall remain as wastes, which can be deposited without creating a hazard potential for subsequent generations (*principle of precaution*).

(2) The following principles shall apply to waste management:

1. Waste quantities and their pollutant contents shall be kept as low as possible (*waste prevention*);
2. Wastes shall be reused as material **or energy**, to the extent that this has ecological advantages, is technically feasible, and that the additional costs so incurred are not disproportionate as compared to other methods of waste management, and that there

ist, die dabei entstehenden Mehrkosten im Vergleich zu anderen Verfahren der Abfallbehandlung nicht unverhältnismäßig sind und ein Markt für die gewonnenen Stoffe *oder die gewonnene Energie* vorhanden ist oder geschaffen werden kann (*Abfallverwertung*);

3. Abfälle die nicht verwertbar sind, sind je nach ihrer Beschaffenheit durch biologische, thermische oder chemisch-physikalische Verfahren sonst zu behandeln. Feste Rück stände sind möglichst reaktionsarm und konditioniert geordnet abzulagern

is a market or that a market can be created for the recycled substances *or energy* (*waste recycling*);

3. Wastes that cannot be recycled shall be treated otherwise by means of biological, thermal or chemical-physical methods, depending on their condition. Solid residues shall be deposited after sorting in a low-reaction and conditioned form.

The waste management law establishes avoiding and recycling waste is not, as is often claimed, an end in itself but that it is rather a basic principle to achieve an ecological waste management according to §1 of the waste management act (AWG):

The Austrian waste management law defines waste material as recyclable/reusable if there are both, a method of treatment and a market for the products.

Residues may be defined as appropriately disposable, if their emissions fail to result in any major environmental impact on air/water/soil through-out centuries.

In other words, a final disposal site can only be a landfill whose material flows have no such environmental impact that would require prior treatment of the residues. Thus, only solid waste can be disposable.

A basic demand must be that wastes which can neither be avoided nor recycled or reused shall be disposed of in a way as to prevent any future environmental burden; i.e. they must neither end up in abandoned polluted areas, nor must a long-term supervision of the disposal site become necessary.

This demand can, however, only be met by the decontamination or post-treatment of residues so that their eluates will not exceed the limiting values for potable water, which means that they certainly will never have a negative impact on the ground water.

In addition, the disposal ordinance of April 10, 1996, coming into force in 2004, prohibits in general to dispose of waste with more than 5% organic carbon. Therefore, in future all household waste will have to be treated before being landfilled.

Waste Situation in Vienna

Table 15.1, Waste Quantities in Vienna, shows the development of the situation of waste over the last ten years:

- The waste minimization process is still in its infancy
- The capacities of resource recovery are still growing
- The amount of secondary raw materials collected during the last ten years has increased from 95,000 t to approximately 313,000 t, i.e. from 13% up to 35% of the total amount of MSW

Table 15.1 Waste Quantities in Vienna

	1989 [t/a]	1991 [t/a]	1993 [t/a]	1994 [t/a]	1995 [t/a]	1996 [t/a]	1997 [t/a]	1998 [t/a]	1999 [t/a]
MSW coll. by ordinary system	476.850	468.806	456.885	440.256	447.756	467.256	469.238	469.118	488.456
Com. Waste similar to MSW	147.321	160.955	112.945	80.268	70.822	66.227	64.271	71.297	89.044
Sep. Coll.: Paper, glass, biowaste	94.753	147.309	216.796	254.797	270.054	275.44	280.366	289.931	312.800
Total amount of waste	**718.924**	**777.070**	**786.625**	**775.321**	**788.632**	**808.927**	**813.878**	**830.346**	**890.300**
MWI-Flözersteig	110.648	84.719	179.828	182.874	189.105	184.820	179.697	180.155	187.023
MWI-Spittelau	9.338	256.353	230.185	249.113	250.91	241.970	258.557	248.632	258.373
Iron scrap from MWIs	-1.500	-7.803	-11.770	-10.753	-10.346	-12.176	-11.349	-10.276	-10.378
Untreated landfilled waste	**480.716**	**283.129**	**155.211**	**86.931**	**76.493**	**96.307**	**90.280**	**106.715**	**126.458**
Waste from landfill to MWIs	—	—	4.497	7.968	21.855	394	2.654	3.359	802

MSW = Municipal Solid Waste

MWI = Municipal Waste Inicinerator

- The amount of untreated landfilled waste has increased during the last five years from 76.000 t to 126.000 t.

The table clearly shows that essential goals of the waste management law have been ignored or circumvented.

- Landfill volume has been wasted and could have been saved by the thermal treatment of waste.
- The landfilled residues have certainly not met the standards of an ecologically oriented waste management plan as far as their disposability is concerned, because they are neither sufficiently inert nor non-polluting for final disposal.

In 1999 the two waste incineration plants treated 445,000 t of waste thermally, nevertheless approximately 126.500 t of wastes were disposed of in the sanitary landfill Rautenweg. Even if both waste incineration plants had provided nearly the theoretically possible maximum operating capacity of 450,000 t p.a., a total of more than 120,000 t of waste would still have had to be landfilled without prior treatment. Moreover, there is a shortage of redundant treatment facilities, which are indispensable in order to guarantee the disposal of exclusively inert material.

Emissions into the Atmosphere Caused by Waste Treatment Processes

Emissions Resulting from Waste Incineration Plants

Incineration is aimed at transforming the material to be incinerated into gaseous incinerator products and mineralized inert solid residues which can be characterized as having a texture comparable to that of the earth crust. This process is aimed at accomplishing various goals:

- decomposition of organic substances under the supervised conditions of a perfectly regulated incineration process
- mineralization of organic substances
- reduction of the volume of material to be incinerated to about a tenth of the original volume which results in
- the concentration of components to be supervised and/or to be recovered
- generation of long-distance energy and electricity from heat gained in the incineration process by the transformation of chemical energy into thermal energy leads to
- decreased consumption of fossil fuels and thus reduction of emissions coming from other incineration plants with higher emission rates.

The incineration of organic components of household waste with the oxygen of the atmosphere generally results in the formation of carbon dioxide and water vapour. Each ton of waste is estimated to result in the emission of approx. 5,000 m³ of flue gas.

This flue gas is mostly composed of nitrogen, water vapour, carbon dioxide, and residual oxygen, but also contains dust particles carrying heavy metals, and toxic pollution gas, which is formed in the incineration process from waste components or from flue gas.

For this reason, the flue gas leaving the waste incineration plants must be carefully cleaned before being blown into the atmosphere. Even before the issuing of legal standards, the Municipality of Vienna decided in the early eighties to equip the waste incineration plants of Vienna with efficient scrubbers; at the time these scrubbers helped to set new and strict emission standards in the former law relating to boiler emission. Being a pioneer in this field, Austria is, even internationally, still among the model countries regarding the emission standards for waste incineration plants. Emission standards are defined as those emission concentrations which must not be exceeded in the operating process of a plant; that is why plant owners make an effort that their emissions remain below the legal requirements at any rate.

The permissible maximum burden of emission that does not cause adverse health effects now no longer serves as the basis for determining the emission standards for waste treatment plants, but has been replaced by those values which can be achieved by using progressive flue gas cleaning technologies.

Therefore, the burden of emission caused by waste incineration plants is far below any limiting values for emission concentrations. Therefore if emission standards are occasionally exceeded, it does not mean that health hazards are entailed, but that the state of the art of scrubbing technologies has not been reached.

The evaluation of the continuously measured emission values of the (MWI) Spittelau measured during 1999, as shown in Table 15.2 shows that the actual emissions were clearly far below the legal standards.

Similarly, the discontinuously measured contaminants, such as hydrogen fluoride, heavy metals, or ammonia from the DeNOx-filtering unit, were also far below the required emission standards.

Also all dioxine measurements (more than 50 measurements) show that the emission figures are clearly below the emission limit of 0.1 ng TE per m³ flue gas at a level of 0.04 ng TE per m³.

Despite the fact that not the supply of electricity and long-distance energy but the inertization of wastes is the prime function of a waste treatment plant, the supply of energy, which is in accordance with the strict emission standards, is of much greater advantage for the environment than the supply of energy from individual stove heating.

Comparing Emissions of the MWT Spittelau with Household Stove Heating

The comparison of emissions from the MWI-Spittelau and household heating is based on the following sources:

- the figures for stove heating efficiencies are drawn from studies conducted by the Institute of Chemical Engineering of the Technical University of Vienna
- the emission factors for stove heating have been researched by the study group "Schadstoffemission/Energieverwertung", dealing with emission of contaminants and recovery of energy; this working group was set up for preparing the energy report of the Federal Ministry for economic affairs
- annual throughput of 240,000 t of waste
- 6,000 m³ of flue gas per t of waste, i.e. 1,440 million m³ of flue gas per year
- thermal capacity of 60 MW and additional 6 MW of electricity, with a total efficiency of about 80 % (the evaluation aspects for electricity and long-distance energy were the same)
- the figure for dust emission was 0,7 mg per m³, based on the average of 1994
- the figures for heavy metals of dust generated in the MWT Flötzersteig served as a guideline for the contents of heavy metals

Table 15.2 Emission Analysis 1999 of the Municipal Waste Incineration Plant Spittelau, Vienna

Emission figures: Half hour mean values [HMV] at 11% O_2 and dry flue gas in mg/Nm³

		Dust	HCl	SO_2	CO	NO_2	Unburned Hydro-carbons	Incin. Waste [t]	Fluegas Volume [Mio Nm³]	Produced Heat [MWh]	Produced Electricity [MWh]
Emission limits*		15.0	15.0	40.0	100.0	100.0	20.0				
January '99	min	0.2	0.1	0.1	16.1	0.9	0.1				
	max	2.6	3.2	12.0	53.7	30.2	2.4	24,281	121.41	46,428	3,999
	φ	0.9	0.5	3.5	29.8	19.0	0.3				
February '99	min	0.8	0.1	0.1	13.4	2.5	0.1				
	max	9.0	0.6	13.8	84.7	216.5 (1)	4.1	22,975	114.87	41,695	3,487
	φ	2.2	0.3	2.7	31.4	20.6	0.3				
March '99	min	0.5	0.1	0.1	8.5	2.1	0.1				
	max	7.8	1.7	17.3	54.3	49.1	9.7	25,220	126.10	41,939	3,548
	φ	1.4	0.2	3.4	25.1	25.1	0.7				
April '99	min	0.5	0.1	0.1	12.1	8.6	0.1				
	max	4.6	1.9	10.1	68.0	54.1	5.3	23,680	118.40	42,210	3,702
	φ	1.1	0.3	3.4	25.6	29.0	0.4				
May '99	min	0.5	0.1	1.4	7.2	1.5	0.1				
	max	11.1	5.9	7.5	77.9	31.7	7.6	21,111	105.55	37,619	3,192
	φ	1.6	0.5	3.7	24.8	25.0	0.5				
June '99	min	0.3	0.1	0.1	8.1	2.0	0.2				
	max	42.6 (2)	1.1	14.1	99.1	52.9	14.4	7,877	39.39	16,198	348
	φ	3.2	0.5	0.4	21.9	16.9	1.0				
July '99	min	0.1	0.3	0.1	4.8	1.8	0.2				
	max	12.8	1.1	14.0	102.0 (3)	148.3 (4)	6.6	13,985	69.92	26,459	203
	φ	1.2	0.6	5.3	21.9	13.4	0.6				
August '99	min	0.3	0.1	0.1	6.3	2.1	0.1				
	max	5.5	0.6	35.0	112.6 (5)	77.2	23.4 (6)	25,637	128.18	44,215	2,613
	φ	1.1	0.4	5.8	23.1	25.0	0.4				

(Contd.)

Table 15.2 (Contd.)

		Dust	HCl	SO_2	CO	NO_2	Unburned Hydro-carbons	Incin. Waste [t]	Fluegas Volume [Mio Nm³]	Produced Heat [MWh]	Produced Electricity [MWh]
September '99	min	0.7	0.0	0.8	0.0	0.0	0.0				
	max	5.2	0.6	14.0	81.9	50.9	18.5	24.591	122.96	42,861	3,934
	ø	2.1	0.2	3.6	28.6	24.5	0.2				
October '99	min	0.8	0.0	0.0	13.6	3.3	0.0				
	max	6.7	0.5	10.7	54.5	53.8	1.0	25.005	125.03	45,879	4,005
	ø	1.9	0.2	3.6	26.7	25.1	0.0				
November '99	min	0.5	0.0	1.0	7.5	5.1	0.0				
	max	25.2 (7)	0.7	15.8	61.5	53.9	13.4	23.013	115.06	40,332	3,459
	ø	3.1	0.2	4.6	27.7	24.9	0.2				
December '99	min	0.6	0.0	0.3	9.3	0.0	0.0				
	max	7.7	0.9	8.2	58.4	51.5	5.1	25.781	128.91	46,118	3,981
	ø	1.8	0.2	3.9	29.7	25.0	0.1				
1999	min	0.1	0.0	0.0	0.0	0.0	0.0				
	max	42.6	5.9	35.0	112.6	216.5	23.4	263.156	1.315,78	471.954	36,471
	ø	1.8	0.3	3.7	26.4	22.8	0.4				

*Emission limits according to the permit for existing plants
ø.... Monthly mean values calculated from HMV
UHC ..... Unburned hydrocarbons
(1) NO_2: 2 HMV - DeNOx burner failure
(2) Dust: 1 HMV - Unexspected waste quality
(3) CO: 1 HMV - too high heat content of waste
(4) NO_2: 1 HMV - DeNOx burner failure
This table is also publised via Internet:
http://www.magwien.gv.at/ma22/emission/oot.pdfl

(5) CO: 3 HMV - Too low heat content of waste
(6) UHC: 1 HMV - Too low heat content of waste
(7) Dust: 2 HMV - Elektro filter failure in line 2

The fact that there are no emission values for some contaminants from stove heating does not mean that these substances are not emitted at all; it only means that differently from waste treatment plants, no such emission values are available for these substances. There definitely is emission of heavy metals when burning oil or coal.

As can be seen from Table 15.3, the emission amount e.g. of the MWI Spittelau is 0,5 t of organic compounds per year, whereas the production of the same quantity of heat in "environmentally sound" stoves fired with wood would have resulted in about 4.400 t of unburned hydrocarbons, to which naturally the total range of PAHs, PCDDs, PCDF and many more belong.

Table 15.3 compares the emissions of the MWI Spittelau with those emissions of stove heating which would produce the same amount of heat as the waste treatment plant.

Table 15.3. Emissions

	MWI-Spittelau		Stove heating			
	Conc. (mg/m³)	Emission (t/a)	Wood	Coal	Oil	Gas
				(tons p.a t/a)		
HCl	2.5	3.600				
HF	<0.01	<0.015				
SO₂	6.2	8.928	44	1560	540	4
CO	33.3	47.950	30.940	23.610	380	290
NOₓ	26.9	38.740	110	160	120	130
org. C	0.5	0.503	4420	1770	60	40
Dust	0.7	1.008	440	630	20	0
	(μg/m³)					
Pb	0.011	0.0158				
Zn	0.057	0.0820				
Cd	0.001	0.0014 *				
As	<0.001	<0.0014				
Ni	0.007	0.0108				
Cr	<0.001	<0.0014				
	(mg/m³)					
Hg	0.004	0.0058*				

* The annual amount of waste incinerated in the MWI Spittelau contains approx. 400 kg of mercury (Hg) and approx. 2400 kg of cadmium (Cd).

Gaseous emissions of dioxines and furanes from waste incineration plants

Without the use of special PCDD and PCDF separators, the PCDD and PCDF emissions of the Viennese waste treatment plants were found to range between 0.5 and 2.0 ng of the so-called international toxicity equivalents (ITE) per m³ of flue gas, i.e. they exceeded the emission standards required for waste incineration plants of 0.1 ng ITE per m³ by five to twenty times.

This means that the required limiting value can only be reached by installing additional scrubbers.

Tests have demonstrated that the required limiting value can in any case be obtained by means of activated carbon filters.

Regardless of this fact, the catalytic converter, with which the MWI Spittelau is equipped and which destroys nitrogen oxides, also reduces the PCDD and PCDF emissions substantially.

The emission concentrations obtained after using the catalytic converter are currently slightly above or below the emission standard. Tests have shown that the expansion of the catalytic converter system is a definite improvement in the destruction of dioxine, and that the emission standard is observed. This experience resulted in the installation of a third catalytic-converting unit.

All measurements of dioxine emission between October 1991 and now have shown emission concentrations of 0.03 to 0.09 ng ITE per m^3 of flue gas, the average value is 0.04 ng ITE per m^3. This shows that the emission standard for dioxine and furane in waste treatment plants can also be met through catalytic destruction.

PCDDs and PCDFs in untreated household waste

PCDDs and PCDFs do, however, not only emerge in waste incineration processes in incineration plants; enormous concentrations of these substances have also been found in untreated household waste prior to incineration. According to an Environment Canada report[6], the content of dioxine in household waste can range between 6 and 70 ng per kg of waste and is likely to be found in leachates of household landfills later on.

Jager J. et al., 1990 have come up with an amount of 183 ng TE per kg in mixed compostable waste, and even 7 to 17 ng TE per kg in biowaste samples.

Johnke and Stelzner 1991, in their studies relating to the German household waste analysis, have found an average dioxine content of 80 ng TE per m^3 by analyzing 70 samples of household waste. The examination of 15 incineration lines in 12 waste incineration plants provided a spectrum of PCDD and PCDF inputs of 11 to 350 ng TE per kg of waste. Assuming that a total of 5 m^3 of flue gases per kg of waste is produced in incineration processes, one can clearly deduct that less dioxine is blown into the atmosphere than is contained in the waste delivered at the incineration plant.

Gaseous emissions in landfills of household waste

Household waste in landfills undergoes various processes of biological decomposition, originating in the course of time and slowly decomposing different organic sediments, methane and carbon dioxide. The final product is a landfill gas consisting of 40 to 60 % of methane, 20 to 40 % of carbon dioxide, and up to 4 % of various other organic trace elements. Contained in these organic elements are odorous substances such as mercaptanes or hydrogen sulphide, as well as the total range of aliphatics or aromatics like benzene, toluene, xylene, etc.

Besides, landfill gas contains a large number of organic and inorganic compounds, which have entered the landfill together with the landfilled waste and are redistributed into the environment through leachates or landfill gas.

Tabasaran and Rettenberger, 1987[3] have demonstrated that the amount of landfill gas produced per ton of household waste ranges between 200 and 250 m^3, with an expected annual formation of 2.5 to 7.5 m^3 of landfill gas per ton of waste in times of maximum development, and that even after a storage period of 50 years this formation of landfill gas has not come to a standstill (Dernbach 1988).

Table 15.4 referring to Eisenmann 1985, provides some examples of trace elements in landfill gas.

Table 15.4. Range of concentration of trace elements in landfill gas

		Range	Peaks
CO_2	%	0 – 85	
H_2S	mg/m³	0 – 200	23000
Merkaptane	mg/m³	0 – 50	200
Hexen	mg/m³	3 – 18	1200
Benzene	mg/m³	0.1 – 10	450
Toluene	mg/m³	0.2 – 600	1700
Xylene	mg/m³	0 – 400	
Ethylbenzene	mg/m³	0.6 – 240	
Cumol	mg/m³	0 – 30	
Acetaldehyde	mg/m³	0 – 150	
Dichlormethane	mg/m³	0 – 200	3000
Vinylidenchloride	mg/m³	0 – 100	
1.2-Dichlorethene	mg/m³	0 – 700	
Vinylchloride	mg/m³	0 – 30	250
Total	mg/m³	0 – 5000	20000
Total Chlorine	mg/m³	10 – 600	
Total Fluorine	mg/m³	1 – 100	
Mercury	mg/m³	up to 0.3	
Arsenic (AsH_3)	mg/m³	?	

The list of landfill gas components shows that the tremendous environmental impact caused by the disposal of untreated household waste is not only due to leachates but also due to landfill gas. In this context it has to be mentioned that methane, which is produced in the landfilling process of untreated household waste, has a 21 times greater greenhouse effect than has the same amount of carbon dioxide.

MWI Spittelau—A Landmark in Vienna

After a year-long discussion with the head of the environmental protection department of the municipality of Vienna, the Viennese artist Friedensreich Hundertwasser understood the urgent need for waste incineration and that the waste incineration plant is a facility for protecting man's environment and agreed to undertake the new architectural design of the plant's facade without accepting any remuneration.

The artist succesfully demonstrated that industrial construction is just as much part of our life as residential buildings. Visual pollution by aggressive and inhumane constructions, he says, is the most dangerous form of pollution because it destroys man's dignity and essential nature.

Hundertwasser's vision is, that this spectacular industrial construction should be an example of a symbiotic harmony between technology, ecology and art, a reminder to society to mend its wasteful ways.

Public Acceptance

The artistic form of the MWI Spittelau is not only a work of art by the famous artist

Friedensreich Hundertwasser that causes an extra cost of 80 million Austrian Schillings (approx. 7 million US$) but has also increased the acceptance of waste incineration among the Viennese population.

Several opinion polls conducted by the Austrian Institute of Empirical Social Investigations IFES (Institut für Empirische Sozialforschung) after the reconstruction of the MWI Spittelau have shown that in Vienna the acceptance of MWI has increased significantly. The two polarizing questions about the attitude towards MWI were answered as shown in Table 15.5.

The two questions were:
- Are you in favour of MWI which uses waste energy for district heating?
- Are you principally against MWI?

Table 15.5. Attitude of the Viennese people towards MWI

	Principally against MWI	In favour of MWI with district heating
1990	4%	73%
1991	3%	79%
April 1992	4%	77%
June 1992	3%	81%

These polls show that opposition against MWI in Vienna lies relatively constant between 3% and 4 % whereas the acceptance among the population has risen from 73% to 81% . From this point of view the effort for the artistic appearance of the MWI Spittelau has not only added a new attraction to Vienna but has also been a good investment in environmental PR.

However 3% to 4% of the population still oppose MWI and make much more noise than the 81% who are in favour of MWI.

REFERENCES

1. Canadian National Incinerator Testing and Evaluation Programme: Environmental Characterization of Mass Burning Incinerator Technology at Quebec City - Summary Report EPS 3/UP/5, June 1988, Environment Canada.

2. Dernbach H. (1988). Entgasungsmaßnahmen bei Altablagerungen. In: *Handbuch der Altlastensanierung*. Economia Verlag Bonn.

3. Eisemann R. (1985). Bewertung von Spurenstoffen in Deponiegas. In: Dokumentation einer Fachtagung "Deponiegasnutzung - Spurenstoffe, neuere Planungen, Entwicklungen und Erfahrungen". BMFT und UBA Berlin.

4. J. Jager, K. Fricke, H. Vogtmann, M. Wilken (1990). Bewertung von organischen Schadstoffen in Kompst. Witzenhäuser Abfalltage.

5. Johnke und Stelzner (1991). Dioxin-und NO_x-Minimierungstechnik. VDI-Seminar in München, 17-18. September.

6. Leitbild Für Die Schweizerische Abfallwirtschaft in: Schrittenreihe Umweltschutz Nr. 51, hg.v. Umweltamt für Umweltschutz, Bern, 1986.

7. Tabasaran O. und Rettenberger G., Grundlagen zur Planung von Entgasungsanlagen. Nr. 4547, 1/1987. In Kumpf W. et al.: Müll- und Abfallbeseitigung Müllhandbuch. E-Schmidt Verlag seit 1964.

8. Vogel G., Entscheidungsgrundlagen Zur Fortschreibung Des Wiener Abfallwirtschaftskonzeptes, hg. Magistrat der Stadt Wien, MA 48 - Stadtreinigung und Fuhrpark. Oktober 1988.

16

Co-firing: Primary Energy Savings and Carbon Dioxide Avoidance

IVO BOUWMANS

TBM/E&I, Jaffalaan 5, 2628 BX Delft, The Netherlands

In the discussions about the reduction of greenhouse gas emissions, following the Kyoto agreements, waste co-firing (or co-incineration) may play an important role. If waste is burnt in a power plant or an industrial plant rather than in a waste incinerator, its calorific value avoids the use of fossil fuels and therefore results in an overall reduction of carbon dioxide emission. However, the merits of this type of incineration, claimed by the owners of the industrial and power plants, may well be overstated, if the waste treatment processes are evaluated as an integrated system rather than as separate installations.

In this chapter, three installations that can process waste are compared: a dedicated waste incinerator, a power plant, and a cement kiln, where both the calorific and the material content of the waste are used. The effects of the incineration of three typical examples of waste (mixed plastic waste, rubber, and sludge from a waste water treatment plant) are evaluated. The calorific value of the waste and the carbon dioxide emissions are compared with those of the primary fuel that would otherwise have been used in the processes.

This integrated approach makes clear that for the system as a whole the net gain may be smaller than often claimed. However, provided that other environmental criteria are met, the substitution value can be substantial and application of waste in a power plant or a cement kiln can have considerable advantages.

Introduction

Climate change caused by the enhanced greenhouse effect is currently considered to be the major global environmental problem. The debate about whether this is connected to the anthropogenic emission of large quantities of carbon dioxide is still in full swing. Even so, at the 1997 meeting of the United Nations Framework Convention on Climate Change in Kyoto, the participants decided on a worldwide reduction of CO_2 emissions. The member states of the European Union agreed on an average reduction of 8% in 2012 compared to 1990. The target for the Netherlands is a 6% reduction for 2010.

One of the options for emission reduction is the use of biomass when fossil fuels would be used otherwise (Ekmann *et al.*, 1998). A special case is the co-firing of waste, where the calorific value of the organic part of the waste is used to generate electricity or to fuel an industrial process. Wastes can not only be incinerated in dedicated incinerators, but also in power plants or industrial processes, in which case it is referred to as "co-firing". An example of the useful application of waste in an industrial process is co-firing in a cement kiln.

From an environmental point of view, it is better if the calorific value of the waste is used as much as possible. For that reason, owners tend to overstate the energetic performance of their installations. The usual claim is that the amount of primary energy saved is equal to the calorific value of the waste that is co-incinerated. This chapter offers a more integrated approach for estimating the merits to the environment of co-firing as a system. The burning of waste in a dedicated waste incinerator is compared with incineration in a coal-fired power plant and a cement kiln. A systems approach is adopted to make a proper evaluation of these processes. This yields more accurate results than an analysis of each process in isolation and makes the model more generally applicable.

The Model

Three Processes

The system under consideration consists of three parallel processes: a waste incinerator, a coal-fired power plant and a cement kiln (see Fig. 16.1).

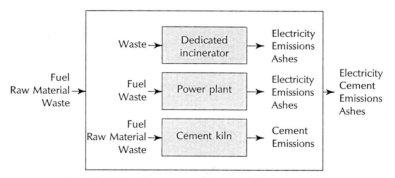

Fig. 16.1 Model system consisting of three processes.

The waste incinerator's only input flow is waste. The heat from the burning waste is used to generate electricity, as is the case in all Dutch waste incinerators. Other output flows are ashes and emissions.

The input flow of the power plant is fuel, which is burnt to generate electricity. In the system considered here, the fuel is coal, part of which can be substituted by waste that meets certain quality criteria. As in the case of the incinerator, the output flows are electricity, ashes and emissions.

In the cement kiln, raw materials are used to make clinker in a furnace where primary fuel is burnt. It is assumed in the model that the fuel used in the kiln has the same calorific value as that used in the power plant, but this is not a model restriction. As in the cases of the incinerator and the power plant, part of the fuel can be substituted by waste that meets the quality criteria. The mineral content of the waste is incorporated in the clinker, reducing the demand for raw materials. The output flows of the kiln are clinker and emissions.

The three processes combine to a system that has three input flows:
- primary fuel
- raw materials
- waste

The combined system has four output flows:
- electricity
- clinker
- ashes
- emissions

Of these, the demand for electricity and clinker, which are external factors, are taken to be constant.

The demand for primary fuel and raw materials, as well as the resulting amounts of ashes and emissions, will change depending on the way waste is distributed over the three processes. The output flows of ashes and emissions are not explicitly modelled here.

Waste Used in Power Plants

First, we will consider the energetic performance of the processes, taking into account the performance of the combined system. Since the total demand of electricity is taken to be constant, any reduction in the energy output of the incinerator must be compensated for by an increase in the electricity production of the power plant.

An amount of waste (M_w, kg) that is brought to the power plant rather than to the incinerator, will generate an amount of energy (E_p, J) equal to

$$E_p = M_w u_w \eta_p, \tag{1}$$

with u_w the calorific value of the waste (LHV, Lower Heating Value, J/kg) and h_p the energy efficiency of the power plant. In the power plant, this will save an amount of primary fuel ($M_{f,saved,p}$, kg) equal to

$$M_{f,\ saved,\ p} = \frac{E_p}{\eta_p u_f} = \frac{M_w u_w}{u_f}, \tag{2}$$

where u_f is the calorific value of the fuel (J/kg). However, the incinerator now generates less energy because of the lower supply of waste. The reduction amounts to

$$E_{i,\ less,\ p} = M_w u_w \eta_i, \tag{3}$$

(where η_i is the efficiency of the incinerator). This energy must now be supplied by the power plant, which takes an amount of primary fuel of

$$M_{f,\ extra,\ p} = \frac{E_{i,less,p}}{u_f \eta_p} = \frac{M_w u_w \eta_i}{u_f \eta_p}. \tag{4}$$

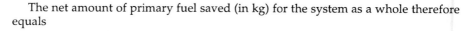

The net amount of primary fuel saved (in kg) for the system as a whole therefore equals

$$M_{system,\ saved,\ p} = M_{f,\ saved,\ p} - M_{f,\ extra,\ p} = M_w \frac{u_w}{u_f}\left(1 - \frac{\eta_i}{\eta_p}\right) \tag{5}$$

or, expressed per amount of diverted waste:

$$\frac{M_{system,\ saved,\ p}}{M_w} = \frac{u_w}{u_f}\left(1 - \frac{\eta_i}{\eta_p}\right) \tag{6}$$

The saving can also be expressed as the amount of carbon dioxide avoided by the use of waste in the power plant, using the carbon intensity (C_f, kg/J), which is the mass of carbon emitted per joule of energy in the fuel:

$$\frac{M_{CO_2,\ avoided,\ p}}{M_w} = u_w\, C_f r\left(1 - \frac{\eta_i}{\eta_p}\right) \tag{7}$$

where r is the molecular mass ratio of carbon dioxide and carbon ($44/12 = 3.67$).

Waste Used in Cement Kiln

Because the demand for clinker is assumed constant, the energy requirement of the cement kiln is constant. Analogous to the power plant case, waste burnt in kilns rather than in incinerators saves an amount of primary fuel equal to the following:

$$M_{f,\ saved,\ k} = \frac{E_k}{\eta_k u_f} = \frac{M_w u_w}{u_f} = M_{f,\ saved,\ p} \tag{8}$$

where E_k is the energy used in the kiln (J) and η_k is the energy efficiency of the kiln. As in the case of waste burnt in the power plant, the incinerator now generates less electricity, so that the energy output of the power plant has to be increased. The extra primary fuel needed to do this is:

$$M_{f,\ extra,\ k} = \frac{M_w u_w \eta_i}{u_f \eta_p}. \tag{9}$$

The resulting net amount of primary fuel saved is

$$M_{system,\ saved,\ k} = M_w \frac{u_w}{u_f}\left(1 - \frac{\eta_i}{\eta_p}\right) = M_{f,\ saved,\ p,} \tag{10}$$

so this is the same as when waste is brought from the incinerator to the power plant ($M_{f,\ saved,\ p}$).

Again, the gain can be expressed as the amount of carbon dioxide avoided:

$$\frac{M_{CO_2, avoided, k}}{M_w} = u_w \, C_f \, r \left(1 - \frac{\eta_i}{\eta_p}\right) = \frac{M_{CO_2, avoided, p}}{M_w} \tag{11}$$

equal to the amount avoided when the waste is used in the power plant.

It follows that both the amount of primary fuel saved and the amount of carbon dioxide avoided when waste is used in either a power plant or a cement kiln rather than in a waste incinerator, depends on the efficiencies of the incinerator and the power plant—not on that of the kiln. The result also contradicts the belief that the amount of energy saved is equal to the energy content of the waste. Moreover, the equality of eqns 5 and 9 shows that from the energy point of view no distinction can be made between application in the power plant and application in the cement kiln.

In countries where power generation by waste incinerators is not standard practice, however, co-firing would lead to relatively larger savings: because η_i (the efficiency of the incinerator) is effectively zero in that case, eqns (5) and (10) then equal eqns (2) and (8), respectively. For the same reason, the amount of carbon dioxide avoided will be larger in those countries.

Example

A combination of process characteristics that is typical for the Dutch situation is used here as an example: the energy efficiencies of the incinerator, the power plant and the cement kiln are 20%, 40% (both electrical efficiencies) and 70%, respectively. The coal that is used in both the power plant and the cement kiln has a calorific value of 31 MJ/kg and a carbon intensity of 25 g/MJ. These values are typical for coal.

Three types of waste will be used as examples: mixed plastic waste (MPW), rubber, and sludge from a waste treatment plant. These represent waste with energy contents higher than, equal to and less than that of coal, respectively. The energy contents of these waste flows are given in table 16.1.

Table 16.1. Energy content of the analysed waste flows.

Waste	Energy (MJ/kg)
MPW	38
rubber	31
sludge	8

Because the amounts of primary fuel saved (both uncorrected and corrected) and the amounts of carbon dioxide avoided are equal for the cases of the power plant and the cement kiln, we may safely drop the indices p and k without the risk of confusion.

Energy

In this section, the energetic gain of co-firing is evaluated for the three sample waste flows.

Table 16.2 gives the amount of energy per tonne of waste in the incinerator, the power plant and the cement kiln ($M_w u_w$ times the energy efficiency of the process).

Table 16.2. Energy per tonne of waste (in GJ).

Waste	Incinerator	Power plant	Cement kiln
MPW	7.6	15	27
Rubber	6.2	12	22
Sludge	1.6	3.2	5.6

The middle column of table 16.3 shows the uncorrected amount of fuel saved, divided by the amount of waste, for the three waste flows. These figures are those often claimed by the owners of power plants and cement kilns as representing the amount of primary fuel saved by using waste.

Table 16.3. Uncorrected and corrected amounts of fuel saved per amount of waste (kg/kg)

Waste	$M_{f,\ saved}/M_w$	$M_{system,\ saved}/M_w$
MPW	1.2	0.61
Rubber	1.0	0.50
Sludge	0.26	0.13

However, the corrected results in the right hand column of Table 16.3 (from eqn 6) show that the actual saving are much less: 50% of the uncorrected values.

The amount of energy saved can be further increased if the power plant is a cogeneration plant, where the surplus heat is used for the heating of buildings. The total energy efficiency of a cogeneration plant can go up to 90%. In the case of the realistic value of 80%, 0.92 tonne of primary fuel would be saved per tonne of MPW, and analogous increases hold for the other waste flows.

If the primary energy saved is expressed per amount of energy in the waste, one joule of waste saves 0.50 J of primary fuel for all waste flows. In the case of a cogeneration power plant, the energy saved would be 0.75 J/J. These figures show that the claim "the energy saved equals the calorific value of the waste" overstates the real value by 100%, or 33% in the case of a cogeneration plant.

Carbon Dioxide

In the light of the international agreements, the amount of carbon dioxide emission avoided becomes an important measure in evaluating the merits of waste co-firing. Here, too, uncorrected values could be given, but we will restrict ourselves to the net results ($M_{CO_2,\ avoided}$), for these give the result for the system as a whole.

Table 16.4 gives the results for the power plant (energy efficiency 40%) and the cogeneration plant (energy efficiency 80%). As in the case of the saved energy, the figures also apply to the situation where the waste is used in a cement kiln.

Table 16.4. Avoided carbon dioxide per amount of waste (kg/kg)

Waste	Power plant	Cogeneration plant
MPW	1.7	2.6
Rubber	1.4	2.1
Sludge	0.37	0.55

At first sight it may seem strange that 1 kg of MPW avoids 2.6 kg of carbon dioxide. However, most of the carbon dioxide's weight (73%) originates from the oxygen in the combustion air.

In 1990, the official figure for the annual emission of carbon dioxide in the Netherlands was 160 Mton (ECN, 2000). The Dutch Government's target for 2010 is 6% lower than this, which would have meant a reduction of about 10 Mton. In the meantime, however, the emission figure has risen some 10% (Keijzers, 2000), making the target even more difficult to reach. Application of suitable waste flows in power plants or cement kilns could help to achieve the goal. The annual amount of combustible waste in the Netherlands is of the order of 5 Mton. However, this number is expected to decrease. In addition, a very large part of the waste can only be burnt in dedicated incinerators. The reasons for this will be dealt with in the next section.

Limitations

The application of waste as a fuel in industrial processes is limited in several ways, because only specific waste flows satisfy the demands made by these processes.

In the first place, the calorific value of the waste must be high enough. However, skimming high-calorific waste from the incinerators would cause the calorific value of the remaining waste to drop, which would result in inferior incinerator performance. This would decrease the overall energy output of the system under consideration.

Secondly, the waste should not have high concentrations of volatile chemicals. In a power plant, for instance, the more aggressive gases formed during the incineration can cause damage to the interior of the boiler. Therefore, when large amounts of waste are co-fired the temperature in the boiler must be lowered, which in turn reduces the plant's energy efficiency for thermodynamic reasons.

Thirdly, the level of toxins in the emissions to the atmosphere should not be too high. While dedicated waste incinerators in the Netherlands are equipped with extended flue gas treatment installations, power plants and cement kilns have lower standards of emission reduction. Sewage sludge flue gas, for instance, may contain heavy metals, mercury, dioxins and furans, acid gases and nitrogen oxides (Werther and Ogada, 1999). Care should be taken to prevent high emission levels, and regulations should be adapted to the case of co-firing.

Finally, the remaining ashes should not contain high levels of contamination. In the case of cement, chemicals from the waste may affect the quality of the clinker. The Dutch cement industry sees itself in the first place as a producer of a high-grade building material, and is reluctant to do anything that could reduce the customers' confidence in the product. In contrast, cement producers in Belgium and France present their process as an ecologically sound way of waste treatment.

These four considerations mean that in practice, plant owners will only withdraw selected types of waste from the dedicated incinerators.

Conclusion

The claim that the amount of primary fuel saved by the co-firing of waste is equivalent to the calorific value of the waste is rendered unjustified by the integrated systems approach explained here. The equivalent holds for claims about the amount of avoided carbon dioxide emissions. In both cases, a correction should be made because of the energy that is not produced in other installations where the waste would have been treated otherwise.

It is beyond doubt that the best way to save primary energy and avoid carbon dioxide emissions is to reduce the amount of waste, electricity and clinker in the first place. The production of electricity and clinker should never become dependent on the 'production' of waste, because then the incentive to prevent waste would disappear. However, a large amount of primary fuel can be saved and carbon dioxide emissions avoided by co-firing, provided the other environmental conditions are met.

REFERENCES

1. ECN (2000); Energie Verslag Nederland 1999 (in Dutch); available at www.ecn.nl.
2. Ekmann, J.M., J. C. Winslow, S. M. Smouse and M. Ramezan (1998); International survey of cofiring coal with biomass and other wastes; *Fuel Processing Technology*, **54**: 171-18.
3. Keijzers, G. (2000); The evolution of Dutch environmental policy: the changing ecological arena from 1970-2000 and beyond; *J. of Cleaner Production*, **8**: 179-200.
4. Werther, J. and T. Ogada (1999); Sewage sludge combustion; *Progress in Energy and Combustion Science*, **25**: 55-116.

17

Baled MSW and Associated Problem, in the Context of Fire Hazard

William Hogland[1], Diauddin R. Nammari[1], Sven Nimmermark[2], Marcia Marques[1, 3] and Viatcheslav Moutavtchi[1]

[1]Department of Technology, University of Kalmar,
Kalmar P.O. Box 905 SE- 39129, Sweden
[2]Department of JBT, University of Agricultural Sciences, Alnarp, Sweden
[3]Rio de Janeiro State University – UERJ, Department of
Environmental Engineering, Brazil

Introduction

The baling technique has been shown to be the most promising storage method for the baling of waste fuels, according to investigations carried out both at laboratory scale and outdoors on a large scale (Hogland 1998, Hogland et al., 1996). The technique uses a computerized baling machine and lifting device, which presses the waste material into a bale and wraps it with a stretched film of polyethylene (PE). A front-end loader is used to transport the bales from the conveyer belt of the baling machine to storage. Previous studies (Ansbjer et al., 1995, Tamaddon et al., 1995a, Tamaddon et al., 1995b, Hogland W. and Tamaddon F. 1995, Hogland W. and Marques M. 1999, 2000) have shown that storage of such bales affords no risk of self-ignition. There is no temperature rise within the bale, with the inside temperature following the temperature of the surroundings. Bales exhibited good stability during storage with almost no energy or mass loss. Being wrapped, they are clean to handle and transport. There is a reduction in volume of about a third during compression. Problems include swelling induced by water ingress and, odour problems resulting from the build up of fatty acids (propionic, butyric and acetic acid).

In this study, storage tests were carried out under high temperature conditions and the physical and chemical processes operating during storage of round and square bales were studied. In addition, incineration tests were carried out to assess the flammability and the concentration of a variety of pollutants in the solid and gaseous parts of the smoke.

Preparing the Bales

Round bales containing household waste and a mix of household waste and organic material (around 30% organic food waste from the food processing company

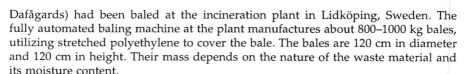

Dafågards) had been baled at the incineration plant in Lidköping, Sweden. The fully automated baling machine at the plant manufactures about 800–1000 kg bales, utilizing stretched polyethylene to cover the bale. The bales are 120 cm in diameter and 120 cm in height. Their mass depends on the nature of the waste material and its moisture content.

The bales were mainly composed of household waste that was collected from both the urban area and the surrounding countryside of Lidköping. A typical composition of the material in the bales is shown in Fig. 17.1. To be able to assess the effect of baling and waste content on the behaviour of the bales, the baling pressure was reduced in some bales and the organic content was increased in others so that it comprised 50% of the total bale content (refer to Fig. 17.2).

Paper and food waste comprised 25% and 30% of the total waste from the countryside and the urban areas, respectively. The waste from the countryside contained a high percentage of garden waste, and a surprisingly high percentage of metals, while non-combustible waste fraction comprised 5-8% of the total. Random samples of the waste were taken for laboratory analysis.

Characterization of Bales

Emissions

A few days after the baling, experiments with six round bales and two square bales (baled at the incineration plant in Västervik, Sweden) were started at the University of Kalmar. With the help of a multi-parameter monitoring system and a linked computer, continuous data measurements were carried out on bales stored at 17 °C and 35 °C.

The characteristics measured (see Table 17.1) were:
- Temperature in the middle of a bale.
- Ambient temperature and the surface temperature of some bales.
- Gas concentrations in each bale (O_2, CO_2 and CH_4).

Table 17.1. Intended storage for the waste bales

| | Round bales | | Square bales |
	Normal pressure by baling	Low pressure by baling	Normal pressure by baling
Storage temperature about 35 °C	Household waste (About 70% household waste + about 30% food waste)	Household waste (About 70% household waste + about 30% food waste)	Household waste
Storage temperature about 17 °C	Household waste (About 70% household waste + about 30% food waste)		Household waste

Over a period of several days, the temperature of the bales slowly increased until it reached ambient room temperature, after which this temperature was maintained. The rate of biological decomposition of the waste inside the bales was very low and after 8-9 months of storage, significant levels of carbon dioxide were recorded in only a few bales (Appendix A. Figures 17.1–17.4). The decomposition of the waste inside the bales caused an intense and unpleasant odour and so the bales had

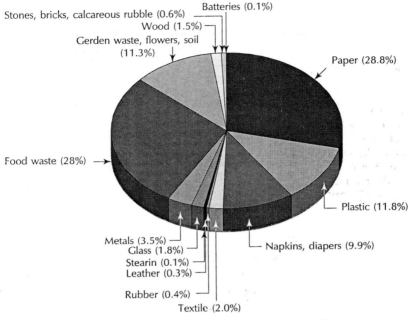

Fig. 17.1. The composition of bales with household waste.

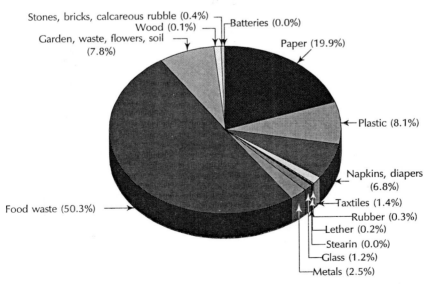

Fig. 17.2. The composition of bales with household waste and extra organic material.

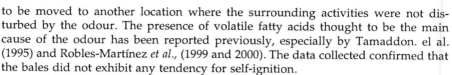

to be moved to another location where the surrounding activities were not disturbed by the odour. The presence of volatile fatty acids thought to be the main cause of the odour has been reported previously, especially by Tamaddon. el al. (1995) and Robles-Martínez *et al.*, (1999 and 2000). The data collected confirmed that the bales did not exhibit any tendency for self-ignition.

The bales were kept at either 17 °C or at 35 °C. The lower temperature (17 °C) was equivalent to the room temperature and the higher temperature was achieved by enclosing the bales in makeshift tents using a heater to heat the air.

As can be seen from the information illustrated on Appendix A (Figures 1–3), the only noticeable change in temperature was in the square bales. Bale Sq2, which was kept at 17°C, exhibited an increase in O_2 consumption during the first month. However, this consumption rate decreased gradually. This indicates an initial burst of degradation followed by a gradual acidification of the bale, which would reduce the rate of CO_2 emission. Hence, on an average it could be said that baling reduced biodegradation.

Leachate and Stormwater

Samples of stormwater runoff taken from the bale storage area and surrounding areas were collected at the Lidköping site. With the help of an automated water sampler, samples were taken at regular intervals during a rain event. No extreme concentrations (see Table 17.2 and Fig. 17.3) were found, although the concentration of zinc, for example, in the collected rainwater was quite high. Some increases in the concentrations of pollutants were observed during the rain event, which might have been caused by dissolution from previously contaminated layers of the surface and subsequent transport in the liquid phase.

Table 17.2. Results from the rain water analysis (Marques M 2000)

Quantity	Unit	Value (median)
COD	mg/L	190
N-total	mg/L	12
P-total	mg/L	0.39
Cu	μg/L	14
Ni	μg/L	10
Pb	μg/L	50
Cr	μg/L	10
Zn	μg/L	120
pH	pH	7.4
Conductivity	mS/m	24

Hg was below the detection limit

Since the bales are wrapped and compacted, they produce leachate in small quantities. This leachate had a high concentration of fatty acids, which are thought to be the main source of odours. This high concentration of acids reduces the pH and hence the activity of methanogenic bacteria, which normally decompose the fatty acids and produce CO_2. Thus it can be said that the bales are truly acidic, which was also found in the study by Robles-Martínez *et al.*, (1999 and 2000). The results of the leachate measurements are shown in Fig. 17.4.

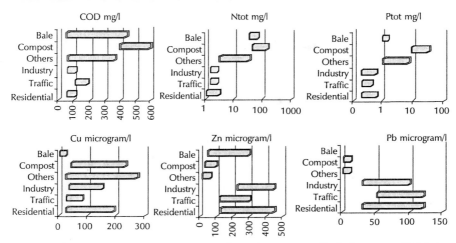

Fig. 17.3. The concentrations in stormwater from the storage of baled waste at Lidköping (bale), compared with data obtained in composting areas (compost) and other areas and roads (at Spillepeng landfill site located on the coast in Malmö in southern Sweden operated by SYSAV AB). Swedish stormwater is recommended for planning in Industry, Traffic and Residential areas with more than 50 inhabitants per hectare (Marques M 2000).

The Burning Tests

Experiments were performed to obtain information on the manner in which bales burn and the pollutants that could be spread into the environment during a possible fire in a bale storage area. Two bales were used in the experiment. They were placed in a covered container and ignited with the aid of cardboard and 10–15 l of diesel fuel. The bales were left to burn for 1 hour before the start of the analysis, thus allowing the flames to stabilize. Because of the compactness and high water content of the bales, it was expected that it would be difficult to set fire to the bales and that the fire would eventually extinguish itself. However, the bales burnt very well. A strong fire developed around the bales, which shed flakes that, in falling, re-invigorated a steadily burning fire. No extreme amount of smoke was released and, according to subjective and non-scientific visual and olfactory perceptions, the smoke from the ordinary bales did not appear to be particularly aggressive or pungent and seemed to be similar to that from a small wood fire. After this initial period was over, the bales were allowed to burn for a further period of 2 h 50 min while data were collected. The smoke from the fire was extracted from the container to analyse the concentrations of the pollutants. During the extraction, the flow rate and the temperature of the gas was monitored. The moisture content of the smoke, as well as the concentrations of Hg, Cd, HCl, HF, HBr and NH_3, were analysed by passing a side-stream through bottles filled with absorbents. Continuous measurements were also carried out on the concentrations of O_2, CO_2, CO, SO_2, NO, NO_x, N_2O, and THC in the smoke. Dust (soot) was also analysed for the concentrations of Hg, Pb, Cd, As, Ni, Cr, Mn, Cu, Co, Sb and V. Samples from the smoke were taken by

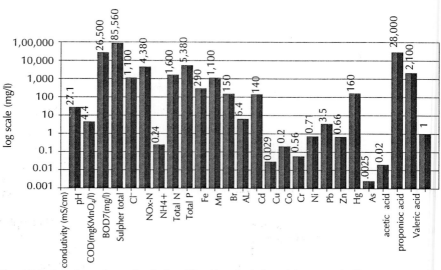

Fig. 17.4. Leachate from the bales. The Y scale is logarithmic and all, except pH and conductivity, have units of mg/L.

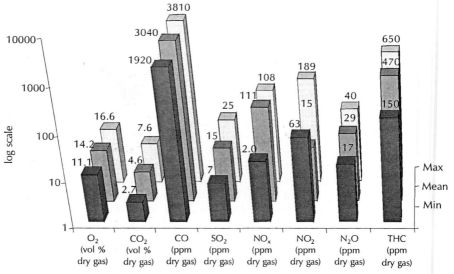

Fig. 17.5 Data collected with continuous measurement for 2 h 50 min smoke flow (mg/m³) was 400 (mean) 360 (min) and 440 (max) while the smoke temperature (°C) was 197 (mean) 142 (min) 247 (max).

passing it through a glass apparatus to measure the concentrations of PAH, PCDD/F and PBrDD/F. The results, averaged over the collection period of 2 h 50 min, are shown in Table 17.3.

Table 17.3. The total of the results obtained for period of 2 h 50 min

Quantity	Unit	Value[1]
O_2	vol (% dry gas)	14.2
CO_2	vol (% dry gas)	4.6
Humidity	vol	8.1
CO	ppm dry gas	3040
NO_x of which NO_2	ppm dry gas	111
SO_2	ppm dry gas	15
Soot	mg/m^{3*}	470
HCl	mg/m^{3*}	99
HF	mg/m^{3*}	0.1
NH_3	mg/m^{3*}	27
Hg	$mg/m^3 *$	8
Cd	mg/m^{3*}	360
TCDD-TEQ**	ng/Nm^{3*}	13
PAH without naphthalene	$\mu g/Nm^3$	3.044 $\mu g/Nm^3$

*m^3 is equivalent to normal dry gas containing 9% CO_2
** Calculated according to Eadons model
[1] The value represents the average over the 2 h 50 min.

Smoke and Particle Analysis

The results from the analysis of smoke are shown in Fig. 17.5–17.7. Throughout these experiments, sampling was carried out twice. The concentration of metals was measured both in the soot and in the filtered material. The whole amount of soot, including filtered material from the filter, was crushed down to a homogenous powder. Then analysis of the metal content was carried out by Atomic Absorption Spectroscopy (AAS).

Samples of Hg, Cd, HCl, HF, HBr and NH_3 were collected by filtering the gases into solutions of appropriate absorbents. All the samples were taken from filtered gas and the soot-examination-instrument, which was placed at the end of the filter stand. The smoke gas analysis was carried out using different devices: for the analyses of CO_2, CO and N_2O an IR-instrument was used, O_2 was measured by using its paramagnetic properties, NO/NO_x by chemical-luminance, and SO_2 by UV instrument and THC flame ionisation.

The results obtained were cross-referenced to values taken from a brief literature survey as well as to the Swedish limits for burning. The values from the literature are summarized in Table 17.4. It was found that, with the exception of Pb and Cd, the trace metal emissions were in accordance with Swedish national limits. All the values obtained were less than those obtained by Hasselrris and Licata (1996) for boiler emissions, otherwise the values were higher.

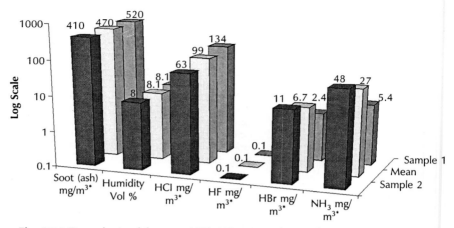

Fig. 17.6 Data obtained for soot, HCl, HF, HBr and NH_3 during manual sampling

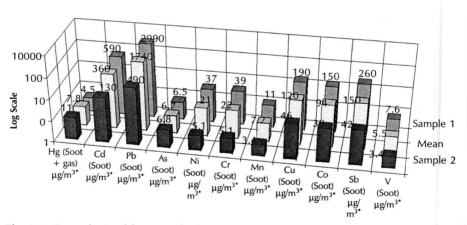

Fig. 17.7 Data obtained for Hg, Cd, Pb, As, Ni, Cr, Mn, Cu, Co, Sb and V, the sum of all the metals was 2.2 mg/m^3.

Table 17.4. Emission of gases and metals

Quantity	Mannien et al. values (1997) 11% O_2 (mg Nm^{-3}) 11% O_2 (%)	Carlsson K (199) Values at 11% O_2 mg/m^3	Authors Hasselrris and Licata (1996) Values at 7% CO_2 μg/Nm^3 Boiler, stack emission	SFS 1997: 692[21]
O_2	12.3-13.7 (%)			6 mg/m^3 Present limit, 6 mg/m^3 Limit as of 31 Dec 2005 10 μg/m^{3}Limit as of 1 Jan 2005 or 1 Jan 2010
CO	44-160 (Mean) 15-35 (BL)			50 mg/m^3 day average 150 mg/m^3 10 min. 95% 100 mg/m^3, 0.5 h in a 24 h period
CH_4	20 – 2			
SO_2	450 – 550			50-200 μg/m^3 Present limit 20-200 μg/m^3 Limit as of 31 Dec. 2005 20-350 μg/m^3 Limit as of 1 Jan 2005 or 1 Jan. 2010
NOx as NO_2	160-180	205(NO) 16(NO_2)		50-110 μg/m^3 Present limit 40-90 μg/m^3 Limit as of 31 Dec. 2005 30-200 μg/m^3 m^3 Limit as of 1 Jan. 2005 or 1 Jan. 2010
N_2O	15			
Particles	2 – 5			Soot 40-90 μg/m^3 Present limit Soot 20-90 μg/m^3 Limit as of 1 Jan. 2005 or 1 Jan. 2010
HCl	30 – 154			Limit for burning 10 mg/m^3 mean 24h 10-60 mg/m^3 half hour period
HBr	3.5			
HF	0.3 – 1.2			1 mg/m^3 mean 24 h 2-4 mg/m^3 half hour period

(Contd.)

Table 17.4. (Contd.)

Quantity	Mannien et al. values (1997) (mg Nm⁻³) 11% O₂	Carlsson K (199) Values at 11% O₂ mg/m³	Authors Hasselris and Licata (1996) Values at 7% CO_2, μg/Nm³ Boiler, stack emission	SFS 1997: 692[21]
NH₃				
Hg	1	0.08 – 0.45 (Hg) gas	475,41	All average taken during testing time of at least of 30 min and a max 8 h, representing the Maximum limit : 0.05mg/m³ New factories 0.1 mg/m³ Existing factories
Cd	18	1.29	1680,2	0.05 mg/m³ New factories, 0.1mg/m³ Old factories
Pb		21.1	18133,2	0.5 μgm³ Present limit 0.5 μg/m³ Limit as of 1st Jan 2005 or 1st Jan 2010
As	0.02 (Ni+As)	(As + Co + Ni + Se + Te) 41	78,0.16	0.05 mg/m³ New factories, 0.1 mg/m³ Old factories
Ni			4240,6	
Cr	0.06 – 0.74 (Cr+Cu+Mn+Pb)		3620,2.3	0.05 mg/m³ New factories,0.1 mg/m³ Old factories
Mn			3235,1	0.5 mg/m³ New factories,1 mg/m³
Cu			8818,54	0.5 mg/m3 New factories,1 mg/m3 Old factories
Co			111,0.3	0.5 mg/m³ New factories,1 mg/m³ Old factories
Sb		(Sb + Pb + Cr + Cu + Mn + V) 32.5	822,0.3	
Sn			800,2	0.5 mg/m³ New factories,1 mg/m³ Old factories
V			257,0.09	0.5 mg/m³ New factories, 1 mg/m³ Old factories
Zn		74.3	909,33	
Tl		(Tl + Cd + Hg) 1.3		0.05 mg /m³ New factories 0.1 mg/m³ Old factories

Organic pollutants

To measure dioxins, a specialized sampling device had to be used, and the samples were later sent for analysis. All the equipment used was made of glass, and was thoroughly cleaned before use.

PBrDD/F

The analysis on PBrDD/Fs showed that the total amount for the brominated dioxin was 6067 ng/Nm^3 in dry gas of 9% CO_2. However, problems with elutriation could have affected the results, while disturbances in the spectrometer increased the uncertainty. The only particular combination that was identified with a high level of certainty was 2378-TBrDD, the concentration of which corresponded to 16 ng/Nm^3 of dry gas in 9% CO_2. More detail is shown in Table 17.5.

Table 17.5. Measurement of the concentrations of dioxins and furans containing bromine

Material	Abundance in experiment: 1.58m^3 dry gas 9% CO_2 Unit ng/ experiment	Pollutants content ng/m^3 (dry gas 9%CO_2)
3Br 278 ClDF	<38*	<24
Sum monoBr triCl Df	<38*	<24
2Br 378 Cl DD	<92*	<58
Sum monoB triCl DD	<92	<58
Sum diBr diCl DF	800	506
23 diBr 78 diCl DD	<15*	<9
Sum diBr 78 diCl DD	<15	<9
Sum triBr monoCl DF	480	304
2378 tetraBr DF	<6000**	<380
Sum tetraBr DF	<8000**	<5063
2378 tetraBr DD	26*	16
Sum tetraBr DD	162	103

  * Elutriation problems might have effected the result
** Some disturbance occurred while the spectrometer was obtaining data, thus increasing
   the uncertainty, thus reducing the possibility of obtaining precise data.

Polyaromatic Hydrocarbons

The total concentration of PAH extracted from the gas was 3.044 µg/Nm^3 in dry gas at 9% CO_2. Fourteen different PAH, from acenaphthylene to dibenzo(a,h) anthracene, were identified. The amount of naphthalene in the extract gas, 826mg/m^3, was included in the total. More details of the PAH content can be seen in Figure 17.8.

A short literature review was carried out to compare the results obtained with those found in the literature. Special emphasis was directed to landfill or other fires. The results are summarized in Table 17.6. The PAH concentrations found were higher than those reported in most of the literature, especially when compared with landfill fires Ruokojärvi et al., (1995), other controlled fires Ruokojärvi P. et al., 2000, or domestic burning (Lohmann et at. 2000). However, the value was within the

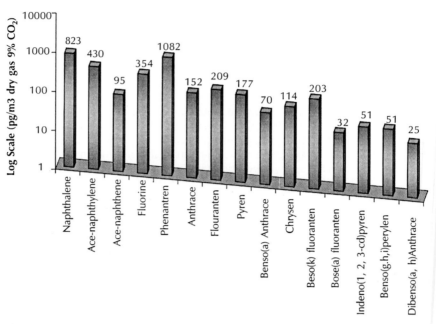

Fig. 17.8 PAH isomers during experiments. The total PAH concentration was 3.044 $\mu g/Nm^3$, excluding naphthalene.

range found by Launhardt *et al.*, (1998) and Mannien H. *et al.*, (1997). The generally high results can be explained by the fact that the bales were burnt inside a closed container. In this circumstance, the levels of PAH would be expected to be higher because there is less mixing with the air. In addition, the temperature of the bales during combustion was lower than incinerator temperatures or landfill fire temperatures (Ruokojärvi *et al.*, (1995) as being 704°C at the ignition point and the centre of the fire).

PCDD/F

The total chlorinated dioxin content varied according to which TEF system was used, with the values (TEQ) being 12.531 ng (I– TEF – 88)/Nm³, 14.091 ng (I – TEF – 99)/Nm³ or 13.858 ng (Eadons)/Nm³. However, these values represent the upper limit since the concentration of 2,3,7,8 – TCDD was 1.5 ng/Nm³ and the total concentrations of TCDD was <32.9 ng/Nm³.

The EU limit for PCDD/F is 0.1 ng I – TE/Nm³ for incinerators, and other emissions into the atmosphere. This maximal value was exceeded. However, the total TEF of PCDD/F was still lower then most of the values obtained in the literature. It is important to note that the comparison has to be taken lightly as waste fuel burnt in incinerators is very efficiently combusted and mixed thoroughly with air. The same can be said for landfill fires and, in this situation, the dilution of the PCDD/F due to the wind should be considered because this reduces the final concentration

Table 17.6. Summary of the results found in the literature for PAHs

Author	City/ Area/experiment	Value ($\mu g/m^3$)
Baker J (1991)	Lake Superior (background air)	0.0038 (mean total)
Bronan D (1991)	Outside and around Stockholm	0.011-0.0025 (total around and outside Stockholm)
Keller C. (1984)	Columbia	0.056 (8 PAH)
Foreman W (1990)	Denver	0.093 (10 PAH)
Jaklin J (1985)	Vienna	0.412
Baek S (1992)		0.0402 (18 PAH)
Halsall C (1993)	Cardiff Manchester London and Stevenage	0.011-0.735
Fromme H (1998)	Berlin, winter and summer for car and subway trains respectively	0.0102 and 0.0287(car) 0.0302 and 0.0675 (subway train)
Ruokojärvi P (1995)	Finland(controlled and uncontrolled– two days) fires–air samples	0.0055 (before controlled fire) 1.670(during day 1 of fire) 1.270 (during day 2 of fire) 0.024 (after fire) 0.816 (uncontrolled fire)
Mannien Helena (1997)	Municipal incinerator, RDF and PDFs Finland Kautta plant	1.15-262 (11 % O_2 dry gas), flue gas including gas phase particles after ESP
Yassa N (2001)	Waste landfills Algiers Metropolitan Area	11 PAHs Down town Algiers winter 0.0083, summer 0.0409 Oued Smar Landfill winter 0.0111 Summer 0.0830
Ruokojärvi P, (2000)	Simulated house fires	0.0064-0.470 (mainly phenanthrene, fluranthene, and pyrene)
Wienecke J (1995)	Hazardous waste incineration	0.04091-0.06684(absorbent silica gel)
Launhardt T (1995)	House heating systems	16 PAH compounds 360-35000 13% O_2
Lohman L (2000)	Diffuse domestic burning in Manchester (A),Clapham(B), Austwick(C), Lancaster(D) in the UK	A.0.026-0.220 B and C : 0.016-0.130 D 0.0063-0.028

in the volume of air sampled. Hence, if more efficient combustion had taken place the PCDD/F concentration would have been less. Figures 9–11 show the results for the concentrations, percentages and the toxic fingerprints of the toxic congeners. 12378-PeCDF was found to have the highest TEF: 35.51% using TEF-88, 31.582 % using TEF-99 and however Eadons model give it a value of 21.19 %, the highest percentage according to Eadons model is 12378-PeCDF 30.242 %. However, OCDD had the highest concentration in ng/Nm³, with a corresponding TEF of 0.39 % TEF-88, 0.027 % TEF-99 and 0 % TEF-Eadon (no toxicity equivalent value is assigned using Eadons model).

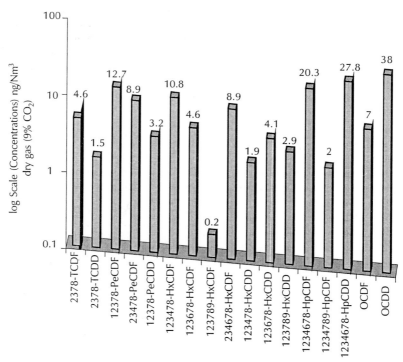

Fig. 17.9. PCDD/PCDF isomeric distribution for the dioxins found during the experiment.

A literature survey was carried out to compare with the results obtained from the burning of bales. It was found that, on an average, the results obtained were within the ranges of some of the values cited for incinerators (Kilgroe J. 1996, Wienecke J. 1995, Buekens A. 1998. However, the values obtained in this work were higher than those for landfill fires (Ruokojärvi P. 1995), simulated household fires (Ruokojärvi P. 2000), domestic burning (Lohmann L. 2000) and household heating systems (Launhardt T. 1998) (see Table 17.7).

Table 17.7. Summary of the results found in the literature for dioxins

Author/Ref	Equipment/process/event	Values
Kilgroe J (1996)	Municipal waste combustors	0.1-20000 ng/dscm n (depending on waste and technology)
	Techniques in control of municipal waste incineration (Table 2) Process (removal of PCDD/Fs from flue gasses by the combination of scrubber, bag filter	Raw gas (untreated) values before treatment by any process

(Contd.)

Table 17.7. (*Contd.*)

Author/Ref	Equipment/process/event	Values
	and activated carbon adsorption)	2.17 ng-TE/Nm^3
	ENTRAINED FLOW	100 ng/Nm^3
	Moving bed	0.3 ng-TE/Nm^3
	Fixed bed	
	Decompositions of PCDD/Fs in	PCDD/F in flue gas only
	flue gases by SCR unit (Table 3)	inlet *(for more information*
	Catalyst	*regarding Space velocity NH$_3$/*
		NO$_x$ molar volume see ref) RT
		= reactor
		temperature
Buekens A (1998)	Pt and Au on Silica-boria	All ng-TE/Nm^3
	alumina composite oxide	• 0.25 (RT 220)
	V$_2$O$_5$-WO$_3$-TiO$_2$	• 0.35 (RT 250)
	Pt on V$_2$O$_5$-WO$_3$-TiO$_2$	• 0.15 (RT 0.15)
	SCR DeNO$_x$	• 2.2 (RT 300)
	SCR DeNO$_x$	• 0.34 (RT 200)
	SCR DeNO$_x$	• 0.01 (RT 200)
	V$_2$O$_5$-WO$_3$-TiO$_2$	• 0.39 (RT 230)
	SCR DeNO$_x$	• 1.64 (RT 280)
	Pt supported	• 0.05 (RT 325)
		• 90 ng- PCDD/Nm^3
		(RT300 -400)
Wienecke J (1995)	Hazardous waste incineration	25 ng TEQ/Nm^3 (about)
	House heating systems (tiled	2-25 pg ITE m^3 with 13% O$_2$
	stove with modern combustion	(untreated wood)
Launhardt T(1998)	technology-designed for wood	38-952 pg ITE/m^3 with 13%
	combustion- and tiled stove with	O$_2$ (for combustion of paper
	old combustion technology-	carton painted wood and
	designed for wood and coal	wood with 2-5 % PVC)
	combustion)	respectively
	House heating systems (Oil	1-2 pg ITE/m^3 and ≤10 pg
	boilers and wood boilers)	ITE/m^3 respectively
Mannien Helena (1997)	Municipal incinerator, RDF and	≤0.01-3.17 ng /Nm^3
	PDFs Finland Kautta plant	≤0.01-0.04 ng ITEQ/Nm^3
Lohmann L (2000)	Diffuse domestic burning in	A : 20-510 fg TEQ/m^3 B, C,
	Manchester (A), Clapham (B),	D : 6.4-110 fg TEQ/m^3
	Austwick(C), Lancaster(D) in the	
	UK	
Ruokojärvi P (2000)	Simulated house fires	1.0- > 7.2* ng-ITEQ/m^3
		(>7.2 due to high PAH
		content)
Ruokojärvi P(1995)	Municipal waste landfill fires in	PCDD/F in working
	Finland	samples taken from air 51-
		427 pg TE/m^3

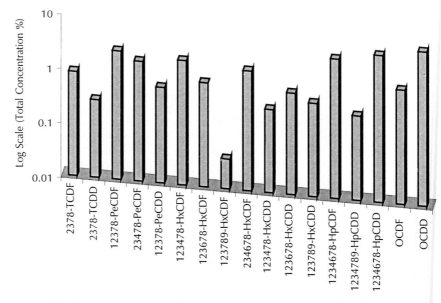

Fig. 17.10. Percentage of the toxic congeners in the analysed smoke.

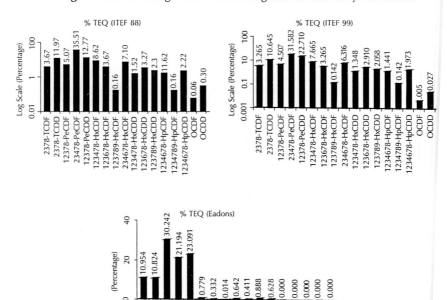

Fig. 17.11. Toxicity fingerprints using ITEF 88, 99 and Eadons model.

Discussion and Conclusion

The risk of self-ignition of stored baled household waste seems to be very small. Although a high content of organic material and/or a high environmental temperature may occur, no increase in temperature inside the bales was found to take place. Accordingly, it should be possible to use the technique in countries with high percentages of organic material in their waste and a hot humid (tropical) climate. Reducing the pressure imposed on the waste during baling, within possible limits for the baling press, does not seem to increase the risk of self-ignition.

The rate of biological processes and oxygen consumption inside the bales seems to be very slow. This might be partly caused by a low rate of oxygen transport due to the high water content of the bales (it is assumed that the rate would be reduced by diffusion in the water phase). It is also considered that the bales, which consist of a fairly compact mass containing a lot of plastic material, present a barrier for gas penetration. The material in these bales was not shredded in a hammer mill, unlike the material used in previous experiments in Sweden, where available oxygen was consumed relatively rapidly. A large amount of organic fatty acids was produced in such bales, which lowered the pH to below levels suitable for the production of methane gas, and no production of methane has been observed so far.

The contamination of rainwater by leachate from the bales does not seem to be very high, according to analyses of wastewater from the storage area in Lidköping. This, of course, depends on the quality of baled waste, and the magnitude of contamination and so is specific to the type of waste in question. Thus in conclusion, it has been observed that the storage of bales will not increase the risk of fire since during storage they exhibit good stability. The bale will only ignite if set fire to and in this case; emissions are expected to be higher than the counterpart fire in a landfill.

Acknowledgements

The Knowledge Foundation, Bala Press AB, Lidköpings Värmeverk AB, Trioplast AB and Renova AB provided financial support. The support of the reference group: Lars Lindskog (BalaPress AB), Bengt-Olof Andersson (Lidköpings Värmeverk AB), Håkan Pettersson (Trioplast AB), Ola Ståleby and Joakim Syrstad (Renova AB) is also acknowledged.

REFERENCES

1. Ansbjer, J., Hogland, W., Tamaddon, F., 1995. Storage of waste fuels with baling technique. ISWA Times No. 3.
2. Baek, So, Goldstone, M. E., Kirk, P. W. W., Lester, J.N., Perry, R., 1992. Concentrations of particulate and gaseous polycyclic aromatic hydrocarbons in London air following a reduction in the lead content of petrol in the United Kingdom. *Science of the Total Environment* **111** (2-3), 169-199.
3. Baker, J., Eisenreich, S., 1991. Concentrations and fluxes of polycyclic aromatic hydrocarbons and polychlorinated diphenyls across the air water interface of lake superior. *Environmental Science Technology* **25** (3), 500-509.

4. Broman, D., Naf, C., Zebuhr, Y., 1991. Long term high volume and low volume of air sampling of polychlorinated dibenzo-para-dioxins and dibenzofurans and polycyclic aromatic-hydrocarbons along a transect from urban to remote area on the Swedish Baltic coast. *Environmental Science Technology* **25** (11), 1841-1850.

5. Buekens, A., Huang, H., 1998. Comparative evaluation of techniques for controlling the formation and emission of chlorinated dioxins/furans in municipal waste incineration. *Journal of Hazardous Material* **62**, 1-33.

6. Foreman, W., Bidleman, T., 1990. Semi volatile organic compounds in the ambient air of Denver, Colorado. *Atmosphere and the Environment* **24a**, 2405-2197.

7. Fromme, H., Oddoy, A., Piloty, M., Krause, M., Lahrz, T., 1998. Polyaromatic hydrocarbons (PAH) and diesel engine emission (elemental carbon) inside a car and a subway train. *Science of the Total Environment* **217**, 165-173.

8. Halsall, C., Burnett, V., Davis, B., Jones, P., Pettit, C., Jones, K. C., 1993. PCBs and PAHs in UK urban air. *Chemosphere* **26** (12), 2185-2197.

9. Hasselrris, F., Licata, A., 1996. Analysis of heavy metal emission data from municipal waste combustion. *Journal of Hazardous Material* **47**, 77-102.

10. Hogland, W., 1998. Baled waste fuel at a thermal power station in Umeå, Northern Sweden. In: Proceeding of the 5th Polish-Danish Workshop on Biofuels, 26-29 November 1998, Starbienino, Poland, pp. 21-28.

11. Hogland, W., Bramryd, T., Persson, I., 1996. Physical, biological and chemical effects on unsorted fractions of industrial solid waste in waste fuel storage. *Waste Management Research* **14**, 197-210.

12. Hogland, W., Marques, M., 1999. Physical, biological and chemical processes during storage and spontaneous combustion of waste fuel. *Resources Conservation Recycling* (in press).

13. Hogland, W., Tamaddon, F., 1995. Seasonal storage of solid waste-fuel at landfill site. In: Proceedings of Sardinia 95, the Fifth International Landfill Symposium, Vol. I. CISA, Cagliari, Sardinia, Italy, pp. 807-814.

14. Jaklin, J., Kernmayr, P., 1985. *International Journal of Environmental Chemistry* 21, 33.

15. Keller, C., Bidleman, T., 1984. Collection of airborne polycyclic aromatic hydrocarbons and other organics with a gas fiber filter-polyurethane foam system. *Atmosphere and the Environment* **18** (4), 837-845.

16. Kilgroe, J. D., 1996 Control of dioxin, furan, and mercury emissions from municipal waste combustors. *Journal of Hazardous Material* **47**, 163-194.

17. Launhardt, T., Strehler, A., Dumler-Gradl, R., Thoma, H., Vierle, O., 1998. PCDD/F -and PAH emission from house heating systems. *Chemosphere* **37** (9-12), 2013-2020.

18. Lohmann, R., Northcott, G. L., Jones, K. C., 2000. Assessing the contribution of diffuse domestic burning as a source of PCDD/Fs, PCBs, and PAHs to the U.K. atmosphere. *Environmental Science and Technology* **34**, 2892-2899.

19. Mannien, H., Peltola, K., Ruuskanen, J., 1997. Co-combustion of refuse derived and packaging-derived fuels (RDF &PDF) with conventional fuels. *Waste Management and Research* **15**, 137-147.

20. Marques, M., 2000. Solid Waste and the Water Environment in the New European Union Prospective. PhD Thesis. Department of Chemical Engineering and Technology, Royal Institute of Technology, Stockholm, Sweden.

21. Miljödepartementet (Ministry of the Environment), 1997. Förordning (1997:692) om Förbränning av Farligt Avfall (Regulation (1997:692) on Incineration of Hazardous Waste). Regulation No.1997:6992. Ministry of the Environment, Sweden (in Swedish).

22. Robles-Martínez, F., Gourdon, R., 1999. Effect of baling on the behaviour of domestic wastes: laboratory study on the role of pH in biodegradation. *Bioresource Technology* **69**, 15-22.

Appendix A

The Measurement of CH_4, CO_2, O_2, ambient temperature, as well as the temperature inside the baled waste fuels and the temperatuare on the surface of the baled waste fuels

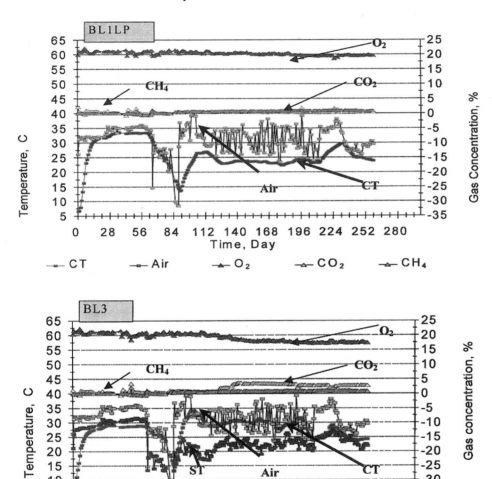

Fig. A.1. Bale BL1LP had low pressure and high organic percentage of waste; Bale BL3 was the same but with normal pressure. Both bales were kept at an average temperature of 35 °C. CT = center of bale temperature, ST = surface of bale temperature, Air = surrounding air temperature

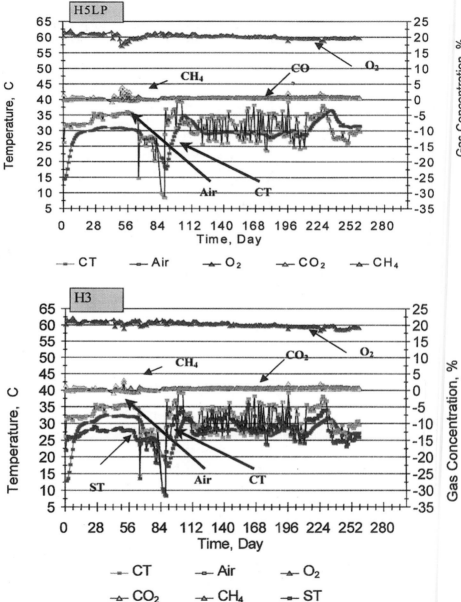

Fig. A.2. Bale H3 was pressed with normal pressure, normal waste content. Bale H5LP contained normal waste; however, it was pressed with low pressure. Both bales were kept at an average temperature of 35 °C, CT = center of bale temperature, ST = surface of bale temperature, Air-surrounding air temperature.

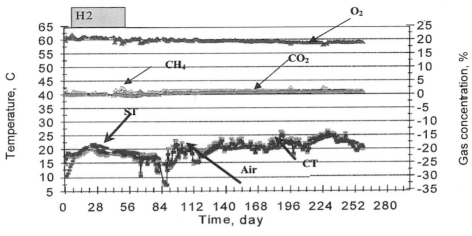

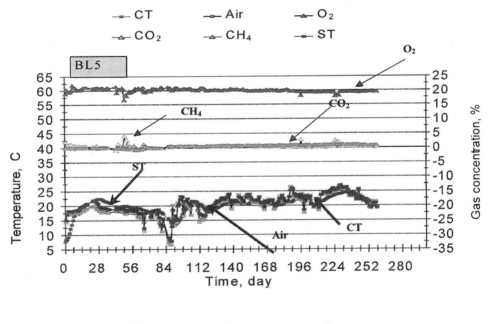

Fig. A.3. Bale H2 was passed with normal pressure, normal waste content. Bale BL5 contained high percentage of organic waste; however, it was pressed with normal Cressure. Both bales were kept at an average temperature of 17 °C, and both bales were uncovered. CT = center of bale temperature, ST = surface of bale temperature, Air =surrounding air temperature.

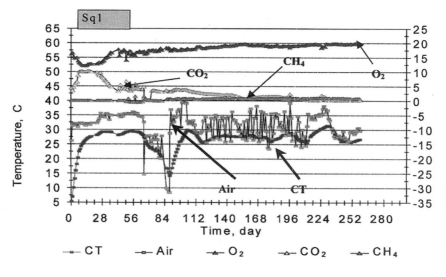

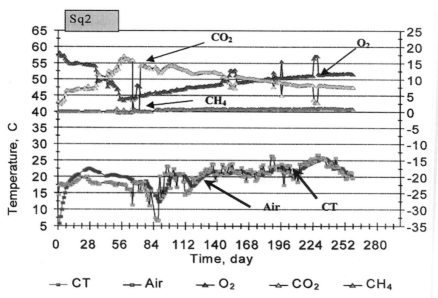

Fig. A.4. Sq1 is a Square Bale kept at an average temperature of 35 °C. Sq2 is also a square bale kept at an average temperature of 17 °C; however, it was uncovered. CT = center of bales temperature, ST surface of bale temperature, Air/AIR = surround air temperature.

23. Robles-Martínez, F., Gourdon, R., 2000. Long-term behaviour of baled household waste. *Bioresource Technology* **72**, 125-130.

24. Ruokojärvi, P., Aatamila, M., Ruuskanen, J., 2000. Toxic chlorinated and polyaromatic hydrocarbons in simulated house fires. *Chemosphere* **41**, 825-828.

25. Ruokojärvi, P., Ettala, M, Rahkonen, P., Tarhanen, J., Ruuskanen, J., 1995. Polychlorinated dibenzo-p-dioxins and –furans (PCDDs and PCDFs) in municipal waste landfill fires. *Chemosphere* **30** (9), 1679-1708.

26. Ruokojärvi, P., Ruuskanen, J., Ettala, M., Rahkonen, P., Tarhanen, J., 1995. Formation of polyaromatic hydrocarbons and polychlorinated organic compounds in municipal waste landfill fires. *Chemosphere* **31** (8), 3899-3908.

27. Tamaddon, F., Bengtsson, L., Hogland, W., 1995. Storage of waste-fuel and related problems. In: Proceedings of R'95 Congress, Switzerland. Volume III, pp. 206-212.

28. Tamaddon, F., Hogland, W., Kjellberg, J., 1995. Retrievable storage of MSW. In: Proceedings of ISWA 25th Anniversary World Congress on Waste Management, 15-20 October 1995.

29. Wienecke, J., Kruse, H., Huckfeldt, U., Eickhoff, W., Wassermann, O., 1995. Organic compounds in the flue gas of a hazardous waste incinerator. *Chemosphere* **30** (5), 907-913.

30. Yassaa, N., Meklati, B. Y., Cecinato, A., Marino, F., 2001. Organic aerosols in urban and waste landfill of Algiers Metropolitan Area: occurrence and sources. *Environmental Science and Technology* **35**, 306-311.

Section IV

Waste to Energy and Climate Change Issues

18

Waste, Energy and Climate Change

Joyeeta Gupta and Frédéric Gagnon-Lebrun

Institute for Environmental Studies, Vrijc Universiteit Amsterdam, The Netherlands

Introduction: The Relationship

Climate change is a global, esoteric, distant and abstract environmental problem. This very characteristic makes it a challenging problem for policymakers. In contrast, wastes are a local, concrete, ever-present and annoying problem for which people can be motivated to take action. Both problems are priorities but at different administrative levels and geographical scales. While the former is a priority for the global community and there are some resources available to deal with it, the latter is essentially a local and unglamorous problem and the resources are often limited. In political theory, one way to address the global problem of climate change is to make links with domestic and local priorities (Gupta and Hisschemöller 1997). In line with that thought process, the thesis of this paper is that a linkage between the two, global and local problems could perhaps lead to harnessing global resources, local populations and politicians to simultaneously address the two major environmental problems.

Before examining the thesis in detail we would like to present some data on the enormity of both problems. The climate change phenomenon refers to the perturbation of the energy balance at the earth's surface due to increasing concentrations of greenhouse gases (GHGs) in the atmosphere. The main greenhouse gases include naturally occurring gases, water vapour, CO_2, CH_4, N_2O and ozone. In addition, there are greenhouse gases developed by humankind: SF_6, HFCs and PFCs. Most of these gases are the result of energy production from fossil fuels as well as agricultural activities and animal husbandry. All these gases have increased substantially in the earth's atmosphere in the last century as a result of human activity. Since 1750, the atmospheric concentrations have increased by 30%, 145% and 15% for CO_2, CH_4, and N_2O respectively (Houghton *et al.*, 1996: 3). This may lead to a mean global temperature rise of 1–3.5 °C by the end of this century, higher than that experienced over the last 10,000 years. Such a rise may lead to changes in the global atmospheric circulation patterns, resulting, inter alia, in shifts in the climatic zones, sea-level rise and increased frequency and intensity of extreme weather conditions.

Another major challenge is the globally recurring phenonmenon of increasing urbanization and industrialisation. Increasing urbanisation poses an enormous challenge in terms of waste management as water, air and soil pollution have become a major health hazard in many countries of the world. Urban populations are increasingly exposed to more kinds and higher levels of industrial pollutants. Human exposure to air pollution causes respiratory and pulmonary diseases, mostly in children while pollution of urban water sources used for drinking, washing and cooking are increasingly threatening human health due to biological pollutants from human wastes and chemical pollution from industrial toxic wastes. Lead poisoning from excessive exposure to lead-based paint and exhaust fumes from leaded gasoline and some industrial activities has also become a major concern mostly in developing countries where traffic is getting denser.

Against this brief background of the seriousness of both problems, this chapter will examine a number of issues. First, it will argue that environmental problems manifest themselves at different geographical levels and scales, and decisions to deal with these problems have to be taken at different levels. Decision-making at these different levels faces interesting challenges (section 2). Second, it will argue that there are many material linkages between the two problems and that in many ways addressing some aspects of one problem also incidentally addresses some aspects of the other problem (see section 3). Third, the paper argues that both the climate change and the waste problems have similar causes in that the root of both problems lies in our system of production, consumption and trade. Hence, if we address the issue of production and consumption patterns, it will become much easier to deal with both problems although we may inadvertently create a new problem (see section 4). Finally, the paper argues that if a political link can be made effectively between the two problems, this may give a boost to addressing both global and local problems.

The Scales of Different Problems

Environmental problems tend to exist at different levels and scales. Problems can manifest themselves at micro and local level, at river-basin (fluvial) level, at national, continental, international and global levels (see Table 18.1; cf. Netherlands Government 1989). We expect that pollution of space will also become a major issue in coming years. Some of these problems occur at different levels simultaneously (e.g. ecotoxicological impacts; photo oxidant formation, etc; Udo de Haes 1996). Each problem can only be adequately dealt with at its own particular political level. The principle of subsidiarity, a key principle in the European Union, implies that problems need to be dealt with at the lowest appropriate level. As far back as in 1973, the First Action Programme for the Environment in the EU stated: "In each category of pollution, it is necessary to establish the level of action (local, regional, national, Community, international) best suited to the type of pollution and to the geographical zone to be protected. Actions likely to be most effective at Community level should be concentrated at that level; priority should be determined with special care". At the United Nations Conference on Environment and Development in 1992, this principle was implicitly accepted and explicitly included in some of the chapters including the chapter on fresh water resources of Agenda 21.

Table 18.1. Scales of problems

Problem	Geographical scale	Policy and political level
Organic and industrial wastes	Local	Local government/ provincial government
Flowing waters in rivers	Fluvial	River basin authority
Acid deposition; long range transboundary movement of hazardous wastes; nuclear accidents	Continental	Regional authorities, e.g. Economic Commission of Europe; International Energy Agency; The European Commission
Pollution of the international seas/oceans; Environmental protection in Antarctica	Oceans/ land outside the national juris- diction of countries	International Treaty, e.g. United Nations Convention on the Law of the Sea
Climate change, Depletion of the ozone layer, Loss of biodiversity	Global	International Treaty, e.g. the United Nations Framework Convention on Climate Change

Clearly climate change is a global problem since the location of the GHG emissions is irrelevant and it is completely uncertain where the impacts will be manifested and how universal the impacts may be. At the other end of the spectrum is the problem of wastes which can be both local and also regional in nature. Having said that, one may argue that the problem of wastes has become very international with the transboundary movement of hazardous wastes which was to some extent dealt with by the negotiation and adoption of the Basel Convention on the Transboundary Movement of Hazardous Wastes in 1989. However, for the purpose of this chapter we focus on the more local waste management issues.

While the principle of subsidiarity is increasingly gaining ground worldwide, there are many practical and political problems in dealing with these issues. Although in the climate change discussion, the whole world has been mobilised into discussing and negotiating the issue extensively, it appears that each country is trying to minimize responsibility for itself. In the initial enthusiasm to deal with the problem (as predicted by Down's 1972 theory on the Issue Attention Cycle), all developed countries were keen to put the issue on the international agenda and to try to deal with it. Since then it has become increasingly clear to politicians and economists that reducing the GHG emissions significantly calls for major societal changes, which at least in the short-term may affect national wealth. This has led to political reluctance to take far-reaching domestic measures in the developed countries. The domestic pressure to take action is also limited because of the abstract, diffuse, uncertain and global nature of the problem. In fact, the problem is caused by the anthropogenic emissions of greenhouse gases by the energy supply, transport, industrial and agricultural sectors but has impacts such as sea-level rise, rising temperatures, shifts in precipitation and extreme weather patterns, frequency and intensity. The exact location of the impacts can be far-removed from the source of the emissions, and the causality is difficult to ascertain. Developing countries have relatively low emission rates compared to those in the North, but they are physically and financially more vulnerable to the adverse effects of climate change. As such there is not much incentive in either the developed countries to take substantial domestic action, nor do developing countries wish to prioritise these issues.

Many in the developed countries argue that since this is a global problem, developing countries also need to commit themselves to action (Clinton 1997). Since developing countries believe that it is the developed countries that should take action first since they have caused the problem, and since the EU and the US could not quite come to terms on the issue, this led to a breakdown in the negotiations in November 2000 in the Hague.

On the other hand, waste management is a local issue and local people are often both the cause and the victims of poor waste management. However, limited resources, management skills and awareness may impede the effectiveness of the implementation of measures. Further, even at the local level, there are scales of operation and there is actual separation between the source of the pollution and its impacts, which often exacerbates the problem of waste management.

Thus both problems are being half-heartedly dealt with; the former globally and the latter mostly in developing countries. However, even in the more developed countries, the principle of NIMBY (not in my back yard) has increasingly led to waste being transported further and further away and sometimes, as mentioned earlier, exported to other countries.

The Material Issue Linkages: Direct

In waste management theory there is a hierarchy of policy responses, with reducing, reusing, recycling, incineration and landfilling following in order from best to worst option. Figure 18.1 shows these policy responses with relation to climate change. This hierarchy is, however, also controversial since it depends on the specific context whether one policy option is better than another policy option. Thus, for example, the quality of the waste streams differ from country to country and appropriate treatment methods also need to differ.

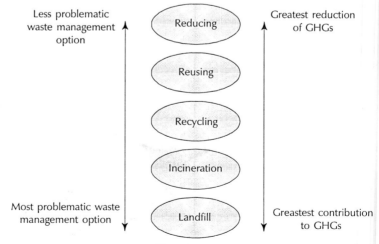

Fig. 18.1. Relation between waste management and climate change.

In many ways the problems of climate change and waste management are closely related. Poor waste management itself leads to enhanced emissions of greenhouse gases. The effectiveness of each policy also varies greatly in terms of reducing GHG emissions. Links exist at each policy response level; we have identified seven links.

The first link is quite obvious: gas emitted in landfills has a composition of about 50% CH_4 and 50% CO_2 (Thorneloe, 2000). When organic matter of solid and liquid waste streams decomposes anaerobically, methane is generated; when it decomposes aerobically CO_2 is generated.[1] It is estimated that solid waste and waste water streams each contribute about 10 % to the total current anthropogenic methane emissions. These emission levels are expected to grow as a result of rising population levels and development (Kruger et al., 2000:). In 1990, it was estimated that the methane emissions were likely to increase by 85% in 2010 and by 163% in 2025. In particular methane emissions from landfills were expected to increase by 50% in 2010 and by 100% in 2025 (IPCC 1990:). The new IPCC report (Kruger et al., 2000) argues that there are two ways of reducing significantly the impact of landfill gas on climate change. The first one calls for a technology that is applied to convert the methane into carbon dioxide. This would contribute directly to combating climate change since methane has a global warming potential 21 times stronger over a 100-year period than that of CO_2. Another technology can also drastically reduce methane emissions by collecting the methane and generating energy from it in the process (Kruger et al., 2000).

A second link is that landfills represent a loss of natural areas (European Environment Agency 1999: 211). If the volume of waste landfilled is reduced, the potential for forested areas would increase, subsequently enhancing the natural intake of CO_2.

A third link is that incineration of wastes can also be used as a source of energy and thus can be an important substitute for other energy sources and reduce GHGs emitted in energy generation from other sources. However, Sheehan (2000) argues that although energy can be produced from incineration of wastes, the production of CO_2 and other toxic emissions and ashes during the incineration outweighs any gains in energy.

Fourth, recycling certainly reduces the quantity of waste going to incineration plants and landfills. However, from the perspective of climate change, recycling may not be as effective as it may appear compared to landfills and incineration plants when they are used to produce energy. Denison (1996) states that the energy produced by incineration corresponds to only one quarter of the energy saved by recycling; when considering the material life cycle as a whole, the energy need for recycled production and recycling collection is substantially smaller than for virgin production and incineration. However, the environmental benefits of recycling may not be that attractive when compared to landfills equipped with CH_4 recovery technologies. In fact, the efficiency of the recycling process in terms of emission reduction is product-specific and depends mostly on the quantity of energy required for collection, recycling, and redistribution compared to the energy needed for virgin production and emissions related to the disposal of the product.

Yet a fifth key link exists between reduction, reusing and recycling of forest products (paper and wood) and GHG emissions. While reducing the use and

---

[1] However, these CO_2 emissions are referred to as neutral as they are part of the natural carbon cycle.

enhancing recycling and reusing of forest products can reduce the amount of waste that has to be dealt with –and thus alleviate the climate change problem; it can also reduce the number of trees cut and thus enhance the natural sequestration of CO_2. Recycling a ton of paper saves about 24 trees which represents an absorption potential of more than 110 kilograms of CO_2 per year (Sheehan 2000). For example, in Jakarta, the Pemulang save about 6 million trees by delivering 378,000 tonnes of waste paper per year to factories (Lardinois and van der Klundert 1993: 13).

Sixth, reusing wastes can also greatly contribute to alleviating climate change. Lardinois and van der Klundert (1993: 18) point out that household compost can be used to fertilize plants in gardens. This substitutes the purchase of fertilisers, which are energy-intensive products for their production process and transportation. In addition, growing plants enhance sinks, which is an additional benefit. Organic material can also be used as animal feed which is, similarly, a substitute to commercial feed also relatively energy-intensive.

Finally, waste does not only release GHGs directly, through aerobic and anaerobic decomposition, but also indirectly through the consumption of energy for dealing with the waste (transportation, treatment of toxic wastes before disposal, cleaning of the products before recycling, etc.). In France, for instance, waste accounted for 15% of total weight freight in 1993, which corresponds to 5% of the total transport sector energy consumption (Ripert 1997).

Table 18.2 presents an overview of the links between solid and liquid waste production and GHG emission for different waste management strategies.

Table 18.2. Linking waste generation and GHG emissions

Policy options	Relation to waste generation	Relation to GHG emissions
CH_4 recovery and conversion of CH_4 into CO_2 in landfills	Reduces fire hazards at landfills	Reduces CH_4 emissions related to decomposition and generates energy that can substitute fossil fuel based energy sources
Reduction of landfilled wastes	Reduces spatial and environmental problems related to landfills	Increases potential for CO_2 sequestration
Incineration	Reduces the space needed for wastes	Generates CO_2 emissions as well as energy that can be used to substitute fossil-fuel based energy sources
Recycling	Reduces waste generated	Reduces methane emissions from landfills and CO_2 emissions from incineration plants
Reusing/recycling of forest products and land use change	Reduces waste generated	Enhances CO_2 sequestration
Reusing	Reduces waste generated	Reduces emissions related to the extraction of resources, the production processes to the disposal of wastes
Local disposal of wastes where possible	Reduces transportation of wastes and associated costs	Reduces GHGs emitted from transportation

This implies that addressing one of these problems could lead to benefits in relation to the other problem; thereby leading to double dividends.

The Material Issue Linkages: Towards Industrial Transformation

There are many ways to respond to a problem. The International Human Dimensions Programme on Industrial Transformation (IHDP; 1999) refers to them as reactive, receptive, constructive and proactive ways (see Fig. 18.3). Reactive approaches are generally end-of-pipe approaches and try to minimize wastes coming out of individual factories at local level. Receptive approaches focus on improving the process and lead to the optimisation of the process. This is mostly done at firm level. Constructive approaches focus on improving the product to be marketed and lead to accelerated changes. This is generally most effective when done at a scale at which the product is marketed in order not to suffer from loss of competitiveness in relation to other products. Finally, pro-active approaches aim at changing the system and society based on a long-term vision for how society should function in the future. This vision is referred to by some as 'Industrial Transformation'. Different scholars interpret it differently. For example, some interpret it as the Factor 4 philosophy which aims at reducing inputs by half while doubling the outputs. Many of the proponents of the Factor 4 philosophy argue that there is need for a Factor 10 philosophy in the West if ecological space is to be made available to the South (von Wiezsäcker 1998). The key point of industrial transformation is that it is about structural change as opposed to incremental change.

The industrial transformation approach tries to close substance cycles. As Frosch (1998) puts it: "The overall idea is to consider how the industrial system might evolve in the direction of an interconnected food web, analogous to the natural system, so that waste minimisation becomes a property of the industrial system even when it is not completely a property of an individual process, plant or industry". However, as Chadwick (1998) argues, it is not so wise to romanticise our notions of nature, since nature is not always an efficient system.

What does the vision of industrial transformation imply? This concept implies the adoption of the principles of eco-efficiency, de-materialization and decarbonization of industrial society, closing material loops globally and/or locally, and minimizing dissipative flows (see Fig. 18.2). This concept can also be used to link waste management with climate change policy and this can be theoretically supported by a number of tools. These tools include substance flow analysis which studies the dynamics and the content of stocks and flows in the industrial system; measuring efficiency through the concept of material intensity for product service (MIPS), ecological rucksacks, life cycle analysis, and environment cum economy models. All these theoretical tools help in analysing the system of production to identify opportunities for reducing the waste flows and increasing the efficiency of energy use (Udo de Haes *et al.*, 1998; Sachs *et al.*, 1998).

Further, industrial transformation goes beyond improved product designs and eco-labelling and ensures the longest durability of goods, promotes sustainable consumption patterns and ensures environmentally-sound reuse of goods. It takes a 'whole system' approach to attain a 'closed substance cycle' or a 'zero waste cycle'.

When materials are used more efficiently in the production stage and consumed more responsibly and in lesser quantity in the consumption stage, less resources are needed in the extraction and material production stages (see Fig. 18.2). This translates into a) reduced energy consumption in the extraction, material production and product manufacturing stages which reduces GHG emissions from fossil fuel combustion and b) reduced quantity of waste produced throughout the life cycle of

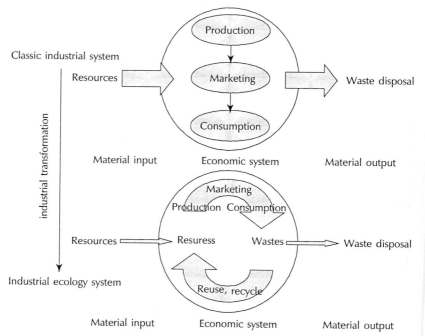

Fig.18.2. Material and energy throughput in the classic industrial system and after industrial transformation.

Source: Based on Hekkert 2000: 18; Udo de Haes *et al.*1998: 64; IHDP 1999: 39.

materials which limits methane and CO_2 emissions from landfills and incineration plants. Since the production and manufacturing stages, which form the industrial sector, accounts for about 40% of the total world primary energy use, the reductions in CO_2 emissions are likely to be significant. Furthermore, about 60% of our energy needs are fulfilled through burning fossil fuels (Hekkert, 2000).

What does this imply in terms of the relationship between climate change and waste management? We would argue that integrated waste management techniques and climate change management both call for industrial transformation. Although such a radical structural and systemic change is desirable from an environmental perspective, it is perhaps not economically and politically feasible. We can thus define a number of intermediate steps to achieve industrial transformation (see Figure 18.3). In line with these steps, reactive approaches to deal with wastes call for dumping, incinerating and landfilling. Reducing, reusing and recycling wastes (referred to as the 3 'R's) can be either receptive, constructive or proactive.

Conclusion

This chapter has argued that the global problem of climate change and the local problem of waste management are not necessarily two distinct compartmentalised problems even though the scales at which they operate and the nature of the problems

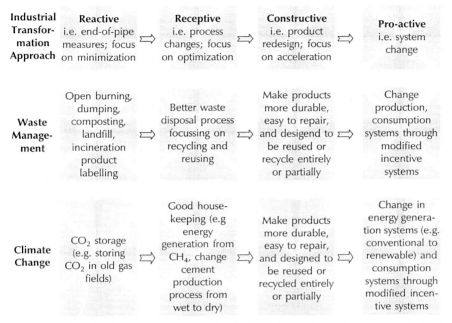

Industrial Transformation Approach	Reactive i.e. end-of-pipe measures; focus on minimization		Receptive i.e. process changes; focus on optimization		Constructive i.e. product redesign; focus on acceleration		Pro-active i.e. system change
Waste Management	Open burning, dumping, composting, landfill, incineration product labelling	⇨	Better waste disposal process focussing on recycling and reusing	⇨	Make products more durable, easy to repair, and desigend to be reused or recycle entirely or partially	⇨	Change production, consumption systems through modified incentive systems
Climate Change	CO_2 storage (e.g. storing CO_2 in old gas fields)	⇨	Good house-keeping (e.g energy generation from CH_4, change cement production process from wet to dry)	⇨	Make products more durable, easy to repair, and designed to be reused or recycled entirely or partially	⇨	Change in energy genera-tion systems (e.g. conventional to renewable) and consumption systems through modified incen-tive systems

Fig. 18.3. The development stages in the industrial transformation approach in the context of waste management and climate change.

appear to be different. There are several direct material/physical links between the two problems which include direct links in the form of GHG emissions from landfills and incineration plants as well as more subtle links. The latter includes emissions from the transport and recycling of wastes and land use changes from landfills and forestry that reduce the sequestration of CO_2.

They are also part of the current paradigm of industrial production. This chapter has also argued that the current paradigm is energy intensive and wasteful and for many visionaries it is important for society to undergo an industrial transformation. Such industrial transformation aims at eco-efficiency, closing substance cycles and de-materialization of society, and draws inspiration from natural systems. Indus-trial transformation, by changing the way we produce and consume goods and services in society, has the potential of addressing both the waste and the climate change problem. Having said that, we would argue that such transformation may indeed create new and unforeseen environmental, economic and social problems.

Given that it is possible to make both direct (incremental, local) and indirect (structural, systemic, global) links between both problems, we would argue that perhaps it is necessary to make political and policy linkages between the two prob-lems. In an earlier paper (Gupta and Hisschemoller 1997), it was argued that cli-mate change being an abstract and esoteric problem, it is difficult to mobilise people to take action given that the benefits of taking action may not always be immedi-ately visible to those taking the action. The paper argued that the only way to accelerate climate change policy is to enhance the participation of stakeholders, to bring the discussion into the local political agenda and to strive towards unanimity. The key to bring the issue on to the local political agenda was to redefine the climate change problem by developing issue linkages to other issues that people could relate to as opposed to something abstract and distant and to allow people to realise that there were immediate benefits for them and that there is a convergence of interest across societies. One way of doing this is by 'glocalisation', matching global issues with local priorities, so that local populations can take action that has visible and immediate local benefits and also has long-term global benefits. This is one reason to link the climate change agenda to the local agenda.

At the same time, we believe that the waste management agenda could be given a considerable push by a strategic link with climate change. Although waste management is generally local or regional in nature and there are clearer benefits to local populations, developing countries are beset with the triple problems of resource shortage, low awareness and education, and problems with handling wastes (Hogland and Marques 2000). The lack of resources and the rush towards industrial growth has put the local populations and ecosystems at risk. In the developed countries too the problem of wastes has not been minimised; it has only been transported to more complex end-of pipe facilities, sometimes in remote places. For both groups of countries it is necessary to reduce the resources used and the wastes generated to reduce costs and to make production more sustainable. However, local municipalities often lack the resources and the power to take action.

Having established different links between climate change and waste management, it appears clear that a set of policies can address both problems. From a policy making and implementation perspective, it seems easier to tackle both problems from a local perspective. However, improving waste management is not commonly included in policies to combat climate change even though several studies proved it to be economically attractive in many cases (Hekkert, 2000). We would argue that there appears to be increasing reason to try to connect these two fields in order to give the global problem of climate change a direct local dimension and to infuse financial and political resources to the local priority of waste management.

REFERENCES

1. Chadwick, Micheal (1998). *Substance Flow Through Environment and Society*. In: *Managing a Material World, Perspective in Industrial Ecology*. P. Vellinga, F. Berkhout, and J. Gupta (eds.). Kluwer Academic Publishers, Dordrecht.

2. Clinton, W.J. (1997). *Remark by the President on Global Climate Change*. Speech at the National Geographic Society, October.

3. Denison, Richard A. (1996). *Environmental Life-Cycle of Recycling, Landfilling and Incineration*, Annual Review of Energy and Environment. In: Sheehan, Bill (2000). *Zero Waste, Recycling and Climate Change*. GrassRoots Recycling Network, October, *http://www.grrn.org/ zerowaste/climate_change.html* (last visited 30/11/2000).

4. Downs, Anthony (1972). Up And Down With Ecology—The Issue-Attention Cycle, *The Public Interest*, Vol. 28, 38-50.

5. European Environment Agency (1999), *Waste generation and management - Environment in EU at the turn of the century (Chapter 3.7)*. *http://themes.eea.eu.int/toc.php/issues/ waste?doc=39050&l=en* (last visited 30/11/2000).

6. Frosch, Robert A. (1998). *Towards the end of waste*. (1998). In: *Managing a Material World, Perspective in Industrial Ecology*.. P.,Vellinga, F. Berkhout and J. Gupta (eds.) Kluwer Academic Publishers, Dordrecht, pp. 38–39.

7. Gupta, J. and Hisschemöller, M. (1997). Issue-linkages: a global strategy towards sustainable development, *International Environmental Affairs*, **9**: (4) 289-308.

8. Hekkert, Marko P. (2000). *Improving Material Management to Reduce Greenhouse Gas Emissions*. Ph.D. thesis, Utrecht University, Faculteit Scheikunde. pp. 193.

9. Hogland, William and Marcia Marques (2000). *Waste Management in Developing Countries*. In: *Solid Waste Management*. B.K. Guha, W. Hogland, and S.G. Stuart (eds.) pp. 2. V.I. Grover, Oxford & IBH Publishing Co. Pvt. Ltd., New Delhi and Calcutta.

10. Houghton, J.T., L.G. Meira Filho, B.A. Callander, N. Harris, A. Kattenberg and K. Maskell (eds.) (1996). *Climate Change 1995: The Science of Climate Change*. Cambridge University Press, Cambridge.

11. IHDP (1999). *Science Plan*. International Human Dimension Programme, Industrial Transformation Project, IHDP Report No.12, Bonn.

12. IPCC (1990). *Climate Change, the IPCC Response Strategies*. Intergovernmental Panel on Climate Change, World Meteorological Organization and United Nations Environmental Program. pp. 58.

13. Kruger, D., T. Beer, R. Wainberg and X. Huqinh (2000). *Solid Waste Management and Wastewater Treatment*. pp. 313 – 327 In: IPCC (2000). *Methodological and Technological Issues in Technology Transfer*, Intergovernmental Panel on Climate Change, Cambridge University Press, Cambridge.

14. Lardinois, I. and A. van der Klundert (1993). Organic Waste: Options for Small-Scale Resource Recovery. *Urban Solid Waste Series 1*, WASTE/TOOL.

15. Netherlands Government (1989). *National Environmental Policy Planning (NEPP)*, Ministry of Housing, Physical Planning and Environment.

16. Ripert, C. (1997). La Logistique et le transport des déchets ménagers, agricoles et industriels. ADEME, Agence de l'environnement et de la Maîtrise de l'Énergie; Ministère de l'Équipement des Transports et du Logement. In European Environment Agency (1999), *Waste generation and management - Environment in EU at the turn of the century (Chapter 3.7)*. *http://themes.eea.eu.int/toc.php/issues/waste?doc=39050&l=en* (last visited 30/11/2000).

17. Sachs, W., R. Loske, M. Linz, et al. (1998). *Greening the North, a Post-Industrial Blueprint for Ecology and Equity*. Zed Books, London.

18. Sheehan, Bill (2000). *Zero Waste, Recycling and Climate Change*. GrassRoots Recycling Network, October, *http://www.grrn.org/zerowaste/climate_change.html* (last visited 30/11/2000).

19. Thorneloe, S., Roquetta, A., Pacey, J. and Bottero C. (2000). *Database of Landfill-Gas-to-Energy Projects in the United States*. MSW Management, March-April, *http://216.55.25.242/msw_0003_database.html* (last visited on 30/11/2000)

20. Udo de Haes, H.A. (ed.) (1996). *Towards a Methodology for Life Cycle Impact Assessment*. SETAC-Europe, Brussels.

21. Udo de Haes, H.A., G. Huppes, and G. de Snoo (1998). *Analytical Tools for Chain Management*. In: P. Vellinga, F. Berkhout and J. Gupta (eds.) (1998). *Managing a Material World, Perspective in Industrial Ecology*. Kluwer Academic Publishers, Dordrecht.

22. von Wiezsäcker, Ernst Ulrich (1998). *Dematerialisation*. In: *Managing a Material World, Perspective in Industrial Ecology*. P. Vellinga, F. Berkhout and J. Gupta (eds.): (1998). Kluwer Academic Publishers, Dordrecht.

19

Waste, Recycling and Climate Change: US Perspective

FRANK ACKERMAN

Global Development & Environment Institute, Cabot Intercultural Center Tufts University, Medford MA, USA

Introduction

Discussion of the causes of climate change usually begins with energy consumption, as it should—but too often ends there as well. It is certainly true that most anthropogenic emissions of greenhouse gases result from the combustion of fossil fuels. Yet we cannot overlook the ways in which other environmental factors, such as waste management, affect emissions. Two reasons can be mentioned. First, there are some significant non-energy sources of greenhouse gases, including the emission of methane from landfills; and second, choices and policies in the realm of waste management have a surprisingly large effect on the ways in which we use energy.

Waste is not only a large contributor to the greenhouse problem; it is also an area where doing the right thing for the environment is politically popular. It is much easier to persuade most people to change the way they handle solid waste than, for example, to get them to drive small, fuel-efficient cars. Thus waste management is a promising area in which to pursue reduction in carbon emissions, and should be part of any comprehensive strategy for climate change mitigation.

In this paper I will present a framework for analysis of the greenhouse impacts of climate change, then offer some estimates of the size of the impacts and finally make approximate calculations of the importance of these impacts for the US. In view of the many uncertainties and approximations that must be made along the way, the numbers that I end up with are not reliable bases for policy making, but hopefully serve to demonstrate that there is something big enough to justify an analysis in greater detail and precision.

Five Categories of Impacts

How does waste management affect greenhouse gas emissions? There are at least five categories of impacts to consider. The first and most obvious is **landfill methane emissions**. Estimates for the US (EPA 1999a) suggest that landfill methane accounts for about 4% of all greenhouse emissions, measured in terms of global warming potential. On a global level the Intergovernmental Panel on Climate Change (IPCC) estimates that landfill methane accounts for 3% of all greenhouse emissions, but may account for 8-10% of all feasible near-term opportunities for emission reduction (IPCC 2001:14).

Most of the organic waste in landfills decays anaerobically, and most of the carbon is gradually released to the atmosphere, about half of it as carbon dioxide and half as methane. The latter is the problem: the same amount of carbon has a global warming potential 21 times greater if it is released as methane rather than carbon dioxide.

This impact is unmistakably caused by modern waste management. When waste ends up as litter, or in small, uncontrolled, uncompacted dumpsites, there are potentially severe problems of sanitation, public health, and aesthetics. But the decay of waste under these conditions is aerobic, releasing virtually all of its carbon as CO_2 rather than methane. As low-income countries develop, they will increasingly move from open dumping of wastes to sanitary landfilling, implying that landfill methane emissions will be a growing problem worldwide in the future.

The other impacts are, for the most part, less visible part of the waste management process. But they are still caused by the decisions we make about waste and materials. Most important is the fact that both recycling and waste reduction lead to **decreased energy use and process emissions in industry**. The IPCC estimates that primary (virgin material) production causes 40 times the greenhouse emissions of secondary (recycled material) production per ton of aluminum. For many other industrial materials, primary production emissions are 4 to 5 times as great as secondary emissions per ton. (IPCC 1996: 670)

Most of this saving reflects the change in industrial energy use. Extractive industries such as mining, and basic materials industries such as metal, paper, and plastics production, are the most energy-intensive branches of industry, using far more energy per dollar of output than later stages of manufacturing. For example, most of the energy required to make an automobile is used to extract raw materials from nature and process them into bulk industrial commodities; much less is used to shape the materials, fabricate car parts, and assemble them. Recycling of raw materials, or using less to begin with, reduces energy use and associated carbon emissions in the most energy-hungry branches of industry.

A third type of impact arises when **energy recovery from waste** displaces fossil fuel consumption. This can occur through incineration; through other energy recovery technologies such as pyrolysis; and through capture of landfill gas. Controlling landfill gas has a double benefit: landfill methane can be substituted for natural gas, a fossil fuel; and combustion converts methane to carbon dioxide, vastly reducing its greenhouse impact. For the same reason, even the simple technique of flaring landfill gas, i.e. burning it without capturing the resulting energy, is of great benefit from a climate change perspective—though burning the gas *with* energy capture is obviously a better idea.

At first glance, it is hard to see how burning waste paper in an incinerator instead of coal in a conventional power plant can reduce atmospheric carbon. Both combustion processes release CO_2 into the air. However, paper comes from trees, which absorb CO_2 from the atmosphere as they grow. Assuming sustainable forestry practices (a controversial assumption, but not one that will be pursued here), the emissions from burning paper will be balanced by the growth of new trees, leading to zero net emissions over the paper life cycle. This assumption is standard in climate change analyses. A parallel assumption can be made, perhaps less controversially, for incineration of other materials of recent biological origin, such as garden waste and food waste.

In contrast, fossil fuels are not renewable on any relevant time scale; their combustion does, therefore, lead to a net increase in atmospheric CO_2. Among ordinary solid wastes, the same is true only for plastics, which are made from fossil fuels. So when a waste-to-energy facility substitutes for a fossil-fuel-burning power plant, the appropriate comparison is between the CO_2 emissions from plastic wastes (only a fraction of the incinerator's feedstock) and the emissions from all of the fossil fuel. There are longstanding debates over the environmental merits of incineration of paper, involving this and many other technical issues (see, for example, Blum *et al.*, 1997 and Grieg-Gran *et al.* 1997).

Varieties of Sequestration

A fourth category of impacts also depends on complex hypotheses about forestry and other environmental policies: paper recycling and reduction may have an effect on **carbon sequestration** in forests. Any decrease in the production of virgin paper means that fewer trees need to be cut down. Hence, depending on assumptions about other factors that affect forest practices, there may be more carbon left standing in the woods.

A recent US EPA study of waste and climate change (EPA 1998a) employed an intricate series of forestry and paper industry models to estimate the sequestration effect. (I was a member of the large research team for that study, though I was not involved in the sequestration analysis.) The forestry models essentially showed that an increase in recycling or source reduction of paper leads immediately to decreased timber harvesting, implying an increase in the volume of wood standing in the forests. Forest owners will gradually respond by reducing their stocks of wood, either by planting less or by using their forests for other purposes. However, this adjustment is slow, due to the time lags involved in planting and growing trees; and even in the long run, the adjustment may be less than complete.

Other models and assumptions could lead to other conclusions. However, the EPA study (EPA 1998a) finds on the one hand that there are rather small energy savings due to paper recycling—that is, the second impact category, as discussed above, is of relatively little importance in this case. On the other hand, it finds that the forest sequestration savings due to recycling or reduction are quite large. On this basis, it finds paper reduction or recycling to be far better than incineration from a climate change perspective.

There are other opportunities for carbon sequestration in waste management and materials use, though they are on a smaller scale than in forestry. Carbon can also be sequestered in wood buildings and furniture, and in paper products. All of us who have not gotten around to cleaning out old file drawers full of forgotten papers are doing our bit to sequester carbon at home and at work.

A final, paradoxical form of sequestration should be mentioned briefly. In this case, I confess that I remain puzzled by the work of my colleagues on parts of the

US EPA study. According to laboratory experiments done by one of the researchers, a noticeable fraction of the carbon in landfilled yard waste and newspaper is never released, but remains sequestered indefinitely in the landfill. The same experiments showed almost no sequestration for landfilled office paper and food waste. For newspaper, landfill sequestration is smaller than the forest sequestration that results from recycling; the best thing to do with newspaper, from a greenhouse perspective, is still to recycle it. For yard waste composting, there is no such alternative to forest sequestration, so it is possible that net carbon emissions are somewhat lower when yard waste is landfilled rather than composted.

This result, which has surprised almost everyone, is based on only one set of laboratory experiments. It will be important to see whether it is confirmed by other researchers. (Unpublished research is currently exploring the possibility that land application of compost leads to increased carbon sequestration in soils, an effect which was omitted in EPA 1998a but should be included in the compost life cycle. With a credit for soil sequestration it is possible that composting will cause slightly lower greenhouse emissions than landfilling.) The effect is not large in any case; no version of the analysis suggests that much progress could be made in reducing greenhouse emissions by changing the management of yard waste. The analysis does, however, cast doubt on past assumptions that composting is a natural strategy for reducing carbon emissions.

The final impact category is **energy required for transportation** of waste materials. If recycled materials are transported far enough, the energy savings from recycling may be offset by the energy consumed in moving the materials. In a worst-case scenario, sending recycled glass by truck from Boston to Denver, about 3000 km, would undo most or all of the greenhouse benefits of recycling. The truck emissions for the journey would roughly negate the carbon reduction achieved in glass production. Note that this is a worst case: emissions are lower for long-distance freight transport by rail or by ship; and emission savings are lower for glass than for most other recycled materials. At the other extreme, recycling aluminum creates such huge per-ton savings in energy and greenhouse emissions that the effects of long-distance transport are insignificant by comparison.

The transportation effect can safely be ignored for recycling in most urban, industrial areas where distances to processing facilities are reasonably short. However, in low-density rural areas far from urban centers the need for long-distance shipping of recyclable materials reduces their environmental benefit, at least from a climate change perspective.

Measuring the Impacts

How large are the greenhouse gas reductions achievable through waste management? Table 19.1 presents the estimates developed in the US EPA study, for nine recyclable materials; Figure 19.1 displays the same information graphically. The table and graph show the change in emissions, relative to landfilling, for each material and each waste management option.

A number of important patterns can be seen in Table 19.1 and Figure 19.1. Almost all the numbers are negative (in the graph, most bars extend to the left of the zero line), indicating that almost everything is an improvement over landfilling

Table 19.1. Changes in greenhouse gas emissions, relative to landfilling

Tonnes of CO_2-equivalent emissions per tonne of material

Material	Waste Management Option		
	Source reduction	Recycling	Combustion
Newspaper	−2.7	−2.5	0.0
Paper	−6.3	−5.4	−2.9
Cardboard	−3.3	−3.0	−0.9
Aluminum cans	−12.0	−15.7	+0.1
Steel cans	−3.4	−2.3	−2.0
Glass	−0.6	−0.4	0.0
HDPE containers	−2.5	−1.5	+0.8
LDPE containers	−3.6	−2.0	+0.8
PET containers	−4.0	−2.5	+0.9

Source: EPA 1998a, Exhibit 8-5, converted to metric tonnes of CO_2-equivalent emissions per metric tonne of material. Figures shown here are the differences between emissions for landfilling and for each of the other waste management options. The landfill scenario used as a baseline assumes that 54% of landfills have methane capture systems, which are 75% effi- cient, implying that an average of 40% of all landfill methane emissions are captured.

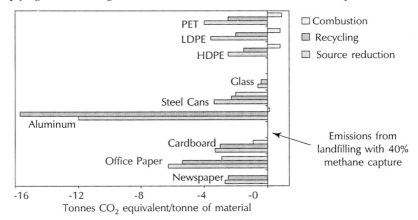

Fig. 19.1 Greenhouse emissions relative to landfilling.

from a climate change perspective. The only significantly positive entries in the table are for incineration of plastics, which gives rise to air emissions. In contrast, plastics are inert in landfills, and do not cause any emissions.

For this study, landfills were assumed to recover 40% of all methane, the ambi- tious new regulatory target (effective as of 2000) that may be above the actual US average. With a lower rate of methane capture, landfilling of paper would look worse, and doing anything else with paper would look comparatively better.

The table shows that incineration is roughly as good as or better than landfilling for non-plastics, but is worse than recycling and source reduction for every material. The surprising climate change benefit from "combustion" of steel cans reflects the fact that incinerators recover and recycle much of the ferrous material they receive.

Combustion of newspaper is no better than landfilling due to the assumption that landfilling of newspaper leads to long-term carbon sequestration. Without that assumption, landfilling newspaper would look worse, and all other newspaper options would look better.

Recycling of all nine materials leads to reduction in greenhouse gas emissions, relative to landfilling. The benefits per ton are greatest for aluminum and smallest for glass. Source reduction is even better than recycling, with the exception of aluminum. The explanation of this puzzle is that source reduction is assumed to replace the existing mix of virgin and recycled aluminum used in the US today, while recycling is assumed to replace purely virgin material. Due to the large difference in energy intensity between virgin and recycled aluminum production, this means that the material being replaced is noticeably less energy-intensive for source reduction than for recycling.

Three Recycling Scenarios

The numbers in Table 19.1 tell us the per-ton effect of recycling on greenhouse gas emissions (under American conditions and the numerous assumptions used in the study). These numbers can then be multiplied by the quantities of recycled material, in tons, to determine the total emission reductions attributable to recycling. Three sets of comparable data on recycling will be useful for the purposes of this calculation, all for 1996:

- The estimated US average level of recycling (EPA 1998b); and
- Recycling in Seattle, Washington and in Bergen Country, New Jersey, two well-documented success stories of American recycling (EPA 1999b).

Table 19.2 shows the levels of recycling of selected materials in these three areas; Figure 19.2 summarizes the extent of paper and non-paper recycling.

Table 19.2. Recycling of selected materials in 1996, in kilograms per person

	US average	Seattle	Bergen
Newspaper, misc. paper	35	165	122
Office/high grade paper	11	24	26
Cardboard	66	185	124
Glass	11	25	26
Plastics	3.6	2.7	12
Ferrous	15	10	32
Aluminum	3.5	4.1	6.8

Sources: US average (EPA 1998b); Seattle and Bergen Country (EPA 1999b).

The next calculation simply consists of multiplying the numbers from Tables 19.1 and 19.2. Table 19.3 shows the reduction in greenhouse gas emissions due to recycling in each of the three cases, in kilograms of CO_2-equivalent emissions per capita; Figure 19.3 again summarizes the impacts from paper and non-paper recycling. The total emission reduction is almost half a ton per person for the US average, and more than a ton for both Seattle and Bergen County. Paper recycling accounts for more than three-fourths of the savings in all cases, and more than 90% in Seattle.

The final calculation presents a further conjecture: how big would the impacts be if all 265 million Americans (the 1996 US population) had recycled at these per

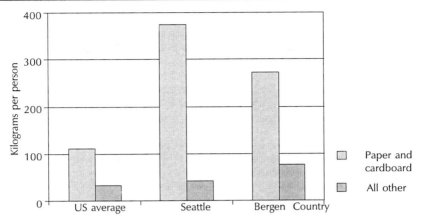

Fig. 19.2 Recycling of Selected Materials, 1996

capita rates? The answer is shown in Table 19.4, both in million tonnes of CO_2-equivalent emissions, and finally as a percentage of US greenhouse gas emissions.

Table 19.3. Greenhouse gas emission reduction per capita due to recycling, 1996

	(kilograms of CO_2-equivalent emissions per capita)		
	US average	Seattle	Bergen
Newspaper, misc. paper	86	413	304
Office/high grade paper	59	128	143
Cardboard	199	556	371
Glass	4.3	10	11
Plastics	7.3	5.4	25
Ferrous	35	22	74
Aluminum	55	64	108
Total	446	1198	1034
Paper as % of total	77%	92%	79%

Sources: Recycling impacts per tonne from Table 19.1 multiplied by quantities from Table 19.2. Newspaper coefficient applied to miscellaneous paper; average of plastics products used for plastics.

Table 19.4. Total emission reduction due to recycling at selected rates, 1996

	US average	Seattle	Bergen
Million tonnes of CO_2-equivalent	118	318	274
Percentage of US total emissions	2.0%	5.5%	4.7%

Source: Calculated from Table 19.3 totals, using 265.2 million US population, and 1582.0 MMTCE US net greenhouse gas emissions (EPA 1999a), both for 1996 for comparability with recycling scenarios.

The existing rate of recycling already saved 2% of nationwide emissions, relative to the baseline of landfilling all discards. If everyone achieved the Bergen Country or Seattle recycling rates, an additional 2.7%-3.5% of total emissions would be saved.

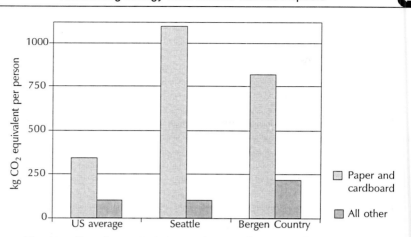

Fig. 19.3 Greenhouse Emission Reductions from Recycling, 1996.

These percentages may sound small. But recall that the Kyoto targets call for roughly 30% reduction in total emissions, relative to a "business as usual" scenario. Thus matching the best recycling programs nationwide could get us something like one-tenth of the way to compliance with the Kyoto Protocol.

Conclusion

More research is needed to solidify the numerous assumptions used in this analysis. The scenarios presented here, with potential savings due to recycling ranging from 2% to 5.5% of national greenhouse gas emissions, are meant to illustrate the approximate magnitude of the effects of waste management, not to provide hard results for planning purposes. Among the crucial areas for further investigation are the actual rate of landfill methane capture, the impact of paper reduction and recycling on forest carbon sequestration, and the puzzling possibility of carbon sequestration in landfills. (Since this last puzzle is unresolved, no calculations have been included here for greenhouse impacts of composting.)

Yet despite these uncertainties, the effect of waste management choices on climate change is large enough that it is well worth studying in greater detail. Paper recycling, and the analysis of the paper life cycle, appear to be of particular importance. It is remarkable to think that we have the potential to achieve one-tenth of the Kyoto targets through an activity that already has widespread grassroots support. Although the data presented here are for the US, similar conclusions apply to other countries with similar waste streams, such as Australia and Canada. (For an earlier version of this analysis developed for Australia, see Ackerman 2000.[1]) The impact of waste on climate change is an important subject to pursue as we develop strategies for greenhouse gas reduction for the twenty-first century.

---

[1] In that analysis, all paper grades were aggregated, leading to slightly higher estimates for carbon savings from paper recycling. The disaggregated treatment presented here is more appropriate when the data is avaliable.

REFERENCES

Ackerman, F. (2000). "Waste Management and Climate Change," *Local Environment* **5** no. 2, 223-229.

Blum, L. *et al.*, (1997). "A Life-Cycle Approach to Purchasing and Using Environmentally Preferable Paper: A Summary of the Paper Task Force Report," *Journal of Industrial Ecology* **1**: (3) 15-46.

EPA (US Environmental Protection Agency) (1999a). *Inventory of US Greenhouse Gas Emissions and Sinks: 1990-97* (EPA 236-R-99-003).

EPA 1999b. *Cutting the Waste Stream in Half: Community Record-Setters Show How* (EPA 530-R-99-013).

EPA 1998a. *Greenhouse Gas Emissions from Management of Selected Materials in Municipal Solid Waste,* (EPA 530-R-98-013).

EPA 1998b. *Characterization of Municipal Solid Waste in the United States: (1997) Update* (EPA 530-R-98-007).

Grieg-Gran, M. *et al.*, 1997. "Towards a Sustainable Paper Cycle: A Summary," *Journal of Industrial Ecology* **1** (3): 47-68.

IPCC (Intergovernmental Panel on Climate Change) 2001. *Summary for Policy Makers of the IPCC WG III Third Assessment Report.*

IPCC 1996. *Climate Change 1995: Impacts, Adaptations and Mitigation of Climate Change* (Cambridge University Press).

Section V

Health Impacts

20

Incineration and Human Health: Characterization and Monitoring of Incinerator Releases and their Impact

Michelle Allsopp, Pat Costner, Paul Johnston and David Santillo

Greenpeace Research Laboratories, University of Exeter,
Prince of Wales Road, Exeter, Ex4 4PS, UK

How Incinerators Effect Health

The impact of waste incinerators on health and their hazardous combustible emissions such as dioxins and PAHs are of great public concern (Ardevol *et al.*, 1999). Research has identified numerous toxic compounds, which are emitted in stack gases and in ashes, as well as many unidentified substances of unknown toxicity (see section 2). Individuals exposed to incineration related hazards, include workers connected with incinerator facilities and populations living in the vicinity. Studies on how exposure to incinerators effects health have focused entirely on these two groups.

A recent publication by the National Research Council (NRC 2000), an arm of the National Academy of Sciences, established to advise the US government, concluded that it was not only the health of workers and local populations that could be affected by incinerators. The NRC reported that even populations living at distances from incinerators were likely to be exposed to some incinerator pollutants.

> *Persistent air pollutants, such as dioxins, furans and mercury, can be dispersed over large regions – well beyond the local areas and even the countries from which the sources first emanate. . . . Food contaminated near an incineration facility might be consumed by people close to the facility or far away from it. Thus, local deposition on food might result in some exposure of populations at great distances, due to transport of food to markets. However, distant populations are likely to be more exposed through long-range transport of pollutants and low-level, widespread deposition on food crops at locations remote from a source incineration facility.*

and,

> *The potential effects of metals and other pollutants that are very persistent in the environment may extend well beyond the area close to the incinerator. Persistent pollutants can be carried long distances from their emission sources, go through various chemical and physical transformations, and pass numerous times through soil, water, or food. Dioxins, furans, and mercury are examples of persistent pollutants*

for which incinerators have contributed a substantial portion of the total national emissions. Whereas one incinerator might contribute only a small fraction of the total environmental concentrations of these chemicals, the sum of the emissions of all the incineration facilities in a region can be considerable. The primary pathway of exposure to dioxins is consumption of contaminated food, which can expose a very broad population. In such a case, the incremental burden from all incinerators deserves serious consideration beyond a local level.

This chapter reviews research on human exposure to pollutants from incinerators and how they affect the health of workers and local populations. A broad range of health impacts have been documented in these two groups, including adverse effects on children living in local populations near incinerators. However, whether the observed associations with pollutants from incinerators are causal is often difficult to confirm.

Types of Research Study

The impacts of incinerators on human health have been assessed after conducting three types of studies. These are human exposure studies, epidemiology studies and finally, risk assessment studies. Exposure studies and epidemiological studies provide the most compelling evidence on health impacts of incineration since they involve generating scientific data directly from the individuals under investigation. On the other hand, risk assessments are theoretical estimations of what health effects may occur based on mathematical calculation.

Exposure Studies

Exposure to compounds emitted from incinerators may occur by inhalation of contaminated air, or by consumption of local agricultural produce or soil that has been contaminated by deposition of airborne pollutants. Workers at incinerator plants may additionally be exposed to contaminated ashes.

To assess the impact on health from exposure to incinerator emissions, reliable methods of assessments are required. One method to assess potential exposure is to monitor levels of contaminants in air from incinerators, and in soils, vegetation and agricultural produce. However, such investigations do not permit the "internal exposure" in humans to be assessed directly (Ardevol *et al.*, 1999). Evaluation of internal exposure requires the quantification of compounds in the human body. With technological advances, it has become possible to monitor the level of certain toxic compounds from incinerators, in the body tissues of humans. This involves the determination of contaminant concentrations in biological samples, such as, blood, urine, hair or breast milk.

Exposure studies analyse biological samples for: (i) chemical pollutants that are released from an incinerator, or, (ii) metabolites (breakdown products) of these chemicals, or, (iii) biomarkers of exposure (which show biological effects of a toxic exposure). The results of the analyses are compared with a control group of unexposed individuals.

A number of studies have been conducted to assess exposure of incinerator workers and populations living near incinerators using the analyses described above.

Epidemiological Studies

Epidemiological studies attempt to establish the state of general health and how it may be related to the intake of pollutants released from incinerators. Information relating to the potentially contaminated people, such as birth and death certificates, disease registries, physicians' reports, self-reported symptoms and illnesses, is used. This is compared with similar information from potentially uncontaminated or less contaminated people. Some of the major challenges used while establishing a cause-and-effect relationship through epidemiological studies were (NRC 2000):

- Identifying suitably exposed populations that were large enough for establishing a useful degree of statistical significance.
- Identifying the many factors that modified the effect (e.g., age, sex, etc.) and/or potentially confounding factors (e.g., smoking, diet, etc.)
- Identifying biases (including reporting biases) in data collection.
- Measuring the frequency of occurrence and concentrations of specific pollutants within the affected population and a potentially unaffected control group.
- Measuring effects that were small, occured infrequently, took many years to appear, and/or occured not in the exposed individuals but in their offspring during infancy, childhood or adulthood.

Epidemiological studies have investigated a variety of health hazards resulting from exposure to incinerator emissions, both in workers and in populations living close to incinerators. Cancer and respiratory effects have been analysed. Human epidemiological studies are limited, despite the widespread concern about potential health effects. These studies are generally more valid than other health studies insofar as exposure to all pollutants emitted from incinerators are explicitly or implicitly accounted for, mirroring the "real" situation (Rowat 1999), although their power is determined by their design and is often limited by a combination of uncontrolable factors (Hu and Shy 2001).

Risk Assessment

Risk assessment attempts to estimate exposure to a particular chemical from the emissions in question and ultimately calculates the probability of ill effects on health from the estimated exposure. Risk assessment is a step by step process which involves the use of mathematical equations to estimate the emissions, transport and transformation of the emissions in the environment, human exposure and, finally, the likelihood of health suffering from this exposure. The use of risk assessment is largely for regulatory decision making.

The process of risk assessment itself is, however, fraught with uncertainties, necessarily over simplifies environmental processes and warrants being viewed with deep scepticism as to whether it can actually be protective of human health (e.g. Johnston *et al.*, 1996). A fundamental problem which risk assessment faces is that estimation of the consequences of pollution is still a poorly understood science. Even for dioxin (TCDD), one of the most intensely studied chemicals, many unknowns remain and, since risk assessment relies on toxicological data to estimate health effects, it can only be as good as the data on which it is based. Limitations in toxicological information contribute to imprecise results generated by risk assessments. Moreover, in the case of a developing foetus, the uncertainty of the toxicological significance of long-term low-dose exposure to pollutants is huge. It is clear that the developing stages of life are the most vulnerable to toxic insult. Risk assessments however are generally based on estimation of risk in adults and may significantly underestimate the potential impact on the foetus and the developing young.

In estimating the probability of impacts on health, many uncertainties appear also at other stages of the risk assessment process. For instance, there is uncertainty in estimating the quantities of emissions, in estimating the transport and transformation of pollutants in the environment and, from this calculation, estimating human exposure. It is indeed extremely difficult, if not impossible, to determine the actual doses involved in environmental exposures. In order to overcome uncertainties in estimations, risk assessors use "conservative" estimates and so assume that overestimating risks overcomes these problems and is therefore protective of public health. However, the notion of "conservative" is ill-defined and in practice raises significant questions concerning exactly how conservative a risk assessment should be, e.g. all uncertain parameters conservatively treated or just a selected few. In this way it becomes apparent that risk assessment not only contains many uncertainties, but it is also a subjective rather than a scientifically objective process. This again calls into question whether regulations derived from risk assessments can be relied upon to be truly protective of human health.

A further concern regarding the estimation of emissions with regard to incinerators is that data are usually based on test burns that are carried out under optimal conditions. It is likely that such data on emissions would underestimate emissions under operational conditions, (Webster and Connett 1990).

Risk assessments on incinerators generally focus on only one or a few substances that are known to be emitted, in particular dioxins and selected heavy metals. However, in reality, emissions from incinerators consist of complex mixtures of hundreds of chemicals, including many unknown compounds with unknown toxicity. Risk assessment omits to take into account health impacts of many of the known chemicals and all the unknown chemicals (Johnston *et al.*, 1996). In addition, in looking at just single chemicals, it does not address the issue of the combined toxicity of the chemical mixtures in stack emissions (Johnston *et al.*, 1998). For instance, the combination of two or more chemicals together may cause an additive, greater than additive (synergistic) or a less than additive (antagonistic) effect.

A further problem in risk assessment is that it is very difficult to determine the most appropriate and sensitive endpoints for detecting the toxicity of chemicals. An adverse effect on the immune system or respiratory system may, for instance, be more sensitive and be instigated at lower chemical concentrations than another sort of health impact. For health risk assessment on incinerators, toxicological endpoints can include both cancer and non-cancer effects on health. Whatever endpoint is chosen, it is thereby accepted as a key metric capable of being used to protect human health in a holistic manner. It is questionable, however, whether the correct endpoints are ever used in risk assessment. In addition, as discussed above, the developing young are more likely to be sensitive to some adverse chemical effects than adults.

To summarize, there is a dauntingly wide spectrum of inadequacies and uncertainties inherent in the process of risk assessment, from the estimation of type and quantity of pollution, to estimates of exposure and health effects. Each of these problems alone can fatally compromise risk assessment procedures. It is particularly important that these limitations are recognized when risk assessment is applied in the formulation, implementation or enforcement of regulations. Risk assessment should be viewed with deep scepticism unless all the areas of uncertainty are explicitly defined (Johnston *et al.*, 1998).

Incinerator Emissions

All waste incinerators are also waste generators—incineration of waste results in output of waste products. This is because physical matter cannot be destroyed, but can only be transformed into new forms. Thus when things are burned, they do not disappear as is the common perception, but merely change their form. Waste products resulting from incineration take the form of stack gas emissions to the atmosphere, bottom ashes (slag) and fly ashes (caught in filters in the incinerator stack) which ultimately are disposed of to landfill sites. Where water is used for cleaning processes in an incinerator, there are also emissions of waste products to water.

It is a popular misconception that the weight and volume of the original raw waste are reduced during incineration. Although it is often stated that the solid residues (ashes) amount to about one third of the initial weight of the raw waste (Pluss and Ferrell 1991), and volume reduction of about 90% is achieved (Williams 1990), neither of these statistics stand up to scrutiny. If all the waste outputs from an incinerator are summed, then the output will exceed the input. The gases present in the flue stack result from the combination of carbon-based materials with oxygen and are usually ignored in calculating the mass of residues, but the combination with oxygen to form CO_2 increases actual weight. Residues from wet gas cleaning systems can generate appreciable volumes of contaminated water and solids. In the case of the statistic concerning volume reduction, this is usually generated by reference to the volume of uncompacted wastes. Landfilled Municipal Solid Waste (MSW), however, is generally compacted to increase stability and prevent water infiltration as well as reduce the volume of the wastes. Compared on this basis, the actual volume reduction achievable is closer to 45% (DoE 1995).

Numerous chemicals are released into the wastes generated by incineration, including hazardous chemicals. For instance, MSW incinerators are typically fed a mixed waste stream and the combustion of such waste leads to hazardous substances originally present within the waste being mobilized into emissions from the incineration plant. While some chemicals remain in their original form, others are changed into new chemical species. For example, heavy metals are not destroyed by incineration but are simply concentrated in the remaining wastes. They can remain in their original form during incineration or may react to form new compounds such as metal oxides, chlorides or fluorides (Dempsey and Oppelt 1993).

The exact nature of the substances released during incineration depends on the composition of the waste that is incinerated. For instance, incineration of chlorinated organic compounds will cause the formation of hydrogen chloride (HCl) and this in turn can contribute to the formation of dioxins. Technical standards that are applied both to the incineration process and to pollution control equipment will also influence the final products of incineration (EEA 2000). However, whatever control technology is applied, all types of incineration result in emissions of toxic substances in ashes and in the form of gases/particulate matter to air. These substances include heavy metals, numerous organic compounds, such as dioxins, and gases, such as nitrogen oxides, sulphur oxides, hydrogen chloride and hydrogen fluoride, together with carbon dioxide. According to the NRC (2000):

> . . . the products of primary concern, owing to their potential effects on human health and the environment, are compounds that contain sulfur, nitrogen, halogens (such as chlorine), and toxic metals. Specific compounds of concern include CO, NO_x, SO_x, HCl, cadmium, lead, mercury, chromium, arsenic, beryllium, dioxins and furans, PCBs, and polycyclic aromatic hydrocarbons. . . .

In many countries over the past few years, new regulatory emission standards have forced the closure or updating of many old incinerators or the building of new ones. Upgraded plants (together with new ones) may be fitted with modern, improved air pollution control technology. For example, out of the 780 incinerators in operation in the UK in the early 1990s (30 for municipal waste, 700 for clinical waste, 40 attached to chemical companies, 6 for sewage sludge and 4 for hazardous waste), only 110 remained after the tightening of emission standards (Murray 1999). Presently there are 12 operating MSW incinerators in the UK. The closure or updating of old incinerators is considered to have led to a substantial reduction of emissions of toxic substances to air.

One study in the Netherlands has also estimated that dioxin emissions to air have been significantly reduced (Born 1996). Murray (1999) states that the most sophisticated German technology developed during the early 1990's has cut atmospheric emissions broadly by a factor of ten. Although this is a significant improvement, the problem of toxic waste products from incineration has not disappeared. In fact, the problem has shifted so that more of the dioxins and other toxic substances generated now appear in the ashes, thereby creating new disposal and pollution problems. The European Environment Agency (EEA 2000) has warned that even if total emissions from incineration are reduced in the future as standards improve "this might be offset with increased incineration capacity". In this regard, it is of great concern that an increase in the use of incineration is being proposed in some European countries. In the UK for instance, following the closure of numerous old incinerators, up to 165 possible new ones have been proposed by the government (DETR 2000).

On the regulatory front, among the various incinerator outputs, stack gas has received the greatest share of attention and is the most highly regulated since the gas and its toxic components are dispersed directly into the open air. However, the other incinerator wastes also contain toxic pollutants and, consequently, pose threats to public health that may be less obvious and/or immediate but are no less real.

The Commission of the European Communities (CEC) adopted a new waste incineration directive at the end of 2000 (EC 2000). The new directive will establish controls on the incineration of most wastes that are not covered by the previous 1994 directive and set limits for releases of some hazardous substances in stack gases and water. All new MSW incinerators built after the directive comes into force must satisfy the limits in the directive within 2 years (i.e. by 28 December 2002), whilst existing incinerators have a period of 5 years to satisfy the criteria. In addition to the EU regulations, various national guidelines for incinerators are also presently in place and these will have to comply with the directive within two years of it coming into force (EC 2000).

Emissions to Air

This section presents data on substances known to be emitted as stack gases from incinerators. Most research on emissions has focused on dioxins and upon the behaviour of a few toxic heavy metals. Data from research upon other emitted chemicals are sparse. In addition, a very large number of the chemicals emitted from incinerators remain unidentified.

Emissions from incinerator stacks to air are discussed below under the following categories: organic compounds; heavy metals; gases and particulates. The EC has included limits for air emissions from incinerators in the new directive for only a few of the compounds falling under these categories. These limits are given in Table 20.1.

Table 20.1 EC Air Emission Limit Values

Substance	Proposed EC Limit (mg/Nm³)
Dioxins	0.1 ng TEQ/Nm³
Mercury	0.05[b]
Cadmium + Thallium	Total 0.05[b]
Sb, As, Pb, Cr, Co, Cu, Mn, Ni, V	Total 0.5[b]
Carbon monoxide	50[c]
SO$_2$	50[c]
NOx	200[c]
HCl	10[c]
HF	1[c]
Particles (total dust)	10

[a] Average values measured over a sample period of a minimum of 6 hours and a maximum of 8 hours.

[b] all average values over the sample period of a minimum of 30 minutes and a maximum of 8 hours.

[c] daily average value

Organic Compounds

Dioxins
Polychlorinated dibenzo-*p*-dioxins (PCDDs) and polychlorinated dibenzofurans (PCDFs) are a group of chemicals often referred to simply as dioxins. There are more than 200 individual congeners (members) of the PCDD/Fs group. The most widely known and most toxic congener is 2,3,7,8-TCDD. It has been described as the most toxic chemical known to mankind and is recognized as a human carcinogen. Dioxins are persistent in the environment, toxic and bioaccumulative (build up in the tissues of living organisms).

The toxicity of individual dioxins and furans varies by several orders of magnitude. Because analytical data may report 17 different congeners as well as totals for homologue groups (i.e. all congeners containing the same number of chlorine atoms), it is often necessary to summarize data so that individual samples can be directly compared. This is generally done by expressing the amount of dioxins present as toxic equivalents (TEQs) relative to 2,3,7,8-TCDD. Until recently, the most common TEQ system used was the international toxic equivalents system (I-TEQ). The TEQ system works by assigning TCDD, the most toxic congener, a toxic equivalence factor (TEF) value of 1. The toxicity of all other congeners is expressed relative to this, such that they are assigned a TEF value between 0 and 1. The I-TEQ of a sample containing a mixture of dioxins is obtained by multiplying the concentration of each congener by its TEF and summing the results.

One important consideration in relation to dioxin emissions from incinerators is that regulations consider only the chlorinated varieties. It has been known for some time that incinerators generate and emit brominated and mixed chloro-bromo substituted dioxins in appreciable quantities (Schwind *et al.*, 1988). These are regarded to be of similar toxicological significance to the chlorinated dioxins, producing a similar array of biological impacts at similar molar concentrations (Weber and Greim 1997). Despite these compounds being highly persistent when associated with fly ash particles, little attention has been directed at evaluation of their significance to human health and there are currently no obligations on the part of incinerator operators to monitor and control these chemicals.

Formation of Dioxins in Incinerators

Dioxins are produced as unintentional by-products of many manufacturing and combustion processes, especially processes that use, produce or dispose of chlorine or chlorine derived chemicals. All types of incinerators produce them. Research has shown that while dioxins can be destroyed in the combustion zone of incinerators, they can be regenerated in the post-combustion zone by processes that are dependent on the temperature profile (Blumenstock *et al.*, 2000, Huang and Buekens 1995, Fangmark *et al.*, 1994). The predominant formation pathway of dioxins has been reported to be *de novo* synthesis (Johnke and Stelzner 1992), and they are also formed from precursors that are either constituents of the waste or are formed by chemical recombination of materials in the waste. The chlorobenzenes and chlorophenols are two such groups (Huang and Buekens 1995).

It has therefore been demonstrated that the process of waste incineration itself creates dioxins. Mass balance calculations show that the total amount of dioxins coming out of an incinerator in the various waste products is greater than the amount going into the incinerator as contaminants in raw waste (Williams 1990; Hansen 2000). This appears still to be the case for modern and updated incinerators operating in the late 1990s, although very few data are available from the scientific literature beyond a single, recent Danish study (Hansen 2000).

In another example from Spain, a mass balance estimate based on measurements from eight operating municipal waste incinerators showed that more dioxins are emitted from the incinerators than were present in the raw waste (Fabrellas *et al.*, 1999). Estimates showed the level of dioxins (PCDD/Fs) input in raw waste to the incinerator amounted to 79.8 g I-TEQ/year. This compared to the total estimated output of flue gases (1-1.2 g I-TEQ/year), fly ashes (46.6-111.6 g I-TEQ/year), and bottom ashes (2-19 g I-TEQ/year). An alternative dioxin mass balance conducted on another Spanish municipal waste incinerator was ambiguous. One test showed a greater dioxin output than input whereas another test showed the reverse (Abad *et al.* 2000). This is not particularly surprising because emissions of dioxins and other substances from individual incinerators are highly variable depending on waste input and combustion conditions. In addition, the precision of such estimates is often not high, encompassing a wide range of values.

Dioxin Inventories and Incineration

During the 1980s and early to mid-1990s, MSW incineration was identified as a major source of dioxins emitted to atmosphere. For example, the Dutch government research organization RIVM estimated that incineration was responsible for about 79% of all dioxins emitted to air in the Netherlands for the year 1991. In the UK, MSW incinerators were estimated to be responsible for about 53-82% of all dioxins emitted to air in 1995, whereas in the US, such facilities accounted for about 37% of total annual emission (Pastorelli *et al.*, 1999). A summary of data from 15 countries, described as a "global" inventory, showed that incineration accounted for about 50% of dioxin emissions to air in 1995 (Fiedler 1999). MSW incineration has been identified as being responsible for the greatest proportion of emissions compared to other types of incineration (Alcock *et al.*, 1998), although from "global" inventory data for 15 countries, Fiedler (1999) noted that all sectors of incineration in 1995 were major emitters in many countries. This included incineration of MSW, hazardous waste, sewage sludge and waste wood, as well as crematoria. Table 20.2 shows the estimated dioxin emissions for different types of incinerators for 1997 in the UK.

Table 20.2. PCDD/F emission estimates for the UK, (numbers in bold type represent estimates calculated from measured emissions, other number 3 are estimates)

Process	1997 Range/Low (g TEQ/annum)	1997 Range/High (g TEQ/annum)
MSW incineration	**122**	**199**
Chemical waste incineration (10 sites)	0.02	8.7
Medical Waste Incineration (5 sites)	0.99	18.3
Sewage Sludge Incineration (5 sites)	0.001	0.37
Cement Manufacture (5 sites)	0.29	10.4
Crematoria	1	35
Domestic Wood Combustion (clean)	2	18
Domestic Wood Combustion (treated)	1	5

Source: Alcock *et al.*, 1998.
Footnote: The estimate for total dioxin emissions from all sources is Range/Low 219 and Range/High 663 g TEQ/annum.

Even in more recent calculations incinerators have been estimated to account for a high proportion of atmospheric dioxins. Hansen (2000) conducted a flow analysis for dioxins in Denmark for 1998-1999. Notwithstanding improvements in technology, municipal solid waste incineration was identified as the single largest source of dioxins to atmosphere, estimated at between 11-42g I-TEQ per year. It is estimated that a further 35-275g I-TEQ of dioxins contained in incinerator residues is disposed of to landfill each year. Hansen (2000) also drew attention to the potential importance of the brominated and mixed halogenated dioxins and estimates that between 2 and 60 g of brominated dioxins are emitted to atmosphere from Danish MSW incinerators per year.

A 1997 publication cited by the Commission of the European Communities (EC 1998) noted that incineration of non-hazardous waste may contribute up to 40% of all dioxin emissions in Europe. Nevertheless, in some European countries, it has been estimated/predicted that the contribution of MSW incineration to national inventories has fallen significantly during the mid- to late 1990s. This is due to closure of old incinerators that which emitted high levels of dioxins to air and the fitting of pollution abatement equipment both to remaining plant and new installations. Estimates suggest that such improvements will have resulted in the significant reduction of dioxin emissions from incinerators to air. For instance, strong downward trends of emissions have been identified in countries with modern technology or rigid legislation (Fiedler 1999). Considering atmospheric emissions alone, Her Majesty's Inspectorate of Pollution (HMIP) and the Department of the Environment (DoE) estimated that the contribution to the total annual emission in the UK would fall from 53-82% in 1995, to around 4-14% in the future. Similarly, the German UBA estimated a contribution of 33% for the years 1989-1990 falling to 3% for the years 1999-2000. These estimated data remain to be confirmed with empirically derived data.

A more accurate estimate of emissions can only be established by continuous monitoring of emissions for extended periods of time. Start-up and shut down periods in the operation of MSW incinerators are particularly prone to result in high dioxin emissions. A study on a Belgian incinerator, using continuous monitoring, was undertaken in an attempt to demonstrate that retro-fitted modern pollution control equipment would prevent excedence of the 0.1 TEQ/Nm^3 regulatory limit at all times. In fact the results revealed that monitoring over a period of 6 hours gave an average emission concentration of 0.25 ng TEQ/Nm^3. Moreover, the average over 2 weeks in the same period gave a result of 8.2 to 12.9 ng TEQ/Nm^3 which was substantially greater and clearly exceeded the regulatory limit (De Fre and Wevers 1998).

The above study shows, in a convincing manner, that taking measurements from individual incinerators under the normal regulatory protocols (i.e. point measurements), can significantly under-estimate the dioxin emissions from incinerators. In this case, point measurement under-estimated the average dioxin emissions by a factor of 30 to 50. The significance of this finding to other incineration facilities is simply not known.

Most mass balance inventories consider only dioxin emissions to atmosphere (Fiedler 1999); the output of dioxins in ash from incinerators is not included. Webster and Connett (1998) stress that the fate of dioxin captured in ash receives insufficient attention. A recent study on a Spanish incinerator showed that stack gas emissions were only responsible for a minor contribution to the total dioxin emitted compared to amounts present in fly ash (Abad *et al.*, 2000). The fact that dioxins formed by incineration have become more concentrated in ashes as air pollution control technologies have evolved, thereby generating other hazards, has already been highlighted above.

Considered together, the sampling methodology commonly employed in emission regulation from incinerators, coupled with failure to consider the dioxin mass balance in an holistic manner, suggest that it is highly probable that many current dioxin inventories greatly underestimate emissions from incinerators.

Performance of Updated and New Incinerators

As indicated above, most dioxin monitoring carried out at incinerators in Europe and reported in scientific literature has been derived on the basis of point measurements rather than continuous monitoring, leading to an underestimate of emissions. This situation seems set to continue under the new EC legislation which specifies compliance monitoring based on only two point measurements per year taken over a period of six to eight hours (EC 1999). This basis for regulation and control, as opposed to continual monitoring is unlikely to describe accurately dioxin emissions from these facilities.

In many cases studies carried out on the basis of point measurements have reported that emissions from some European incinerators fall within the new EC limit of 0.1 ng I-TEQ/m^3. For instance, a series of monthly to two-monthly point measurements taken between 1994 and 1997 from a newly constructed German

MSW incinerator were below the specified limit (Gass *et al.*, 1998). Two point measurements, taken within a day of each other, subsequent to initial testing of a newly constructed MSW incinerator in Venice were also below the 0.1 ng I-TEQ/m^3 limit (Pietro and Giuliana 1999). A study on a German hazardous waste incinerator, based on 11 long-term monitored samples taken between 1998-9, showed that emissions were well within the 0.1 ng I-TEQ/m^3 limit (Mayer *et al.*, 1999).

Not all studies, however, have returned data indicating compliance with the 0.1 ng I-TEQ/m^3 regulatory limit. For example, point measurements taken at 1 to 4 monthly intervals, January 1997 to April 1999, from 8 Spanish MSW incinerators revealed that 2 incinerators failed to comply (Fabrellas *et al.*, 1999). Emission values were 0.7 and 1.08 ng I-TEQ/m^3. In Poland, analysis of stack emissions from 18 new or updated medical waste incinerators in 1994-7 found that almost half had emissions below 0.1 ng TEQ/m^3, but others exceeded the limit (Grochowalski 1998). For 5 of the incinerators, the limit was considerably exceeded with concentrations ranging from 9.7 to 32 ng TEQ/m^3. As discussed previously, a Belgian incinerator exceeded the EC regulatory limit when emissions were measured by continuous monitoring (De Fre and Wevers 1998).

It is important to note that the scientific literature reporting emission levels from new and old incinerators presently operating in many countries, including developing countries is extremely limited. One study of dioxin emissions from the ten incinerators reported to be operating in Korea (Shin *et al.*, 1998) noted a wide variation between different incinerators. Emitted levels ranged from 0.07 to 27.9 ng TEQ/Nm^3 of dioxin in the stack gases.

Even fewer data have been published on incinerators burning wastes other than MSW. Nonetheless, one study in Japan, reported point measurements on nine industrial waste incinerators (Yamamura *et al.*, 1999). Dioxin emissions were below 0.1 Ng I-TEQ/Nm^3 for two of the incinerators but were above this level (0.13 to 4.2 Ng I-TEQ/Nm^3) for six of them. Tests on cement kilns in the US that were operated using coal as fuel were found to emit 0.00133 to 3.0 ng I-TEQ/dscm (dry standard cubic metre) (Schrieber and Evers 1994). In the US, a further study reported on dioxin emissions from mobile incinerators (Meeter *et al.*, 1997). The on-site remediation of soils at hazardous waste sites by such incinerators is carried out where sites contain compounds that are regarded as difficult to destroy. Data collected primarily from trial burns of 16 incinerators showed that 10 of the incinerators failed to meet the proposed US Environmental Protection Agency (EPA) standard of 0.2 ng TEQ /dscm. The authors commented that a significant number of mobile incinerators used in these applications could have problems meeting the proposed future EPA limit.

Other Organic Compounds

With a very few exceptions, very little research has been carried out on the other organic chemicals known to be emitted to air from incinerators. Of the compounds, which have been studied, the focus has largely been directed at higher molecular weight compounds rather than the less persistent volatile organic compounds that are known to be emitted (Leach *et al.*, 1999). Compounds for which data have been reported include polycyclic aromatic hydrocarbons (PAHs) and several groups of highly toxic chlorinated compounds including the polychlorinated biphenyls (PCBs), the polychlorinated naphthalenes (PCNs), the chlorobenzenes and the chlorophenols.

PCBs: This group consists of 209 different individual congeners. Around half this number has been identified in the environment. PCBs are persistent, toxic and bioaccumulative. Accordingly, like the dioxins they have a tendency to build up in the fatty tissues of animals and humans where they can persist almost indefinitely. The more highly chlorinated PCB congeners are the most persistent and account for the majority of those found as environmental pollutants. PCBs have become globally ubiquitous chemicals, and they are even found at highly elevated concentrations in the tissues of animals living in what have traditionally been regarded as pristine environments. Arctic marine mammals, such as whales, seals and polar bears have been studied and the presence of PCBs together with other organochlorine contaminants confirmed (see Allsopp *et al.*, 1999). PCBs are known to exert a wide range of toxic effects on health including reproductive, neurological and immunological effects. They are suspected of causing many deleterious health effects both in wildlife and in humans (see Allsopp *et al.*, 1997, Allsopp *et al.*, 1999). Some PCB congeners also cause "dioxin-like" effects on health since they are structurally similar chemicals.

PCBs produced as industrial chemicals were mainly used for insulation in electrical equipment. Production of PCBs has almost totally ceased worldwide, although there are reports of it continuing in Russia. At least one third of the PCBs that have been produced are estimated to have entered the environment (Swedish EPA 1999). The other two thirds remain in old electrical equipment and in waste dumps from which they continue to leach into the environment. Although this is by far the major source of PCB pollution in the environment today, some PCBs are also produced as by-products of incineration and certain chemical processes involving chlorine.

PCBs are known to be formed in incinerators (Blumenstock *et al.*, 2000, Wikstrom 1998, Sakai *et al.*, 1996, Fangmark *et al.*, 1994) and are present in stack gases released to the atmosphere (Miyata *et al.*, 1994, Wilken *et al.*, 1993, Magagni *et al.*, 1991). Data on levels of PCBs in stack gases are, however, somewhat sparse in scientific literature. A study on MSW incinerators in Japan in 1992 found that emissions of the highly toxicologically significant coplanar PCBs varied considerably between different incinerators (Miyata *et al.*, 1994). The mean level (1.46 ng TEQ/m^3) was greater than the guideline (0.5 ngTEQ/Nm^3) for newly constructed incinerators in Japan. The study concluded that waste incinerators were a source of PCB contamination in humans, food and environment.

PCNs: A group of chlorinated compounds that are also persistent, bioaccumulative and toxic. When originally produced they were used in similar applications to PCBs which eventually superseded them. PCNs are known to be produced as unintentional by-products of thermal processes involving chlorine, including incineration and metal reclamation (see: Falandysz and Rappe 1997). PCNs have similar properties to dioxins and PCBs and many of them have high toxicity even at small doses (see: Abad *et al.*, 1999, Abad *et al.*, 1997).

PCNs have been found to be present in the stack gas of MSW incinerators. The concentration of PCN (mono- to octa-chlorinated) varied from 1.08 to 21.36 ng/Nm^3 in five MSW incinerators in Spain, while levels of dioxins varied from 0.01 to 5 ng ITEQ/Nm^3 (Abad *et al.*, 1999). In addition, PCN congeners exhibiting dioxin like toxicity have been identified in the emissions from municipal waste incinerators (Falandysz and Rappe 1997, Takasuga *et al.*, 1994).

PCNs from incineration and other combustion sources are present at detectable levels in wildlife and these processes may contribute a significant loading of these highly toxic and persistent chemicals to the environment (Falandysz and Rappe 1997, Falandysz *et al.*, 1996) in addition to the environmental burden resulting from historical manufacture.

Chlorinated Benzenes: Chlorinated benzenes are also formed in incinerators (Blumenstock *et al.*, 2000, Wikstrom *et al.*, 1998, Fangmark *et al.*, 1994), as are the chlorinated phenols (Wikstrom *et al.*, 1999). It has been shown that these chemicals are released in stack gases (Wilken *et al.*, 1993). The production of hexachlorobenzene (HCB), the fully substituted form of benzene, is of particular significance. HCB is persistent, toxic and bioaccumulative. It is toxic to aquatic life, land plants, land animals and humans and has been used extensively as a pesticide and as seed dressing. Recent research indicates that HCB can contribute significantly to the dioxin-like toxicity caused by organochlorine chemicals in human milk (van Birgelen 1998). It is listed by the IARC as a Group 2B carcinogen, *i.e.* it is possibly carcinogenic to humans and also appears to be a tumour promoter. HCB may damage the developing foetus, liver, immune system, thyroid, kidneys and central nervous system. The liver and nervous system are the most sensitive organs to its effects (ATSDR 1997, Newhook and Meek 1994).

Halogenated Phenols: 14 chlorinated, 3 brominated and 31 mixed bromo-chloro-phenols have been identified in MSW incinerator flue gas (Heeb *et al.* 1995). These chemicals are of considerable importance since dioxins can be formed by condensation reactions of two halogenated phenol molecules. The concentrations of mixed brominated and chlorinated phenols found in the raw combustion off-gas (4nmol/Nm^3; 1.2ug/Nm^3) and stack gas (1 nmol/Nm^3; 0.5ug/Nm^3) exceeded typical raw gas concentrations of the dioxins (0.2nmol/Nm^3; 0.1ug/Nm^3) in MSW incineration plant.

Brominated and Mixed Halogenated Dioxins: In addition to chlorinated dioxins and furans, numerous other halogenated compounds will be formed during incineration, including brominated and mixed chlorinated-brominated dioxins and furans.

Polychlorinated dibenzothiophenes: PCDBTs are sulphur-containing compounds that are structurally very similar to dibenzofurans. A sulphur atom substitutes for the oxygen atom found in the furan moiety of dibenzofuran structure. Little is known about their toxicology, but due to their structure they are suspected to be toxic. PCDBTs have been detected in the stack gas of waste incinerators (Sinkkonen *et al.*, 1991).

PAHs: A group of compounds which are produced as by-products of incomplete combustion of organic substances. Some are persistent, toxic and bioaccumulative; others are carcinogenic. PAHs are emitted by incinerators in stack gases (Yasuda and Takahashi 1998, Magagni *et al.*, 1991). Waste composition, temperature and excess air during the incineration process determine the quantity of PAHs emitted by a given facility. High emissions of PAHs have been shown to occur during start-up of incinerators (see Yasuda and Takahashi *et al.*, 1998). Measurements of total PAH incinerator emissions reported in one study were 0.02 to 12 µg/Nm^3 (see: Marty 1993).

VOCs: Few studies have been conducted on the vast array of other chemicals emitted from waste incinerators. However, one study has been undertaken specifically to identify and quantify volatile organic compounds (VOCs) in the stack gas of a

MSW incinerator (Jay and Stieglitz 1995). This study identified a total of around 250 different VOC compounds, for which concentrations ranged from 0.05 to 100 µg/m³. The list includes highly toxic and carcinogenic compounds such as benzene and the substituted phenols, together with other known toxic compounds such as phthalates. Data on the environmental and toxicological significance of many of the VOCs emitted are very limited, but VOCs are known to contribute to ozone formation in the lower atmosphere (see below).

Organic compounds emitted by incinerators are generally monitored on the basis of a group parameter which sums the total amount present in a sample of the flue gas: Total Organic Carbon (TOC). In the study reported by Jay and Stieglitz (1995), the 250 compounds identified were found to account for about 42% of the TOC. The remaining 58% were shown to consist of aliphatic hydrocarbons of unknown identity.

Leach *et al.*, (1999) noted that processes which generate large quantities of VOCs are of environmental significance since, mixed with nitrogen oxides and exposed to sunlight, they aid in the formation of photochemical oxidants (ozone and peroxyacyl nitrates), with deleterious impacts upon ambient air quality. The new EC limit for total VOC (expressed as carbon) is 10 mg/Nm³ as a daily average.

Heavy Metals

Heavy metals are emitted from all types of incinerators. Many heavy metals are known to be toxic at low concentrations and some are persistent and bioaccumulative. Heavy metals enter the incinerator as components of various materials in the raw waste. The process of incineration leads to their being concentrated by a factor of up to 10 in the waste residues (ashes) as the volume of waste is reduced through combustion (Buchholz and Landsberger 1995). A proportion of these toxic trace metals is emitted in the stack gases of incinerators to atmosphere. The major proportion is generally present in fly ash and bottom ash with the exception of mercury, where the greater proportion is vented via the flue stack.

Each metal has its own major source in the raw waste. Mercury is present due to the disposal of batteries, fluorescent light bulbs and paints (Carpi 1997). Cadmium is present in paints, PVC plastics and the pigments used to colour plastics. Lead is present in batteries, plastics and pigments (Valerio *et al.*, 1995, Korzun and Heck 1990), and antimony is present in flame-retardants (van Velzen and Langenkamp (1996) used in plastic items.

On a global scale, incineration contributes significantly to atmospheric emissions of many heavy metals, as shown in Table 20.3 (EEA 2000). Within the EU, figures for 1990 estimated incineration to be responsible for 8% (16t/yr) of all cadmium emissions and 16% (36t/yr) of mercury emissions. Emissions of chromium and lead amounted to 46 tonnes and over 300 tonnes respectively (EC 1998). A variety of flue gas treatment systems have been devised in order to reduce stack emissions of heavy metals (EEA 2000). Stack gas data for hazardous waste incinerators indicate that the fabric filter removal efficiencies (with the metals retained in the ash arisings) are in the order of 95% for most metals except mercury.

The EEA (2000) note that control of mercury emissions constitutes a special problem in incineration. Almost 100 % of the elemental mercury present in waste is emitted *via* the stack gases because it does not adsorb to filter dusts or ashes.

Elemental mercury comprises about 20-50% of the total mercury emitted. The remainder is in the form of divalent mercury which may be predominantly mercury chloride ($HgCl_2$). After emission to the atmosphere, divalent mercury, which is water soluble, may be deposited close to the incinerator. On the other hand, elemental mercury may be transported for long distances by atmospheric currents before it is eventually converted to the divalent form and subsequently deposited (Carpi 1997).

Despite the acknowledged significance of the fate of toxic heavy metals present in the waste-streams, published data on the concentrations of heavy metals in stack emissions appear to be very limited. Nonetheless, according to an emissions inventory in the Netherlands, stack emissions of cadmium and mercury were reduced considerably between 1990 and 1995 from MSW incinerators as a result of modernisation (Born 1996). During this period, the contribution to the total Dutch emissions of cadmium reduced from 44 to 13% and mercury from 53 to 11%. The reduction of atmospheric emissions (assuming the data are reliable) means that metals retained in the facility by pollution control devices will be retained in fly ash residues.

Table 20.3. Worldwide Atmospheric Emissions of Trace Metals from Waste Incineration

Metal	Emissions (1000 tonnes/year)	Emissions (as a % of total emissions from all sources)
Antimony	0.67	19.0
Arsenic	0.31	3.0
Cadmium	0.75	9.0
Chromium	0.84	2.0
Copper	1.58	4.0
Lead	2.37	20.7
Manganese	8.26	21.0
Mercury	1.16	32.0
Nickel	0.35	0.6
Selenium	0.11	11.0
Tin	0.81	15.0
Vanadium	1.15	1.0
Zinc	5.90	4.0

Particulate Matter

Minute particles of matter suspended in the air, often called particulates, are present as a result of both natural and human activities. Those of natural origin are derived from wind blown soil particles, sea salt, dusts from volcanic eruptions, spores from fungi and pollen grains from plants. Those from human activities are the result of combustion processes, such as coal-burning, incineration and vehicle exhaust. As a broad generalization, natural particulates are generally larger in size (>2.5 µm) than the finer particulates formed from combustion processes (<2.5 µm), (QUARG 1996, COMEAP 1995, EPAQS 1995). It is these finer particulates, known as "respirable particles," which are of great concern in relation to human health. Particulate pollution is implicated in the worsening of respiratory illnesses such as asthma, and increasing premature mortality from respiratory and heart diseases. This is because the respirable particulates are small enough to be inhaled into the extremities of the lung airways, whereas larger particles are prevented from reaching the deep airways by the respiratory system's protective mechanisms. Those particulates sized < 0.1 µm, termed ultrafine particles, are of greatest concern with regard to adverse effects on human health.

Incineration gives rise to emissions of particulates (EC 1998). Poorly controlled incineration plants can emit high levels of particulate matter and contribute to local environmental problems. Modern incinerators emit lower levels, but data suggest that the particulates emitted are fine in size and therefore could be contributing to adverse health effects (EC 1998). Indeed, the majority of particles formed from combustion processes, including all types of waste incineration, are ultrafine particles that are less than 0.1 μm in size. Even the most modern MSW incinerators do not have technology that prevents the release of ultrafine particles. Collection efficiencies for respirable particles (less than 2.5 μm) are between 5 and 30 % using current bag filter technology. For particles less than 1 μm in size, which includes all ultrafine particles, most will pass through incinerator filtration systems unabated. Furthermore, there are indications that some of the modern pollution abatement equipment installed in incinerators, particularly ammonia injection systems, which attempt to reduce oxides of nitrogen, may actually increase the emissions of the finest, most dangerous particles (Howard 2000).

At present, there is only limited information on the chemical composition of particulates. Emissions from incinerators include, for example, particles formed of mineral oxides and salts from the mineral constituents in the waste (Oppelt 1990). Heavy metals and organic chemicals such as dioxins, PCBs and PAHs can adhere onto the surface of the particles. Metals may absorb in a number of different forms including metal oxides, soluble salts and metal carbonates. The chemical nature of particulates, for instance, the form of metal or the type of other potentially toxic chemical adhered to the particle surface, may ultimately influence the effects on health resulting from exposure (QUARG 1996, Seaton 1995, Marty 1993).

Ultrafine particles have been found to be highly chemically reactive, even when they originate from material which itself is not reactive. This is due to their minute size and consequent high surface area. Research has shown that a proportionally higher number of surface atoms are present as the particle size decreases. This leads to their surface becoming highly charged and therefore chemically reactive. Ultrafine metal particles have been shown to be especially chemically reactive (Jefferson and Tilley 1999).

MSW incinerators typically have a mixed waste input containing heavy metals and halogenated organic compounds. They emit ultrafine metal particulates. Since these particles are especially reactive, it can be argued that MSW incinerators will therefore produce a more toxic ultrafine particulate aerosol than, for example, a coal-fired power station (Howard 2000). In this regard, emissions from incinerators may have long-term consequences for the health of the general public.

The new EC directive on incineration of waste does not give any specific limits for PM 2.5, that is respirable particles less than 2.5 µm. In this way the directive ignores the particulate pollution from incinerators which is of most relevance to public health. The directive does specify a limit for total dust emissions 10 mg/m^3 from incinerators. Data published in the 1980s gave emissions of particulate from UK MSW incinerators ranging from 18-4105 mg/m^3 (Williams 1990), and from US hazardous waste incinerators ranging from 4-902 mg/m^3 (Dempsey and Oppelt 1993). A recent study on MSW incinerators in Sweden reported emissions of 0.003 to 64 mg/m^3. Four out of 21 Swedish incinerators exceeded the EC limit on dust emissions (Greenpeace Nordic 2000).

Inorganic Gases

Inorganic acidic gases, notably hydrogen chloride (HCl), hydrogen fluoride (HF), hydrogen bromide (HBr), sulphur oxides (SOx), and nitrogen oxides (NOx) are formed and emitted by incinerators. These gases arise as a consequence of the elements chlorine, fluorine, bromine, sulphur and nitrogen being present in waste (Williams 1990). NOx are also formed as a result of the direct combination of nitrogen and oxygen, a process that is accelerated at high temperatures.

HCl is emitted in greater quantities from incinerators than from coal-fired power stations. This is due to chlorine in the waste, notably in the form of plastics such as PVC (Williams 1990). The new EC directive sets a limit (daily average value) of 10 mg/m^3 for HCl and 1 mg/m^3 for HF (EC 2000). A recent study of 21 Swedish MSW incinerators reported that emissions from 17 of them exceeded the EC limit, often to a substantial degree (Greenpeace Nordic 2000). The average release from the 21 incinerators was 44 mg/Nm^3 with a range of 0.2-238 mg/Nm^3.

Oxides of nitrogen (NOx), including nitrogen dioxide (NO$_2$), and oxides of sulphur (SOx), including sulphur dioxide (SO$_2$), are emitted from industrial combustion processes including all types of incinerators. These gases can also influence the pH of rain, making it acidic. Over time, acid rain can have deleterious effects on soil and water quality, and adversely affect ecosystems. Like exposure to particulate air pollution, exposure to NOx and SOx is also linked to adverse effects on respiratory health of individuals with pre-existing respiratory disorders. For instance, research has shown associations between increased air pollution levels of SO$_2$ and increased premature deaths in individuals who had pre-existing respiratory or cardiovascular illness. Similarly an association is evident with increased hospital admissions in individuals with pre-existing respiratory illness such as asthma or chronic obstructive pulmonary disease. Studies have also shown associations between exposure to NO$_2$ and worsened symptoms of respiratory illness, although the data are not conclusive (Ayres 1998).

NOx and SOx emissions also result in the formation of particulates, known as secondary particulates. The formation of secondary particulates occurs as a consequence of these gases undergoing chemical reactions in the atmosphere. They originate from the chemical oxidation of sulphur and nitrogen oxides in the atmosphere to acids, which are subsequently neutralised by atmospheric ammonia. The particles formed include ammonium sulphate and ammonium nitrate. These particles, which are generally soluble in nature, can persist in the air for long periods of time. A less abundant type of secondary particle is ammonium chloride, which originates from HCl gas. Like primary particles, secondary particles can have a wide variety of other potentially toxic organic compounds adsorbed onto their surfaces such as PAHs and dioxins (QUARG 1996, COMEAP 1995, EPAQS 1995). Similarly, secondary particulates are also thought to have adverse impacts on human health (e.g. see EC 1998).

NOx emission limits are included in the new EC directive. The limit (daily average value) for nitrogen monoxide and nitrogen dioxide, expressed as nitrogen dioxide, is 200 mg/m^3 (for existing incineration plant with a capacity exceeding 6 tonnes per hour, or new incineration plant). A recent study of 12 Swedish MSW incinerators documented emissions ranging from 1.2 – 236 mg/Nm^3. 4 of the 12 exceeded the EC limit.

The EC directive on incineration of wastes also includes a limit (daily average value) for sulphur dioxide of 50 mg/m^3. A recent report on 10 Swedish incinerators found that emissions ranged from 1.2 to 236 mg/Nm^3. Of the 10 incinerators, 9 of them had emissions that exceeded the EC limit (Greenpeace Nordic 2000).

Other Gases

Carbon dioxide (CO_2) is emitted by incinerators. Municipal waste contains around 25% by weight of carbon and this is released as CO_2 when waste is burned. Approximately one tonne of CO_2 is produced per tonne of waste incinerated. CO_2 is a greenhouse gas that affects climate change and releases have to be kept as small as possible (EEA 2000). There is no EC limit on emissions of CO_2 from incinerators.

Carbon monoxide is also released from incinerators. It is potentially toxic and is also a greenhouse gas. Research suggests that increases in CO levels in the air may be linked to health impacts in certain susceptible individuals with pre-existing heart disease (Ayres 1998). A recent study on Swedish incinerators found that of the 15 incinerators which recorded emissions, 10 exceeded the new EC limit of 50 mg/Nm^3 (Greenpeace Nordic 2000). Emissions ranged from 2.6 to 249 mg/Nm^3.

Emissions to Water

Incinerators emit wastes to water from cleaning equipment. Published scientific data on these emissions is very limited. Wastewater from wet exhaust gas cleaning contains heavy metals, the most significant in terms of quantity emitted and toxicity being lead, cadmium, copper, mercury, zinc and antinomy. Wastewater from wet slag removal equipment contains high levels of neutral salts and also contains unburned organic material from the residue (EEA 2000).

Emissions to Ashes

Ashes from waste incineration generally contain the same pollutants as air emissions, but may differ in concentration and composition (EEA 2000). Fly ashes and bottom ashes contain dioxins and heavy metals although, as for air emissions, little is known about many other compounds present in fly ash.

Organic Compounds

Information about the contents of organic compounds in bottom ashes is scarce, with the exception of dioxins for which there are some data (EEA 2000).

Dioxins

Dioxin emissions from incinerators to air and water have decreased in recent years due to improvements in pollution control equipment. However, it is difficult to tell whether the total emissions of dioxins from incinerators have declined at the same time. It is highly probable that while emissions to air via stack gases have decreased, emissions to ashes have increased. Indeed, it has been proposed that the total dioxin emissions from incineration probably have not been reduced greatly in recent decades (Wikstrom 1999). A theoretical assessment of the total emissions from a MSW incinerator in Sweden also found that a reduction of dioxins emitted in flue gases would result in an increase in concentrations in ash (GRAAB 1996). Thus, the total dioxin emissions from the plant would remain the same, regardless of improvements in air pollution abatement technology.

There are relatively few data regarding dioxins in fly ashes and bottom ashes because many installations are not obligated to control them (Fabrellas *et al.* 1999, Greenpeace Nordic 1999). A theoretical assessment of emissions from an incinerator in Sweden suggested that 97% of the total dioxin emissions from an incinerator would be emitted in ash. This is in close agreement with direct measurements from an incinerator (Spittelau) in Austria, which showed that 99.6% of the total dioxin emissions were in ash residues (Greenpeace Austria 1999). A study on a Spanish incinerator also noted that only a minor proportion of dioxin emissions is through stack gases, the majority being in ashes (Abad *et al.*, 2000). In addition to chlorinated dioxins, it is also likely that other halogenated dioxins and furans are present in ashes, as in flue gases, such as brominated and mixed chlorinated/brominated compounds. A study on medical and MSW incinerator fly ashes found results suggesting that iodinated dioxins are also likely to be present (Kashima *et al.*, 1999).

With regard to levels of dioxins in incinerator emissions, the highest levels have been found in fly ash. Levels characteristically range from parts per trillion (ppt) to parts per billion (ppb), (EEA 2000). Research on eight MSW incinerators in Spain found mean levels in fly ash between 0.07 and 3.5 ng I-TEQ/g (ppb) (Fabrellas *et al.* 1999). Another study on a MSW incinerator in Spain reported levels which fell within this range from two measurements, which were 0.37 and 0.65 ng I-TEQ/g (ppb) (Abad et al. 2000). Particularly high levels were reported for one Spanish incinerator in 1997 (41 ppb TEQ) although levels in 1999 were lower (Stieglitz *et al.* 1999).

Lower concentrations are apparent in bottom ash samples, typically ppt levels (EEA 2000). For instance, mean values for 3 MSW incinerators in Spain were 0.006, 0.013 and 0.098 ng I-TEQ/g (ppb), (i.e. 6, 13 and 98 ppt TEQ), (Fabrellas *et al.* 1999). Similarly, levels in bottom ashes from five MSW incinerators in Bavaria, Germany ranged from 1.6 to 24 ppt TEQ (Marb *et al.*, 1997). Ash from 18 new or updated medical waste incinerators in Poland sampled in 1994-7 had substantially higher levels of dioxins, ranging from 8-45 ppb TEQ (Grochowalski 1998).

Based on limited sampling, Abad *et al.* (2000) noted that, although the highest concentrations of dioxins are present in fly ash, the high production of bottom ash in incinerators means that the annual output of dioxins in bottom ash is comparable to that of fly ash. However, a study of eight MSW incinerators in Spain calculated that the overall output of dioxins was higher for fly ash (Fabrellas *et al.* 1999). The total yearly output of dioxins from 8 MSW incinerators reported to be operating in Spain, based on point measurements, were as follows: flue gases, 1-1.2, fly ashes, 46.6-111.6 and bottom ashes, 2-19 g I-TEQ/y (Fabrellas *et al.*, 1999).

As mentioned in the previous section, dioxin inventories most often underestimate emissions from incinerators because emissions in ashes are not included in calculations. A report on output of dioxins from Swedish incinerators has proposed that the Swedish EPA have grossly underestimated total incinerator emissions by underestimating emissions in ashes (Greenpeace Nordic 1999).

Other Organic Compounds

As previously discussed in this report, emissions of organic compounds to stack gases are multitudinous and fly ashes are similarly laden with numerous compounds. The EEA (2000) note that fly ash contains concentrated organic compounds, such as PAHs and soot as well as chlorinated organic compounds. PCBs are known to be present in fly ash (Sakai *et al.*, 1996). PCBs were reported to be detected in fly ash of hospital and MSW waste incinerators (Magagni *et al.*, 1994), and in sewage sludge incinerator fly ash and bottom ash (Kawakami *et al.*, 1998). The level of PCBs in fly ash from sewage sludge incinerators was 7.1 ng/g with the proportion of PCBs to dioxins being similar to that found in MSW incinerators. PCNs have also been identified in incinerator fly ash (Schneider *et al.*, 1998).

A study on fly ash from MSW incinerators identified 72 different phenolic compounds in the ash including many unknown ones (Nito and Takeshita 1996). Most of the compounds were hydroxy compounds of PAHs, polychlorinated PAHs, PCBs and dioxins. The study noted that some of these halogenated hydroxy compounds may be persistent and toxic and that their toxicities should be evaluated because they will be leached from fly ash into the environment after disposal in landfills. Another study identified many new kinds of aza-heterocyclic hydrocarbons (azaarenes and other basic compounds in fly ash (Nito and Ishizaki 1997). These compounds are produced by incomplete combustion and this study confirmed that incinerators are a source of them. The study identified 63 and 18 kinds of azaarenes from two different fractions of fly ash respectively. Of these compounds, quinoline, alkylquinoline, benzoquinoline, benzacridine, azapyrene, azabenzopyrene, phenylpyridine, biphenylamine and their isomers comprised the majority. Of concern is that many of them are known to be carcinogenic or mutagenic compounds. Leaching of such compounds from fly ash into landfills would release these toxic chemicals into the environment.

Heavy Metals

Both fly ash and bottom ash residues from incinerators contain many heavy metals. Fly ash generally has higher metal concentrations than bottom ash if the large, unburned metal fragments from the bottom ash are excluded (Bucholz and Landsberger 1995). Table 20.4 shows concentrations of heavy metals detected in fly ash and bottom ash from two Spanish MSW incinerators (Alba *et al.*, 1997) and table 20.5 shows concentrations detected in ashes from a US incinerator (Bucholz and Landsberger 1995). The concentrations of heavy metals in incinerator ashes are very high compared to background levels in the environment. For instance, if concentrations in bulk ash (combined fly + bottom ash) are compared with average concentrations of heavy metals found in soil globally, it is clear that bulk ash contains elevated amounts of many metals (Bucholz and Landsberger 1995). In addition, the process of incineration greatly enhances the mobility and bioavailability of toxic metals compared with raw municipal waste (Schumacher *et al.*, 1998). Consequently, there is greater potential for leaching of metals into the environment from ashes dumped in landfills than from ordinary waste (see section 2.4.1).

A study on incinerator ashes from veterinary college incinerators in which animal carcasses are burned found that levels of metals varied considerably between incinerators (Thompson *et al.*, 1995). Generally levels of metals in the ashes were much lower than levels found in MSW incinerator ashes. One exception was zinc, which was at a similar level. It was noted that burning of plastics in the waste may contribute to lead and zinc content in the ashes.

Given that incinerator companies are not required under national laws in many countries routinely to monitor ashes, published data on heavy metal levels in ashes

Table 20.4. Ranges of elemental abundance in MSW incinerator ashes and in soil. All concentrations are in mg/kg unless otherwise specified.

Element	Fly ash	Bottom ash	Soil
Ag	46-55.3	17.5-28.5	0.1
Al	3.19-7.84%	6.20-6.68%	7.1%
As	269-355	47.2-52.0	6
Br	3830-3920	676-830	5
Cd	246-266	47.6-65.5	0.06
Co	11.3-13.5	65.2-90.3	8
Cr	146-169	623-807	100
Cu	390-530	1560-2110	20
Hg	59.1-65.0	9.1-9.7	0.03
In	1.50-1.67	0.45-0.71	0.07
Mo	14-26	100-181	2
Pb	3200-4320	2090-2860	10
Se	6.7-11.2	<2.52	0.2
Sn	470-630	300-410	10
Th	2.85-3.21	4.31-4.86	5
Ti	3300-6300	7500-18100	5000
V	27-36	46-137	100
Zn	13360-13490	6610-6790	50

Source: Buchholz and Landsberger (1995).

Table 20.5. Minor and trace element concentrations in MSW incinerator residues

Element	Fly ashes (mg/kg dry residue)	Bottom ashes (mg/kg dry residue)
Cr	365 ± 18	210 ± 8
Zn	9382 ± 208	2067 ± 9
Pb	5461 ± 236	1693 ± 22
Ni	117 ± 2	53 ± 3
Cu	1322 ± 90	822 ± 4
As	<50	<50
Cd	92 ± 2	<12.5
Hg	0.29 ± 0.03	<0.035

Source: Alba *et al.,* (1997).

are sparse. A study on hazardous waste incinerators conducted the US revealed that the metals which most frequently exceeded regulatory limits were arsenic, nickel and lead (Dempsey and Oppelt 1993).

The Solution: Reduce, Re-use and Recycle and Phase Out Incineration

A lack of landfill space, tighter regulations restricting the quantity of waste going into landfills together with environmental problems from old landfills have driven municipalities in many countries to look for new methods of handling waste. Presently, 60% of waste generated throughout countries in the European Union goes into landfills (Hens *et al.*, 2000). This situation is made worse by the growing amount of waste being generated. For example:

- total waste production in the EU rose by nearly 10% between 1990 and 1995 and a further 20% increase has been predicted by 2010 (EEA 1999).
- In Estonia, Slovenia, Lithuania, Slovakia, Bulgaria, Hungary, Czech Republic, Romania and Poland, economic growth may lead to a doubling of municipal waste generation by 2010 (EEA 1999).
- In Asia, municipal waste from urban areas is predicted to double by 2025 (World Bank 1999).

One of the chosen methods for dealing with the current waste crisis is incineration, for it reduces the volume of solid waste, thereby lessening the burden on landfills. However, incinerators are not the solution to the waste problem. They are symptoms of non-existent and/or ill-conceived policies for the management of material resources. In a world of shrinking resources, it is irrational to let valuable resources "go up in smoke," and doubly so when the smoke is known to carry persistent and other hazardous chemicals. Incineration cannot be regarded as a sustainable technology for waste management and has no place in a world striving to change towards zero discharge technologies.

Incineration has already been banned by the government of Philippines, a move primarily instigated by public opposition to incineration. The Philippines is the first country in the world to ban incineration on a national scale. The Philippine Clean Air Act of 1999 specifically bans the incineration of municipal, medical and hazardous wastes and recommends the use of alternative techniques (for municipal waste) and non-burn technologies. Waste reduction, re-use and recycling are being promoted. The Clean Air Act mandates a three-year phase out period for existing medical incinerators, and during this time, limits hospitals to only incinerate infectious waste.

Problems of Incineration

Environment and Health

No matter how modern an incinerator it inevitably results in the release of toxic emissions to air and the production of toxic ashes and residues. This leads to contamination of the environment and to potential exposure of animals and humans to hazardous pollutants. Many hazardous compounds are released from incinerators including organic chemicals such as chlorinated and brominated dioxins, PCBs and PCNs, heavy metals, sulphur dioxide and nitrogen dioxide. Furthermore, innumerable substances are emitted which are of unknown toxicity. The entire impact on human health of exposure to the whole mixture of chemicals emitted from incinerators is unknown. However, several studies imply that individuals who work at waste incinerators and who live near incinerators have suffered from increases in the rate of mortality as well as many diseases and effects that diminish the quality of their lives. Although other studies report no such associations, it is important to remember that against the numerous confounding factors of lifestyle and experimental design, it may be very difficult to detect and quantify impacts on exposed populations (Hu and Shy 2001). Moreover, a prestigious scientific body has recently expressed *substantial* concern about the impacts of incinerator-derived dioxin releases on the health and well-being of broader populations, regardless of the implementation of maximum achievable control technology (NRC 2000).

Economics

The economics of waste management in general, and in particular incineration, are extremely complex and are outside the scope of this report. It has been noted that incineration is a technology of the previous industrial era and is only economically feasible if much of its cost is externalised i.e. borne by the general public. Pollution control constitutes the major proportion of the cost, but using such technology to reduce the toxic emissions to the air cannot help but redistribute them back to deposits in the ash.

A recent trend has been to generate electricity from burning waste in MSW incinerators. This can only be seen as a by-product of incineration and not a contributor to sustainable energy production. Indeed, incinerators burning MSW, are inefficient energy producers, with only 20% of the energy generated by the waste usually being captured. Murray (1999) has described incineration as inefficient both as a disposal option and as an energy generator. It leads not to material conservation and hazard reduction but to material destruction and hazard creation.

In the UK, a situation has arisen whereby contracts with incinerator operators lock local authorities into long term commitments to provide huge amounts of waste each year. This works against waste prevention, re-use and recycling since local authorities would have to pay financial penalties to incinerator owners if waste was reduced and diverted to re-use/recycling schemes.

Sustainability

The Convention for the Protection of the Marine Environment of the North East Atlantic (the OSPAR Convention, formed from the amalgamation of the former Oslo and Paris Conventions) entered into force in March 1998 and covers 15 States of the North East Atlantic Region and the European Commission. At the OSPAR meeting held in Sintra in June 1998, Ministers agreed on a clear commitment for the cessation of release of hazardous substances within one generation (by 2020). In essence, the commitment means that a target has been set for the cessation of discharges, emissions and losses of hazardous substances and their substitution with non-hazardous alternatives. In practice this means a shift away from dirty technologies towards clean production and zero emission strategies. Incinerators can never comply with the zero emissions strategy or be classed as a clean production technology.

Current EU Policy and Waste Management

Waste policy in the EU widely accepts the hierarchy of waste management to be (in order of priority): waste prevention – re-use – recycling – thermal decomposition with energy recovery (i.e. incineration with energy recovery). In spite of this general consensus, and a growing coherence of this hierarchy in policy lines of individual EU member states as a consequence of EU-Directives, the majority of waste in Europe is either landfilled or incinerated. Importantly, these are the methods which also entail the highest and most serious environmental and health risks (Hens *et al.*, 2000).

A move towards a waste policy aimed at reducing health effects must put more emphasis on prevention and re-use. The current EU waste policy is not founded upon public health protection. Fortunately the available data show how waste management even coincide with the hierarchy proposed by the EU (Hens 2000). For example, waste prevention is deemed to be the most important (no waste equals no health effects), followed by re-use and recycling. Despite this, the lack of consideration of the environment and human health is clearly visible in EU policy. For instance, regulations put in place for incineration by the EU together with national limits on this issue, are based more on what is technically achievable rather than on health and environmental protection goals.

Although emission limits set in the new EU directive have resulted in the closure and upgrading of some older incinerators in European countries, the policy itself is already outdated with regard to the OPSPAR agreement to phase out the releases of all hazardous substances within one generation. The EU directive is based on the conception that small releases of hazardous substances are acceptable. This is the conventional (though misguided) approach which proposes that chemicals can be managed at "safe" levels in the environment. However, it is already widely accepted, that there are no "safe" levels of many environmental chemical pollutants such as dioxins, other persistent, bioaccumulative and toxic chemicals and endocrine disruptors. In addition, the abandonment of the principle is increasing in political circles. For instance, with regard to incineration, the UK environment minister, Mr. Michael Meacher, recently recognised the nature of the problem when in answer to a parliamentary question he replied:

Q440 . . . "I repeat that emissions from incinerator processes are extremely toxic. Some of the emissions are carcinogenic. We know scientifically that there is no safe threshold below which one can allow such emissions" (cited in Howard 2000).

Despite the commitment by the OSPAR Convention for the cessation of releases of all hazardous substances by year 2020, a recent trend for plans to build new incinerators by the government in the UK and other European countries continues.

The Way Forward: Adoption of the Precautionary Principle and Zero Emissions Strategy

Adoption of the Precautionary Principle

The precautionary principle acknowledges that, if further environmental degradation is to be minimized and reversed, precaution and prevention must be the overriding principles of policy. It requires that the burden of proof should not be laid upon the protectors of the environment to demonstrate conclusive harm, but rather on the prospective polluter to demonstrate no likelihood of harm. The precautionary principle is now gaining acceptance internationally as a foundation for strategies to protect the environment and human health (McIntyre and Mosedale 1997, Santillo et al. 1999).

Current regulation for incinerators is not based on the precautionary principle. Instead it attempts to set limits for the discharge of chemicals into the environment which are designated as "safe". In the current regulatory system the burden of proof lies with those who need to 'prove' that health impacts exist before being able to attempt to remove the cause of the problem and not with the polluter themselves (Nicolopoulou-Stamati *et al.*, 2000). Based on knowledge regarding the toxic effects of many environmental chemical pollutants, which have accumulated over recent decades, a more legitimate viewpoint is that *chemicals should be considered dangerous until proven otherwise.*

We have now reached a situation, and indeed did some time ago, where health studies on incineration have reported associations between adverse effects on health and residence near, or employment at, incinerators. These studies are warning signs which should not result in government inactivity, but should lead to decisions being taken which implement the precautionary principle. There is already sufficient evidence that human health suffers from environmental contamination to justify phasing out of the incineration process as a substantial contributor. To wait for further proof from a new generation of incinerators from an already harmful and dirty technology would be to disregard the environment and human health.

Adoption of Zero Discharge

The aim of "zero discharge" is to halt environmental emissions of all hazardous substances. Although it is sometimes discussed as being simplistic or even impossible, it is a goal whereby regulation can be seen as resting places on the way to achieving it (Sprague 1991).

Zero discharge necessitates the adoption of clean production techniques both in industry and agriculture. It is essential that the change to clean production and material use should be fully supported by fiscal incentives and enforceable legislation.

The principle of clean production has already been endorsed by the Governing Council of the UNEP and has received growing recognition at a wide range of international fora. For instance, the adoption of the one generation goal for the phase out of releases of all hazardous substances by the OSPAR Convention in 1998 necessitates instigating clean production technology under a zero discharge strategy.

In terms of waste management strategies, incineration is a dirty technology that can never fulfil the criteria of zero discharge. The way forward for waste management, in line with a zero emissions strategy and hence towards sustainability, lies in waste prevention, re-use and recycling.

Implementation of Reduce, Re-use and Recycle

We live in a world in which our resources are generally not given the precious status by industry and agriculture which they deserve. In part, this has led to the creation, particularly in industrialized countries, of a "disposable society" in which enormous quantities of waste, including "avoidable waste" are generated. This situation needs to be urgently changed so that the amount of waste produced both domestically and by industry is drastically reduced.

Ways to help waste reduction include the use of economic instruments and environmental taxes. The use of these measures is supported by the EC and a number environmental taxes are already in place in several European countries (Steenwegen 2000). However, far more action is presently required to stimulate the change needed for much more waste reduction to become a reality.

Current levels of recycling in European countries vary considerably. For instance, the Netherlands recycles 46% of municipal waste whereas the UK manages just 8%. Intensive re-use and recycling schemes could deal with 80% of municipal waste. It is recognized that fiscal measures can play a considerable role in encouraging re-use and recycling schemes whilst discouraging least desirable practices such as incineration and landfill (Steenwegen 2000).

Measures to be taken in the drive towards waste reduction, re-use and recycling, and therefore towards lessening the adverse effects on health from waste management should include:

- The phase out of all forms of industrial incineration by 2020, including MSW incineration. This is in line with the OSPAR Convention for the phase out of emissions of all hazardous substances by 2020.
- Financial and legal mechanisms to increase re-use of packaging (e.g. bottles, containers) and products (e.g. computer housings, electronic components).
- Financial mechanisms (such as the landfill tax) used directly to set up the necessary infrastructure for effective recycling.
- Stimulating markets for recycled materials by legal requirements for packaging and products, where appropriate, to contain minimum amounts of recycled materials.
- Materials that cannot be safely recycled or composted at the end of their useful life (for example PVC plastic) must be phased out and replaced with more sustainable materials.
- In the short term, materials and products that add to the generation of hazardous substances in incinerators must be prevented from entering the waste stream at the cost of the producer. Such products would include electronic equipment, metals and products containing metals such as batteries and florescent lighting and PVC plastics (vinyl flooring, PVC electrical cabling, PVC packaging, PVC-*u* window frames etc) and other products containing hazardous substances.

and more generally:

- Further the development of clean production technologies which are more efficient in terms of material and energy usage, produce cleaner products with less waste and which ultimately can operate in a "closed loop" configurations to serve the needs of society in a more equitable and sustainable manner;
- Implement fully the Precautionary Principle, such that, in the future, we may be better able to avoid problems before they occur. The continuation and further development of scientific research has a fundamental role to play in identification of potential problems and solutions, but we must be ready to take effective precautionary action to prevent environmental contamination and degradation even in the face of considerable and often irreducible uncertainties.

REFERENCES

1. Alba N., Gasso S., Lacorte T. and Baldasano J.M. (1997). Characterization of municipal solid waste incineration residues from facilities with different air pollution control systems. *J. of the Air and Waste Management Assoc.*, **47**: 1170-1179.

2. Abad E., Caixach J. and Rivera J. (1997). Dioxin like compounds from MWI emissions: assessment of polychlorinated naphthalenes presence. *Organohalogen Compounds*, **32**: 403-406.

3. Abad E. Caixach J. and Rivera J. (1999). Dioxin like compounds from MWI emissions: assesment of the presence of polychlorinated naphthalenes. *Chemosphere*, **38** (1): 109-120.

4. Abad E., Adrados M.A., Caixach J., Fabrellas B. and Rivera J. (2000). Dioxin mass balance in a municipal waste incinerator. *Chemosphere*, **40**: 1143-1147.

5. Alcock R., Gemmill R. and Jones K. (1998). An updated UK PCDD/F atmospheric emission inventory based on a recent emissions measurement programme. *Organohalogen Compounds*, **36**: 105-108.

6. Allsopp M., Santillo D. and Johnston P. (1997). *Poisoning the Future: Impact of Endocrine-Disrupting Chemicals on Wildlife and Human Health*. Greenpeace International.

7. Allsopp M.,. Santillo D., Johnston P. and Stringer R. (1999). *The Tip of the Iceberg: State of Knowledge of Persistent Organic Pollutants in Europe and the Arctic*. Greenpeace International. ISBN 90-73361-53-2.

8. Angerer J., Heinzow D.O. Reimann W., Knorz W. and Lehnert G. (1992). Internal exposure to organic substances in a municipal waste incinerator. *Int. Arch. Occup. Environ. Impact Assess. Rev.* **8**: 249-265. (Cited in NRC 2000).

9. Ardevol E., Minguillon C., Garcia G., Serra M.E., Gonzalez C.A., Alvarez L., Eritja R. and Lafuente A. (1999). (Environmental tobacco smoke interference in the assessment of the health impact of a municipal waste incinerator on children through urinary thioether assay. *Public Health*, **113**: 295-298.

10. ATSDR (1997). ATSDR's toxicological profiles on CD-ROM. U.S. Department of Health and Human Services, Public Health Service, CRC Press Inc, Boca Raton.

11. Ayres J.G. (1998). Health effects of gaseous air pollutants. In: *Air Pollution and Health. Issues in Environmental Science and Technology*, R.E. Hester and R.M. Harrison (eds.). The Royal Society of Chemistry. ISBN 0-85404-245-8.

12. Blumenstock M., Zimmermann R., Schramm K.W. and Kettrup A. (2000). Influence of combustion conditions on the PCDD/F-, PCB-, PCBz and PAH-concentrations in the chamber of a waste incineration pilot plant. *Chemosphere*, **40**: 987-993.

13. Born J.G.P. (1996). Reduction of (dioxin) emissions by municipal solid waste incineration in the Netherlands. *Organohalogen Compounds*, **27**: 46-49.

14. Brereton C. (1996). Municipal solid waste – incineration, air pollution control and ash management. *Resources, Conservation and Recycling*, **16**: 227-264.

15. Buchholz B.A. and Landsberger S. (1995). Leaching dynamics studies of municipal solid waste incinerator ash. *J. of Air and Waste Management Association*, **45**: 579-590.

16. Buekens A. and Huang H. (1998). Comparative evaluation of techniques for controlling the formation and emission of chlorinated dioxins/furans in municipal waste incineration. *J. of Hazardous Materials*, **62**: 1-33.

17. Carpi A. (1997). Mercury from combustion sources: a review of the chemical species emitted and their transport in the atmosphere. *Water, Air, and Soil Pollution*, **98**: 241-254.

18. Chandler A.J., Eighmy T.T., Hartlen J., Hjelmar O., Kosson D.S., Sawell S.E., van der Sloot H.A. and Vehlow J. (1997). Studies in Environmental Science 67: Municipal solid waste incinerator residues. The International Ash Working Group (IAWG). Published by Elsevier 1997.

19. Chang N-B., Wang H.P., Huang W.L. and Lin K.S. (1999). The assessment of reuse potential for municipal solid waste and refuse-derived fuel incineration ashes. *Resources, Conservation and Recycling*, **25**: 255-270.

20. COMEAP, Committee on the Medical Effects of Air Pollutants (1995). Non-biological particles and health. Department of Health, UK. London: HMSO.

21. De Fre R. and Wevers M. (1998). Underestimation in dioxin inventories. *Organohalogen Compounds*, **36**: 17-20.

22. Dempsey C.R. and Oppelt E.T. (1993). Incineration of hazardous waste: a critical review update. *Air and Waste*, **43**: 25-73.

23. DETR (2000). Waste Strategy 2000, England and Wales, Part 1, Part 2. Published by Stationary Office Ltd. ISBN 010 146 932 2.

24. DoE/WO (1995) *Making waste work: A strategy for sustainable waste management in England and Wales*. UK Department of the Environment White Paper, CM3040, The Stationery Office, London.

25. EC (2000). Directive 2000/76/EC of the European Parliament and of the Council of 4 December 2000 on the incineration of waste. Official Journal of the European Communities L 332: 91-111.

26. EEA (1999). Environment in the European Union at the turn of the century.

27. EEA (2000). *Dangerous Substances in Waste*. Prepared by: J. Schmid, A.Elser, R. Strobel, ABAG-itm, M.Crowe, EPA, Ireland. European Environment Agency, Copenhagen, 2000.

28. ENDS (2000a). Regulatory foul-ups contributed to Byker ash affair. *Environmental Data Services Report 304 (May)*: 17-18.

29. EPAQS, *Expert Panel on Air Quality Standards*, (1995). Particles. Published by HMSO. ISBN 0 11 753199 5.

30. Fabrellas B., Sanz P., Abad E. and Rivera J. (1999). The Spanish dioxin inventory: Proposal and preliminary results from municipal waste incinerator emissions. *Organohalogen Compounds*, **41**: 491-494.

31. Falandysz J. and Rappe C. (1997). Specific pattern of tetrachloronapthalenes in black cormorant. *Chemosphere*, **35** (8): 1737-1746.

32. Falandysz, J., Strandberg, L., Bergqvist, P.-A., Strandberg, B. and Rappe, C. (1996).

33. Chloronaphthalenes in stickleback *Gasterosteus aculeatus* from the southwestern part of the Gulf of Gdansk, Baltic Sea. *Organohalogen Compounds*, **28**: 446-451.

34. Fangmark I., Stromberg B., Berge N. and Rappe C. (1994). Influence of postcombustion temperature profiles on the formation of PCDDs, PCDFs, PCBzs, and PCBs in a pilot incinerator. *Environmental Science and Technology*, **28** (4): 624-629.

35. Fiedler H. (1999). National and regional dioxin and furan inventories. *Organohalogen Compounds*, **41**: 473-476.

36. Fleming L.N., Abinteh H.N. and Inyang H.I. (1996). Leachant pH effects on the leachability of metals from fly ash. *J. of Soil Contamination*, **5** (1): 53-59.

37. Gass H.C., Jager E., Menke D. and Luder K. (1998). Long term study for minimization of the PCDD/PCDF – emissions of a municipal solid waste incinerator in Germany. *Organohalogen Compounds*, **36**: 175-178.

38. GRAAB (1996). Tekniskt underlag dioxinier, MU 96:10.

39. Greenpeace Austria (1999). *Waste incinerating plants in Austria*. Vienna, August 1999.

40. Greenpeace Nordic (1999). *Piles of Dioxin: Dioxin in ashes from waste incinerators in Sweden*. *Greenpeace*, Nordic, November 1999.

41. Greenpeace Nordic (2000). *Hot Air: Will Swedish Incinerators Satisfy the EU?*

42. Grochowalski A. (1998). PCDDs and PCDFs concentration in combustion gases and bottom ash from incineration of hospital wastes in Poland. *Chemosphere*, **37** (9-12): 2279-2291.

43. Hansen E. (2000). Substance flow analysis for dioxins in Denmark. Environmental Project No. 570 2000. MiljØprojeckt. (Danish Environmental Protection Agency).

44. Harada, M. (1997). Neurotoxicity of methylmercury; Minamata and the Amazon. In: *Mineral and Metal Neurotoxicology*. M. Yasui, M.J. Strong, K. Ota, and M.A. Verity, (eds.). CRC Press Inc., ISBN 0849376645.

45. Heeb N.V., Dolezal I.S., Buhrer T., Mattrel P. and Wolfensberger M. (1995). Distribution of halogenated phenols including mixed brominated and chlorinated phenols in municipal waste incineration flue gas. *Chemosphere*, **31** (4): 3033-3041.

46. Hens L., Nicolopoulou-stamati P., Howard C.V., Lafere J. and Staats de Yanes (2000). Towards a precautionary approach for waste management supported by education and information technology. In: Health Impacts of Waste Management Policies. *Proceedings of the seminar "Health Impacts of Waste Management Policies"*, Hippocrates Foundation, Kos, Greece, 12-14 November 1998. P. Nicolopoulou-Stamati, L. Hens and C.V. Howard. (eds.). Kluwer Academic Publishers.

47. Howard C.V. (2000). Particulate aerosols, incinerators and health. In: Health Impacts of Waste Management Policies. *Proceedings of the seminar "Health Impacts of Waste Management Policies"*, Hippocrates Foundation, Kos, Greece, 12-14 November 1998. Eds. P. Nicolopouiou-Stamati, L.Hens and C.V. Howard. Kluwer Academic Publishers.

48. Hu, S.-W. and Shy, C.M. (2001). Health effects of waste incineration: a review of epidemiological studies. J. Air and Waste Manage. Assoc. 51: 1100-1109.

49. Huang H. and Buekens A. (1995). On the mechanisms of dioxin formation in combustion processes. *Chemosphere,* **31** (9): 4099-4117.

50. Jay K. and Stieglitz L. (1995). Identification and quantification of volatile organic components in emissions of waste incineration plants. *Chemosphere,* **30** (7): 1249-1260.

51. Jefferson D.A. and Tilley E.E.M. (1999). The structural and physical chemistry of nanoparticles. In: (R.L. Maynard and C.V. Howard (eds.)). *Particulate Matter: Properties and Effects Upon Health,* BIOS Scientific Publishers Ltd., Oxford, UK. pp 63-84, ISBN 1-85996-172X. (Cited in Howard 2000).

52. Johnke B. and Stelzner E. (1992). Results of the German dioxin measurement programme at MSW incinerators. *Waste Management and Research,* **10**: 345-355.

53. Johnston P.A., Santillo D. and Stringer R. (1996). Risk assessment and reality: recognsing the limitations. In: *Environmental Impact of Chemicals: Assessment and Control.* M.D. Quint, D. Taylor and R. Purchase (eds.). Published by The Royal Society of Chemistry, special publication no. 176, ISBN 0-85404-795-6 (Chapter 16: 223-239).

54. Johnston P., Stringer R., Santillo D. and Howard V. (1998). Hazard, exposure and ecological risk assessment. In: *Environmental Management in Practice,* Volume 1: Instruments for Environmental Management. B. Nath, L. Hens, P. Compton and D. Devuyst (eds.). (pp. - 169-187) Publ. Routledge, London. ISBN 0-415-14906-1:

55. Kashima Y., Mitsuaki M., Kawano M., Ueda M., Tojo T., Takahashi G., Matsuda M., Anbe K., Doi R. and Wakimoto T. (1999). Characteristics of extractable organic halogens in ash samples from medical solid waste incinerator. *Organohalogen Compounds,* **41**: 191-194.

56. Kawakami I., Sase E., Tanaka M. and Sato T. (1998). Dioxin emissions from incinerators for sludge from night soil treatment plants. *Organohalogen Compounds,* **36**: 213-216.

57. Korzun E.A. and Heck H.H. (1990). Sources and fates of lead and cadmium in municipal solid waste. *J. of Air and Waste Management Association,* **40** (9): 1220-1226.

58. Leach J., Blanch A. and Bianchi A.C. (1999). Volatile organic compounds in an urban airborne environment adjacent to a municipal incinerator, waste collection centre and sewage treatment plant. *Atmospheric Environment,* **33**: 4309-4325.

59. Magagni A., Boschi G. and Schiavon I. (1991). Hospital waste incineration in a MSW combustor: chlorine, metals and dioxin mass balance. *Chemosphere,* **23** (8-10): 1501-1506.

60. Magagni A., Boschi G., Cocheo V. and Schiavon I. (1994). Fly ash produced by hospital and municipal solid waste incinerators: presence of PAH, PCB and toxic heavy metals. *Organohalogen Compounds,* **20**: 397-400.

61. Marb C., Hentschel B., Vierle O., Thoma H., Dumler-Gradl R., Swerev M., Schädel S. and Fiedler H. (1997). PCDD/PCDF in bottom ashes from municipal solid waste incierators in Bavaria, Germany. *Organohalogen Compounds,* **32** : 161-166.

62. Marty M.A. (1993). Hazardous combustion products from municipal waste incineration. *Occupational Medicine,* **8** (3): 603-619.

63. Mayer J., Rentschler W. and Sczech J. (1999). Long-term monitoring of dioxin emissions of a hazardous waste incinerator during lowered incineration temperature. *Organohalogen Compounds,* **41**: 239-242.

64. McIntyre, O. and Mosedale, T. (1997). The Precautionary Principle as a norm of customary international law. Journal of Environmental Law **9(2)**: 241.

65. Meeter P., Siebert P.C., Warwick R.O., Canter D.A. and Weston R.F. (1997). Dioxin emissions from soil burning incinerators. *Organohalogen Compounds,* **32**: 441-443.

66. Mitchell D.J., Wild S.R. and Jones K.C. (1992). Arrested municipal solid waste incinerator fly ash as a source of heavy metals to the UK environment. *Environmental Pollution,* **76**: 79-84.

67. Miyata H., Aozasa O., Mase Y., Ohta S., Khono S. and Asada S. (1994). Estimated annual emission of PCDDs, PCDFs and non-ortho chlorine substituted coplanar PCBs from flue gas from urban waste incinerators in Japan. *Chemosphere,* **29** (9-11): 2097-2105.

68. Murray R. (1999). *Creating Wealth from Waste,* 171pp. ISBN 1 898309 07 8.

69. Newhook, R. and Meek, M.E. (1994). Hexachlorobenzene: evaluation of risks to health from environmental exposure in Canada. Environmental Carcinogenesis and Ecotoxicology Reviews- *J. of Environmental Science and Health,* Part C 12(2): 345-360.

70. Nicolopoulou-Stamati P., Howard C.V. Parkes M. and Hens L. (2000). Introductory Chapter: Awareness of the health impacts of waste management. *Proceedings of the seminar "Health Impacts of Waste Management Policies",* Hippocrates Foundation, Kos, Greece, 12-14 November 1998. P. Nicolopoulou-Stamati, L. Hens and C.V. Howard (eds.). pp. 2-25. Kluwer Academic Publishers.

71. NIOSH (1995). (National Institute for Occupational Safety and Health). 1995. NIOSH Health Hazard Evaluation Report. HETA 90-0329-2482. New York City Department of Sanitation, New York. U.S. Department of Health and Human Services, Public Health Service, Centres for Disease Control and Prevention, National Institute for Occupational Safety and Health. (Cited in NRC 2000)

72. Nito S. and Takeshita R. (1996). Identification of phenolic compounds in fly ash from municipal waste incineration by gas chromatography and mass spectrometry. *Chemosphere,* **33** (11): 2239-2253.

73. Nito S. and Ishizaki S. (1997). Identification of azaarenes and other basic compounds in fly ash from municipal waste incinerator by gas chromatography and mass spectrometry. *Chemosphere,* **35** (8): 1755-1772.

74. Oppelt E.T. (1990). Air emissions from the incineration of hazardous waste. *Toxicology and Industrial Health,* **6** (5): 23-51.

75. Pastorelli G., De Lauretis R., De Stefanis P., Morselli L. and Viviano G. (1999). PCDD/PCDF from municipal solid waste incinerators in Italy: an inventory of air emissions. *Organohalogen Compounds,* **41**: 495-498.

76. Pietro P. and Giuliana D.V. (1999). Atmospheric emissions of PCDD/PCDFs from the municipal solid waste incinerator of Fusina (Venice). *Organohalogen Compounds,* **40**: 469-472.

77. Pluss A. and Ferrell R.E.Jr. (1991). Characterization of lead and other heavy metals in fly ash from municipal waste incinerators. *Hazardous Waste and Hazardous Waste Materials,* **8** (4): 275-292.

78. QUARG (1996). Airborne Particulate Matter in the United Kingdom. *Third Report of the Quality of Urban Air Review Group (QUARG),* May. ISBN 0 9520771 3 2.

79. Rowat S.C. (1999). Incinerator toxic emissions: a brief summary of human health effects with a note on regulatory control. *Medical Hypotheses,* **52** (5): 389-396.

80. Ruokojärvi P., Ruuskanen J., Ettala M., Rahkonen P and Tarhanen J. (1995). Formation of polyaromatic hydrocarbons and polychlorinated organic compounds in municipal waste landfill fires. *Chemosphere,* **31** (8): 3899-3908.

81. Sakai S., Hiraoka M., Takeda N. and Shiozaki K. (1996). Behaviour of coplanar PCBs and PCNs in oxidative conditions of municipal waste incineration. *Chemosphere,* **32** (1): 79-88.

82. Santillo, D., Johnston, P. and Stringer, R. (1999). The Precautionary Principle in Practice: a mandate for Anticipatory Preventative Action. In: "Protecting Public Health and the Environment: Implementing the Precautionary Principle" Raffensberge, C. and Tikne, J. (Eds.), Island Press: 36-50.

83. Sawell S.E., Chandler A.J., Eighmy T.T., Hartlen J., Hjelmar O, Kosson D., Van der Sloot H.A. and Vehlow J. (1995). An international perspective on the characteristisation and management of residues from MSW incinerators. *Biomass and Bioenergy,* **9** (1-5): 377-386.

84. Schneider M., Stieglitz L., Will R. and Zwick G. (1998). Formation of polychlorinated naphthalenes on fly ash. *Chemosphere,* **37** (9-12): 2055-2070.

85. Schreiber R.J. and Evans J.J. (1994). Dioxin emission results from recent testing at cement kilns. *Organohalogen Compounds,* **20**: 373-376.

86. Schuhmacher M., Granero S., Xifro A., Domingo J.L., Rivera J. and Eljarrat E. (1998). Levels of PCDD/Fs in soil samples in the vicinity of a municipal solid waste incinerator. *Chemosphere,* **37** (9-12): 2127-2137.

87. Schwind, K._H., Hosseinpour, J., (1988). Brominated/chlorinated dibenzo-p-dioxins and dibenzofurans. Part 1: Brominated/chlorinated and brominated dibenzo-p-dioxins and dibenzofurans in fly ash from a municipal waste incinerator. *Chemosphere,* **17** (9): 1875-1884.

88. Seaton A. (1995). Particulate air pollution and acute health effects. *The Lancet,* **345**: 176-178.

89. Shane B.S., Gutenmann W.H. and Lisk D.J. (1993). Variability over time in the mutagenicity of ashes from municipal solid-waste incinerators. *Mutation Research,* **301**: 39-43.

90. Shin D., Yang W., Choi J., Choi S. and Jang Y.S. (1998). The effects of operation conditions on PCDD/Fs emission in municipal solid waste incinerators: stack gas measurement and evaluation of operating conditions. *Organohalogen Compounds,* **36**: 143-146.

91. Sinkkonen S., Paasivirta J., Koistinen J. and Tarhanen J. (1991). Tetra- and pentachloro-dibenzothiophenes are formed in waste combustion. *Chemosphere,* **23** (5): 583-587.

92. Sprague J.B. (1991). Environmentally desirable approaches for regulating effluents from pulp mills. *Wat. Sci. Techno.,* **24**: 361-371.

93. Steenwegen C. (2000). Can Ecological taxes play a role in diminishing the health impacts of waste management? In: *Health Impacts of Waste Management Policies. Proceedings of the seminar "Health Impacts of Waste Management Policies",* Hippocrates Foundation, Kos, Greece, 12-14 November 1998. P. Nicolopoulou-Stamati, L.Hens and C.V. Howard. (eds.). Kluwer Academic Publishers.

94. Stieglitz L., Hell K., Matthys K., Rivet F. and Buekens A. (1999). Dioxin studies on a MSW-incinerator. *Organohalogen Compounds,* **41**: 117-120.

95. Swedish EPA (1998). *Persistent Organic Pollutants: A Swedish Way of an International Problem.* ISBN 91-620-1189-8.

96. Takasuga, T., Inoue, T., Ohi, E. & Ireland, P. (1994). Development of an all congener specific, HRGC/HRMS analytical method for polychlorinated naphthalenes in environmental samples. *Organohalogen Compounds,* **19**: 177-182.

97. Thompson L.J., Ebel J.G.Jr., Manzell K.L., Rutzke M., Gutenmann W.H. and Lisk D.J. (1995). Analytical survey of elements in veterinary college incinerator ashes. *Chemosphere,* **30** (4): 807-811.

98. Valerio F., Pala M., Piccardo M.T., Lazzarotto A., Balducci D and Brescianini C. (1995). Exposure to airborne cadmium in some Italian urban areas. *The Science of the Total Environment,* **172**: 57-63.

99. Van Birgelen, A.P.J.M. (1998). Hexachlorobenzene as a possible major contributor to the dioxin activity of human milk. *Environmental Health Perspectives,* **106**(11): 683-688.

100. Van den Hazel P. and Frankort P. (1996). Dioxin concentrations in the blood of residents and workers at a municipal waste incinerator. *Organohalogen Compounds,* **30**: 119-121.

101. Van Velzen D. and Langenkamp H. (1996). Antimony (Sb) in urban and industrial waste and in waste incineration. European Commission EUR 16435 EN.

102. Weber, L.W.D. and Greim, H. (1997). The toxicity of brominated and mixed-halogenated dibenzo-p-dioxins and dibenzofurans: An overview. *J. of Toxicology and Environmental Health,* **50**: 195-215.

103. Webster T. and Connet P. (1990). Risk Assessment: A public health hazard? *J. of Pesticide Reform,* **10** (1): 26-31.

104. Webster T. and Connett P. (1998). Dioxin emission inventories and trends: the importance of large point sources. *Chemosphere,* **37** (9-12): 2105-2118.

105. Wikstrom E. (1999). *The role of chlorine during waste combustion.* Department of Chemistry, Environmental Chemistry, Umea University.

106. Wikstrom E. Persson A. and Marklund S. (1998). Secondary formation of PCDDs, PCDFs, PCBs, PCBzs, PCPhs and PAHs during MSW combustion. *Organohalogen Compounds,* **36**: 65-68.

107. Wilken M., Boske J., Jager J. and Zeschmar-Lahl B. (1993). PCDD/F, PCB, chlorbenzene and chlorophenol emissions of a municipal solid waste incinerator plant (MSWI) – variation within a five day routine performance and influence of Mg(OH)$_2$-addition. *Organohalogen Compounds:* 241.

108. Williams F.L.R., Lawson A.B. and Lloyd O.L. (1992). Low sex ratios of births in areas at risk from air pollution from incinerators, as shown by geographical analysis and 3-dimensional mapping. *International J. of Epidemiology,* **21** (2): 311-319.

109. Williams P.T. (1990). A review of pollution from waste incineration. *J. of the Institute of Water and Environmental Management,* **4** (1): 2634.

110. World Bank (1999). The International Bank for Reconstruction and Development/THE WORLD BANK. "What a Waste: Solid Waste Management in Asia," Urban Development Sector Unit, East Asia and Pacific Region, Washington, D.C., June 1999.

111. Yamamura K., Ikeguchi T. and Uehara H. (1999). Study on the emissions of dioxins from various industrial waste incinerators. *Organohalogen Compounds,* **41**: 287-292.

112. Yasuda K. and Takahashi M. (1998). The emission of polycyclic aromatic hydrocarbons from municipal solid waste incinerators during the combustion cycle. *J. of Air and Waste Management,* **48**: 441-447.

Section VI

Social Impacts

21

Socially-responsive Energy from Urban Solid Wastes in Developing Countries

CHRISTINE FUREDY[1] AND ALISON DOIG[2]

ITDG: The Schumacher Centre for Technology and Development, Boutron Hall, Bourton-on-Dunsmore, Rugby, Warwickshire CV23 902, UK

Introduction

That technological development should serve the purposes of poverty alleviation and, in particular, that small scale technologies introduced to communities should bring benefits to the less wealthy has long been a tenet of appropriate technology. (Schumacher 1973; McRobie 1981).

Nevertheless, most technological projects and undertakings in cities of developing countries experience problems in incorporating social perspective. In the majority of cases, the community and it needs are not the starting point for technological imports. Technologies are developed by engineers and are introduced with support from an aid agency, the national government or the private sector. A project proposal may have a section referring to social acceptance or adaptation, or the 'education' needed to attain co-operation with the technology. Such nods towards social concerns are a far cry from the ideal of wedding technology to the needs of the poor.

The argument made here is that to co-ordinate social objectives and technological innovation one can rarely start with the application of a pre-selected technology. Rather, some basic questions have to be asked about the community's resources, technology options, and potential beneficiaries. The decision on what technology will be most appropriate should follow such an assessment. In most cases, socio-economic considerations must take priority over technological ones. This argument is made with particular reference to ideas for capturing the values inherent in urban solid wastes.

The social aspects of WtE projects are a concern of ITDG.[3] This chapter illustrates an approach that gives initial emphasis to social and economic considerations in designing and implementing small or intermediate technologies in cities of developing countries. Project work in Nairobi on briquettes from charcoal dust is ued to illustrate the methodology.

---

[1] Professor Emerita, Urban Studies, York University, Toronto, Canada.

[2] ITDG, Rugby, U.K.

[3] Formerly called International Technology Development Group.

The Allure of Energy from Urban Solid Wastes

The possibility of recovering value from solid wastes to create energy has a long history (Barnard and Kristoferson 1985). With the collection and disposal of solid wastes growing steadily more difficult for cities of developing countries, it is not surprising that WtE continues to appear, in principle, an attractive concept. Attempts at introducing WtE in these cities, however, have had little success.

It is not the purpose here to examine the technological difficulties of WtE with respect to urban solid wastes, but such difficulties serve to reinforce the social concerns about what is 'appropriate.'

It is broadly accepted in international solid waste management that municipal solid wastes in most cities of developing countries are too low in calorific value, have too much organic material, and are subject to too much seasonal variation in moisture content to permit conversion to energy by incineration (see UNEP-IETC 1996; Merritt chapter 2 in this book). Most proposed WtE plants have assumed that the feedstock would consist not of regular municipal solid wastes, but special streams of urban wastes that have a high content of combustible material. In notable cases such assumptions have proved to be wrong, as recyclable materials such as newsprint, office paper and plastics are usually diverted from the municipal system to recyclers. Landfill gas extraction, certainly a possibility where municipal wastes have high amounts of organic matter, as is the case in many cities, requires expensive infrastructure built into a designed disposal site, and there are few examples of effective gas extraction for energy production (UNEP-IETC 1996).

The few experiments with WtE incineration plants or other processes such as pelletisation of solid wastes for fuel have been plagued with planning and design problems. Government ministries and aid agencies that receive private sector proposals for funding of innovations, such as pelletisation, often lack the expertise to assess the proposals, which may not contain details of the waste streams where the technology is to be applied, information on the local demand for energy, or reliable cost-benefit studies. National ministries devoted to promoting alternative energy technologies may be more concerned with funding initiatives than assessing them. Lack of transparency and corruption may play a part in funding inappropriate ventures.

Most failures have had physical or technological deficiencies, but they have also, importantly, lacked social insight. They have not been planned together with the local stakeholders and possible beneficiaries. Statements about benefits to needy residents have been pro forma, lacking research to understand the behaviours, values and needs of local actors. Hence, even without the technical problems, it is hard to argue that support to past urban WtE in developing countries has been warranted.

Yet, even given the past performance of WtE projects, the principle of extracting the maximum value from wastes, thus reducing the solid wastes requiring treatment and disposal, remains important, especially in view of ever-increasing energy needs, and the desire to limit greenhouse gases. Successes in rural areas with small scale energy production suggest that ways can also be found in towns and cities to serve the goals of poverty alleviation, waste management, and environment improvement. The breakthrough in WtE in urban areas will depend more on waste generator co-operation and on social insight, than on advances in technology.

The Social Context of Solid Waste Reuse in Developing Countries

Conventional WtE projects in urban areas have not been geared to poverty reduction. Socially-responsive approaches aim to utilize untapped or under-used waste materials to alleviate resource needs of the urban poor. These initiatives seek first to understand how wastes are currently used.

Many poor people depend upon gathering and using urban wastes to meet basic needs and for livelihoods. In addition to the manufactured recyclable materials that are traded in large quantities in most cities of developing countries, urban organic wastes are also accessed and reused: green and food wastes are used as fodder and feed, and decomposed organics are applied to fields (Furedy 2001). Organic wastes are extensively used for fuel, not just in rural areas. Dung from urban cattles is an important fuel in Northern India (Barnard and Kristoferson 1985), and women and children waste pickers gather wood, coconut shells, cinders and coal dust both for household fuel and for sale to households and small enterprises (Furedy 1990, 1984).

There are several questions that should be carefully researched to ensure a socially-responsive approach in technology choice:

1. Are the 'wastes' designed to be exploited truly going to waste? What reusable wastes in the area are accumulating? Answering such questions requires waste stream analysis together with observations of informal waste recovery and reuse, livelihood and stakeholder analysis. In general, it is not thought desirable to interfere with existing livelihoods derived from waste re-use; that is, poor people should not lose work or access to resources on account of the introduced technology.

2. If wastes are being used, are there ways to increase the efficiency and income-earning potential of this use, with affordable technology, or by training in accessing raw materials or in marketing?

3. Are there alternative ways of using wasted of underutilized resources that would serve poverty alleviation better? With respect to urban solid wastes, this might mean examining whether composting of solid wastes, or treating selected wastes for animal feed are simpler, more efficient, and more viable routes to waste recovery linked to the livelihoods of the poor.

An organization devoted to appropriate technology is now attempting to make these questions central to small scale WtE undertakings.

Socially-Responsive Small Scale Energy Production from Urban Wastes

The ITDG is an NGO (founded by Dr. E.F. Schumacher in the 1960s) that specializes in helping people to use technology for practical answers to poverty. The organization became interested in WtE because, although urban areas have higher rates of energy use than rural ones, the urban poor do not have adequate access to electricity or other forms of clean energy. Moreover, they cannot readily collect biomass from their surroundings, and so are forced to use commercial fuels such as charcoal or kerosene. A high proportion of the income of households and small enterprises is spent in this way.

ITDG recognizes urban solid waste management problems and that some of the solid wastes requiring disposal have the potential for use in energy production. At the same time, the organization has noted that thousands of livelihoods in these cities are intimately linked to the collection, recovery and recycling of waste materials, whether in the informal economy, in linked formal sector enterprises, or among public sector employees.

Taking a socially-responsible view, ITDG has argued that "if sufficient consideration is given to the whole waste management system, particularly to the role of the poor as energy consumers who are also dependent upon waste for livelihood activities, energy access and solid waste management can be improved" (Doig 2001).

ITDG's project on Urban Waste Management for Small Scale Energy Production (part of an international project supported by UK Dept. for International Development Knowledge and Research) is investigating the opportunities for applying WtE technologies to support urban livelihoods[4]. Case studies are being developed in five countries (Senegal, Kenya, Nepal, Sri Lanka and Cuba). The case studies are considering waste streams, the stakeholders involved in waste mamangement livelihood opportunities from WtE, and different WtE technology options. In all cases, the focus is on low-value wastes that are not being extensively reused.

General Methodology of ITDG WtE Projects

The WtE projects are initiated with data gathering on the waste streams of the city. This includes characterization of broad waste streams (the sources, quantities and composition of different streams) and an assessment of 'low-value wastes', that is, ones that are not reused to any extent. These are the wastes that end up at disposal sites in significant amounts. (For wastes that are difficult to track at dumps, such as coal, charcoal and sawdust, point sources have to be investigated). Such an assessment is crucial for ITDG projects. Once it is established that there *are* significant quantities of wastes available, other questions can be addressed.

The energy needs assessment is the next step. This is designed to gain a broad picture of the energy supply and demand of the target groups, at national and city levels, and also the specific locality where a new technology may be implemented. The information analysed includes: the types and distribution of energy consumed; access of proposed beneficiaries to fuel; the markets for fuels; household expenditure on fuel against income; convenience of use by fuel type; fuel preferences of different groups.

This detailed analysis must be related to the WtE technologies under consideration. This means attempting to identify and quantify the energy service needs that could be met by the WtE technologies. Sound research on energy access and needs is needed to judge whether any intervention will benefit needy members of the community.

The livelihoods and stakeholder analysis combines an understanding of the major stakeholders in waste management and energy provision in the city with the analysis of livelihoods of the poorer communities in the study. Livelihood here refers to the means of gaining a living, including income-earning activities (whether formal or informal) and also the activities in which people engage to ensure long-term security.

---

[4] The project manager at ITDG is Smail Khennas. Authors wish to thank him for information and comments.

It includes people's access to assets and financial resources, their legal status, and rights to government services.

ITDG bases its livelihoods analysis on the CARE Household Livelihood Security framework (Sanderson 1999), which judges factors such as livelihood strategies, the rules and regulations pertaining to households, and vulnerability to natural disasters or significant social and economic changes.

The health impacts study gathers information on waste-related injuries and diseases, as well as assessing the probable health effects of introducing new techniques of waste reuse.

When a WtE technique appears applicable, there is a detailed assessment of it in the light of the local data including technical/physical requirements, health aspects, financial feasibility, and markets for the energy produced.

This model addresses the assets (social, human, economic, natural and physical) of the affected communities, and considers their capacity for adopting relevant changes. In addition, it also considers the wider context, the policies and institutional structures, which will influence the success of interventions. It helps to identify the non-technical inputs required to introduce the WtE technology, such as financing, capacity building, changes in policy, or local institutional development to support the intervention. The markets for the products of the new technology are investigated and the impact on various actors who will be affected by its introduction.

Although the application of this method is limited by the availability of background data and the capacity to conduct some of the wide-ranging research required, the methodology itself represents a substantial advance in small scale technology development.

Pilot Work on Briquettes from Charcoal Dust in Nairobi

Background

Nairobi was considered a suitable city to explore WtE production because it is experiencing drastic fuel shortages, as well as suffering a crisis in solid waste management. It is a city where there is some interest in organic waste recycling as the United Nations Centre for Human Settlements is supporting composting from market wastes through community groups.

Background research was conducted on the nature of solid wastes and the stakeholders for energy and waste management, including those primarily involved in waste recovery.

Access to Energy by the Poor

The main fuel used by urban small enterprises is charcoal. Commercial cooking is a very important source of livelihood for the urban poor; it accounts for 52% of energy-using activities. Other main activities are metal working (28%), backing (8%) and welding (4%). Firewood is of secondary importance, followed by kerosene. The entrepreneurs who own kiosks (commercial cooking) use sawdust, firewood and sometimes charcoal. A major source is residues from carpentry workshops. The main energy used by households for cooking in the urban poor areas is also charcoal; kerosene and firewood are secondary fuels.

The poorer families in the slums where monthly incomes are less than KES 3000 (US\$ 45), however, can hardly afford charcoal or kerosene. The survival strategy is to use charcoal dust and firewood. Often they reduce their cooking by eating in kiosks.

The wood for charcoal production is brought from rural areas and processed within city limits. Charcoal vendors are strategically placed in residential and industrial areas. Large amounts of charcoal dust accumulate at these dealers' shops. Each sack of charcoal (0.12 m³) generates about 10 litres (0.01 m³) of charcoal dust. It is estimated that about 10% of all the charcoal dust is disposed as waste, amounting to approximately 15000-20000 tons of dust per year.

Focus on Charcoal Dust Briquettes

Of the wastes that might be used for low-cost energy, charcoal dust was judged to be high preference for technical development.

Table 21.1 summarizes the pros and cons of the wastes considered in the Nairobi case.

Table 21.1. Waste-to-energy Options Assessed in Nairobi

Waste	Use/User	Preference/Priority
Waste paper	Briquettes: households	Medium preference: already in use and in limited production
Charcoal dust	Briquettes: households, commercial cooking, hotels	High preference: already in small commercial production and use
Vegetable waste, human excreta	Biogas: households	Low preference: not yet being used
Combustible waste	Heat recovery from incineration: small enterprises	Low preference: only in limited cases
Waste industrial oils	Small enterprises	Low preference, currently used for wood preservation in timber houses
Cooking oil from restaurants	Small enterprises	Low preferebce: not in use

Adapted from ITDG survey 2001

Charcoal dust briquetting is currently practised by women in Nairobi by combining charcoal dust with mud or other binders to create a low efficiency fuel. This is mostly for their own consumption, but they may also sell small quantities. It must be noted that this fuel is used by very low-income households, whose only alternatives for fuel are lower efficiency biofuels or no fuel for cooking at all. Originally, the charcoal dealers would gives this dust free, but with increased demand, they now sell it at Ksh. 10 (approx. 16 cents) for a 20-litre tin of dust. The dust when moulded into briquettes is adequate to provide for a week's cooking.

ITDG identified ways to assist these women, in particular by improving the technology for briquetting by introducing a simple low-cost hand press for production of briquettes and also by improving the ratio of charcoal dust to binder, to create a higher efficiency and cleaner fuel. The project aims to increase the opportunities for these women to create small enterprises and improve their marketing. The hope is to provide training to about 100 individuals, and that the improved technique will spread locally.

Monitoring Energy Projects

Most WtE projects in developing countries have not undertaken systematic socio-economic research in target areas, along the lines illustrated in the ITDG work. Even if the original planning was deficient in social awareness, however, monitoring can help to reorient projects, as long as there is willingness to considerably modify or even cancel the initiative if the monitoring shows that the process has negative consequences for the community's needy persons-that is, they will suffer difficulties in accessing energy or other resources upon which they depend for their livelihood and daily living.

In general, laboratory testing of the waste processing technology should be a pre-condition for application. If there are serious questions about the safety of a process, it should not be promoted, even if social conditions seem appropriate. For example, fuel briquettes or pellets made from mixed municipal wastes should not be promoted without prior testing, as the occurrence of much plastic and some toxic materials in municipal solid wastes throughout the world means it is highly likely that such fuel cannot be burned in homes and small industries. The standards for the safety of an energy process or product should, if anything, be higher for less developed countries, where supervision of the production over a long period of time is likely to more lax than in counties with high levels of education and training.

Conclusion

The potential for energy from solid wastes is not as great as has been assumed by some in the past. For the developing countries, bioconversion of wastes (mainly through compositing) is more feasible for small scale projects than WtE. However, there are specific options that are worth exploring. The most promising small scale projects are those that capture single-source or separate, relatively uncontaminated wastes such as charcoal, cinders, lumber-yard sawdust, organic market wastes, and food wastes from factories, hotels and restaurants. The ITDG is making headway in small scale projects with designated wastes, as illustrated here.

The next phase of the ITDG project is to implement WtE interventions in at least two of the case study countries. The implementation will not only entail the introduction of a technology, but the establishment of markets, financial, and institutional support, and other mechanisms to support an enterprise based on WtE. There will also be a detailed impact assessment of the interventions, using sustainable livelihood approaches. The aim is not only to develop a few individual enterprises, but to further knowledge and understanding of such small scale, socially-responsive WtE interventions.

This work illustrates opportunities to apply socially-responsive approaches to WtE technologies. Indeed, concern for the needs of local communities should be a priority in these projects. Advance in small scale WtE applications is more likely to come about through changes in waste generator behaviour, whereby suitable wastes are secured as separate streams, and through socially-aware planning, than via technological breakthroughs. In these ways, solid waste disposal requirements can be reduced and poorer residents can enhance their access to cheaper fuel or energy.

REFERENCES

1. Barnard, G. and Kristoferson, L. (1985). *Agricultural Residues as Fuel in the Thrid World.* London: Earthscan.

2. Doig, Alison. (2001). "Urban waste management for small scale energy production." Unpublished report. Rugby: ITDG.

3. Furedy, Christine. (2001). "Reducing health risks of urban organic solid waste use." *Urban Agriculture Magazine,* **1** (3) March: 23-25.

4. Furedy, Christine. (1990). *Social Aspects of Waste Recovery in Asian Cities. Environmental Sanitation Reviews* series, No. 30. Bangkok: Environmental Sanitation Information Centre.

5. Furedy, Christine. (1984). "Survival strategies of the urban poor: scavenging and recuperation in Calcutta": *GeoJournal,* **8:** 129-136.

6. ITDG. (2001). "Fuel use survey in Nairobi." Unpublished report.

7. McRobie, George. (1981). *Small is Possible.* New York: Harper and Row.

8. Sanderson, David. (1999). "Briefing notes on household livelihood security in urban settlements". CARE International UK. Unpublished.

9. Schumacher, E.F. (1973). *Small is Beautiful.* London: Vintage.

10. (UNEP-IETC) International Environmental Technology Centre, United Nations Environment Programme. (1996). *International Source Book on Environmentally-Sound Technologies for Solid Waste Management.* Edited by Larry Rosenberg and Christine Furedy. Osaka: UNEP/International Environmental Technology Centre.

Section VII

Marketing Issues of Waste to Energy Technologies

22

The Market for Small-Scale Waste Gasification/Pyrolysis Projects— Preliminary Scoping Study

Sarah Knapp

M.E.L. Reasearch Ltd, Birmingham, UK

Preface

This report presents the findings of a market analysis for small waste gasification plant in the UK. The project was commissioned by the Energy Technology Support Unit (ETSU) at AEA Technology, Harwell UK, and funded by the UK Department of Trade and Industry under its New and Renewable Energy Programme. The work was carried out by M.E.L. Research Ltd. UK in the autumn of 1998.

The overall objectives of the study were to assess the potential merits of establishing a demonstration plant in the UK, and to identify the size, profile and characteristics of the potential market.

The work drew upon the views of the potential suppliers and buyers of the technology. The views expressed in this report are however those of M.E.L. Research and are not necessarily shared by any other party.

Since 1996 the Energy Technology Support Unit (ETSU) noticed a growth in interest in small waste to energy plant, namely gasification/pyrolysis technologies. Enquiries had come from both local authorities and waste management contractors. In general, what was being sought was a technology that could convert paper and other high calorific value (CV) fractions from Materials Recycling Facilities (MRFs), industrial wood waste and commercial and packaging waste, with the overriding plan to use these technologies as part of an integrated waste management strategy. Others saw an opportunity to use relatively clean material to produce power for the Non Fossil Fuel Obligation (NFFO) market.

During the last fifteen years, major efforts have been made by technology suppliers to develop controlled waste gasification/pyrolysis processes in reactor systems for the high CV fraction of household and commercial waste. These technologies have been surveyed in a recent ETSU publication, ETSU/B/RR/00434/REP and other work such as the 1997 Juniper Report '*The European Market for Pyrolysis and Gasification of Waste: A Technology and Business Review*' authored by Dr. Kevin Whiting. Both reports show that emerging new technologies have the capability to deal effectively with refined municipal waste feedstocks in the 20 - 50 (Kilo tonnes per annum) bracket.

The main feature of these technologies is that the feedstock is converted by heat into an intermediate gas (which is then cleaned before burning), and a stabilized organic fraction. The environmental aspects of such processes are attractive due to effective odour control, as the process takes place in an enclosed vessel, avoidance of insects, rodents and birds as well as minimal waste water production and a high degree of hygiene and stabilisation. This being the case these technologies may lead to environmentally acceptable routes to energy from waste that is cost effective at smaller scales than conventional combustion technology. As such this type of project may be seen as an acceptable opportunity to generate substantial quantities of renewable electricity.

In adopting an integrated 'systems' approach to waste management, it is clear that although recycling can often deal effectively with paper, plastics, glass, cans and putrescible elements of the municipal waste stream, small-scale gasification/ pyrolysis plants may provide another tool in the waste manager's kit to deal with other elements, such as wood, paper that cannot be recycled and other high CV elements from trade sources. Thus, the route to landfill—an ever-decreasing resource may be avoided. With the arrival of the EU's Landfill Directive and the necessity to divert biodegradable waste from landfill, the need to find alternative routes for processing waste has become more pressing.

The application of such processes is likely to be necessary therefore to assist in reaching the national targets outlined in the *UK Government Waste Strategy 2000* paper. In the long run, the achievement of core national strategic waste management goals, and targets such as recovering value from 40% of the municipal waste stream, will require the implementation of schemes such as those considered in this scoping study.

Key Aims

This was a scoping study, and as such, involved carrying out a broad but comprehensive market research exercise, with the aim of identifying the likely demand and drivers for the purchasing of small gasification/pyrolysis plant. The research was broken down into appropriate market segments and sub-segments e.g. local authority and inter-authority or waste management company partnerships.

The core aims were:
- *to establish the approximate size and profile of the UK market for 'small waste to energy' projects based on gasification/pyrolysis;*

  The project assumed a feedstock in the 20 - 50 ktpa bracket. As a result not all local authorities would have sufficient municipal waste collected arisings of the quantities and types of waste required to maintain a plant of such size. Supplementary feedstock collected by local authorities may include paper rich commercial or trade waste, and possibly other items. For example, wood waste is a large component of General Industrial Waste and may only be found in sufficient quantities in larger towns. However, some rural areas may have access to large quantities of waste wood due to forestry activity. It was hoped that detailed research of the waste components handled by local authorities would establish typical niche market buyer profiles of small gasification/pyrolysis plants, such as unitary authorities in urban areas, or, of equal importance, it may suggest that smaller plant sizes should be considered.

Once these typical buyer profiles were established, estimates of the resource available for this type of plant could be established, on the basis of local authority municipal waste arisings statistics, available through the 1997/98 National Municipal Wastes Management survey. Estimates of utilizable resource for typical Waste Disposal Authorities and Unitary Authorities were made, taking into account recycling rates and materials currently diverted. If appropriate these figures were aggregated to the national level to produce estimates of the overall quantities potentially diverted, if the plant was fully utilised in all appropriate circumstances.

- *to determine what, if any, barriers to the take up of this concept pre-existed.*

Barriers would include local authority political factors impinging on the siting of this type of plant. Some local authorities may identify NIMBY factors as important public pressures preventing the siting of such a plant. Other authorities may be strongly committed to the diversion of waste from landfill, owing to the need to 'recycle' more and generate energy from waste as a better environmental option. On a more fundamental level there may be a lack of sufficient levels of feedstock or insecurity of supply; alternatively, authorities may be already tied into long-term contracts. Other perceived barriers may be specific to technological investments, uncertainty of required standards of performance or contracting options for ensuring long-term supply. In rural areas the distances involved in collection may impinge on financial as well as operational viability of plant as well as economies of scale. The mode of recycling in the local authority may impinge on the feasibility, as will present costs of waste disposal.

Approach to the Survey

Review of Literature

In order to cover all aspects of this scoping study it was necessary to gather sufficient background information, by way of a thorough review of literature, on work to date. Papers reviewed included assessment and evaluation reports of the processes and technology; reports by technology suppliers and papers found on the 'web'. These documents provided in-depth information, from different points of view, covering social, economic, legislative, technical and political aspects.

Contact with Suppliers

The purpose of contacting suppliers of the technology was to identify key technical and performance specifications of the plant including: feedstock; security of supply; contamination and segregation. Before contacting the suppliers a questionnaire was designed which would be implemented as a telephone survey. The questionnaire covered all issues that were felt might be raised by local authorities and waste management contractors.

Contact with other Relevant Organizations

These included the Environment Agency, store chains and Friends Of The Earth. Organisations that could, in different ways, effect the feasibility of establishing this type of technology in the waste management market.

The Environment Agency was contacted to gain a definitive response to where this technology stood in terms of Packaging Recovery Notes (PRN). This factor would prove important to any company with responsibility for packaging waste. In simplistic terms under the Producer Responsibility Obligation, companies over a specified turnover have to obtain PRNs to show that they are taking responsibility for packaging waste that they produce. If the technology were able to provide PRNs then this would make it a viable route for waste packaging reprocessing.

Reasons for contacting stores lay in establishing their opinion about gasification/ pyrolysis and whether they would consider this technology as an option for managing their packaging waste.

As a powerful lobbying body it was necessary to find out the position that FOE would be taking towards this technology. In the past they have shown opposition towards mass burn incinerators, if they took a similar stance towards these smaller scale plants then they would prove a ready made opposition.

The Survey Sample

Once the background information had been collected, it was then possible to move onto selecting the survey sample and designing the survey accordingly.

The survey sample covered all of the Waste Disposal Authorities (WDAs) and Unitary Authorities (UAs) in England and Wales (numbering 40 and 102 respectively) and 57 companies operating landfill sites. The sample needed to include examples of each type of authority to examine the bearing of the assumed barriers to the setting up of a facility. Landfill site operators were targeted due to the increasing amount of legislation aimed at the operations at these sites which may be giving companies incentive to diversify their waste management strategies.

Two slightly differing survey questionnaires were designed: one for WDAs/ UAs the other for landfill site operators. They were designed with simplicity in mind in order to gain maximum response. They took the form of a 'faxback' sheet that would take the respondent a matter of minutes to complete. The purpose of the questionnaire was to gain a picture of:

- How useful this technology would be to the authority/company over the next ten years;
- plus as a waste management tool; what feedstocks would be available;
- and what the perceived barriers to installing this technology were.

Alongside the questionnaire, the survey sample were sent an 'Information Sheet'. This drew together information gained from the preliminary stages of the project and pointers from the IEA Bioenergy/CADDET 'Advanced Thermal Conversion Technologies for Energy from Solid Waste' report. It was intended to provide answers to some of the initial questions that might arise.

Additional data on: quantities of waste handled; quantities of waste diverted via recycling and current contractual obligations were gained using the UK municipal wastes management survey returns and follow-up telephone calls.

The Views of the Technology Suppliers

Technology suppliers were contacted to establish the current position of the technology and the feasibility of its application to the management of Municipal Solid Waste (MSW). In total six suppliers were contacted.

A set of questions were drawn up, designed with the aim of gaining sufficient information for writing an 'Information Sheet' that would be sent out to the target sample WDAs/UAs and landfill sites, at a later date. The questions covered: size of feedstock; unsuitable feedstocks; preferred calorific value of feedstock; problems perceived with mixing municipal with household waste; effects of seasonal variation; preferred quantities of waste; size of space needed for the plant; initial costs of setting up a plant; running costs; pay back period; required length of contract; emission levels; where the technology stands in terms of NFFO and PRNs; whether the technology was proven in terms of a MSW feedstock.

Views

Size of feedstock

All respondents commented that the system does require size reduction, with acceptable sizes ranging from 50 to 90 mm³. One respondent commented that the more granular the feedstock the better, although larger plant would accept 150 mm³ sized feedstock. Another respondent had sourced and tested plant capable of processing MSW and other wastes to meet this criterion.

Unsuitable waste streams

There were varying answers to this question. One respondent commented that volatile components are needed to aid the process and that, therefore, metal and glass could cause problems as would waste streams that minimise temperature thus preventing fusion. This respondent also felt that the variable water content of MSW would have a detrimental effect on the process, too much water vapour will wet the spark plugs and also reduce the CV of the gas. This was not considered to be a problem by another respondent who had carried out tests on drying feedstock beforehand using heat from the process itself, although very wet waste can prove uneconomic to process as it needs large amounts of heat to enable it to be processed. Glass and metal were not felt to cause problems by any of the other respondents as long as they were crushed to the suggested size. In addition, the glass and metal could be diverted from the process in order to achieve recycling targets. One respondent considered that tires could only be used if they were mixed in with a cocktail of other wastes. Another commented that those elements containing a high level of volatile metals (mercury, arsenic) might not meet the latest EU emissions standards. Similarly chemical wastes may cause problems. All respondents felt that the process could handle PVCs although a very high PVC content in the feed would perhaps require the exhaust gas remediation system to be reviewed.

Preferred calorific value

All respondents commented that the process could operate on a range of CVs. The higher the better but engines will work on gas with a CV as low as 5MJ/m³. One respondent commented that from the economic aspect an actual or wet CV of 8000 KJ/kg would provide a self-sustaining operation with virtually no auxiliary fuel required. Below this value the quantity of auxiliary fuel increases and affects the economics of this or any other process.

Problems associated with mixing municipal and household waste

One of the three respondents had not worked in this area, and as such felt that it was an unproven area and that there would be numerous problems. The other three respondents did not foresee any problems with this mix of waste as long as the waste streams were sorted, segregated and treated accordingly, e.g. putrescibles dried, metal and glass recyclables removed and the feedstock crushed to the appropriate size.

Effects of seasonal variation

As with the comments above the respondent who had not had hands-on experience of working with MSW felt that any variations in the feedstock would cause problems. The other respondents did not agree. One commented that plant operators would need to set operating parameters and make provision for buffering and storage. Whilst another commented that variations in waste quality are readily accepted by the process and the control system automatically compensates for this.

Preferred quantities of waste

This would be dependent on the size of the plant and could range from 12Kg to 5 tonnes per hour.

Size of space required for the plant

This would be dependent on design and the type of process. These plants can be as small as a dustbin. There are transportable facilities that are similar in size and design to a 40ft ISO container. Gasification plants fed with sewage sludge take up very little space. The processing plant for MSW would be more complicated. It would probably need to be set up in a building and so would take up much more space. One respondent was able to say that a 30,000 tpa unit including heat recovery and power generation would require an area of 20m × 30m. This would not include space for a MRF or bulk storage if they should be required.

Initial costs

These would need to include costs of land, buildings, process equipment and cleaning equipment, specific to application, waste streams and feedstock preparation. Estimated costs ranged from £350 thousand up to £3 million.

Running costs

These would need to include connection charges, utilities, electrical power, chemicals for cleaning gas/water, costs of infrastructure, manpower. The running costs may be offset by energy savings e.g. utilising the energy generated by the plant; sale of excess energy were the project to gain a Non-fossil fuel obligation NFFO contract; revenues from gate fees; and, the sale of the end residues. One respondent commented that the system should be self-sustaining. The heat used to dry materials and the energy used to run the plant and any other machinery. This respondent felt it was better to dedicate the energy to other plant as he had experienced difficulties with gaining NFFO contracts. A more definitive answer was given as £25 per tonne of waste processed.

Pay-back period

Those respondents familiar with working with MSW stated that a five-year pay-back was forecast, although one respondent felt that this would be dependent on whether the heat could be utilized to dry the feedstock. The other respondent put this figure at eight years.

Length of contract

This would need to be commensurate with the life span of the plant and was considered by all to be in the region of 10 to 15 years in line with waste management contracts. For smaller plant shorter contracts of seven to ten years were felt to be acceptable.

Emission levels

One respondent felt that this would depend on the feedstock. The other three stated that emissions to the atmosphere were within the mass burn parameters and would already meet proposed EU emission levels for the treatment of both non-hazardous and hazardous waste.

Position of the technology in terms of NFFO

One respondent was not sure. The other three had no doubts that it was recognized.

Position of the technology in terms of PRN

Two of the respondents were unsure. The other two believed it was eligible. One commented that as this technology is an energy from waste process it would therefore qualify as a waste reprocessor and be entitled to a percentage of the quantity of waste processed under the PRN system. The respondent anticipated that this would be identical to that enjoyed by the mass burn incinerator companies.

Is the technology proven

One felt that the technology is not proven at present, one felt that it was proven on a pilot scale the other two felt it was already proven.

Comments

Due to the need for reducing size of the feedstock an additional plant would be required for crushing.

Although it has been stated that the process can deal with any waste, it would appear that the process operates more effectively and thus economically if certain elements of MSW are removed. For example, the process can deal with crushed glass and metal, but works more efficiently if these elements are removed where possible and perhaps recycled. The same is true of putrescible waste, which would need to be dried before processing or diverted for composting. Certain other elements, such as plastics would require the exhaust gas remediation system to be reviewed, as is the case for mass burn. There does not appear to be any problems envisaged in mixing municipal and household wastes or with seasonal variation as long as there is provision for buffering and storage and operating parameters can take this into consideration.

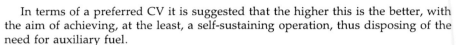
In terms of a preferred CV it is suggested that the higher this is the better, with the aim of achieving, at the least, a self-sustaining operation, thus disposing of the need for auxiliary fuel.

This type of facility, using MSW feedstock, would take up a sizeable area. Additional space would also be needed for the crushing plant, storage/buffering and material segregation. This would then move into the realms of the need for planning permission and the beginnings of public concern, e.g. concerns over increased weight of traffic on local roads and NIMBYism etc.

The technology suppliers were unable to provide exact figures for initial set-up and running costs, although it did appear that an efficiently run facility could be self-sustaining. These figures would be dependent on the size and type of facility and the CV of the feedstock. Pay back period and required length of contract queries produced fairly standard responses and need no further comment. In terms of PRN each facility would be assessed by the Environment Agency and accredited only if it met certain requirements. The same is true, to a certain extent, of NFFO contract, which have to be won.

The technology suppliers were adamant that emissions from these types of facilities would not be problematic, although if exhaust gas remediation systems need to be fitted, this would add an additional cost.

It would appear, from the responses received, that problems are foreseen by those companies not working with MSW, whilst this is not the case for those who are working with MSW. This is no doubt to be expected, as they do not have equal experience or the same agendas.

Contact with Other Relevant Organizations

Background

In order to clarify certain points, other relevant organisations needed to be contacted. These included the Environment Agency, store chains and Friends of the Earth. Contact was made by telephone and in each case the answer to a single question was sought. In the case of the store chains this proved more difficult than expected.

Contact with the Environment Agency

The Environment Agency (EA) was contacted to determine where gasification/ pyrolysis facilities stood in terms of PRNs. If these facilities were eligible for this system it would be an added attraction for them. The EA stated that, as long as plants fitted accreditation criteria, PRNs would apply.

Contact with Store Chains

Stores were contacted to establish their position towards gasification/pyrolysis in terms of their environmental policy, and whether they would consider using this waste management option for obtaining PRNs for their packaging waste. As this option had not been considered by many of those contacted this became a slightly more difficult task than anticipated. In the event only three stores felt able to make a response:

- Store A - Packaging is currently recycled in-house, although consideration would be given to this option, as more information became available.

- Store B - Currently looking in to setting up own in-house gasification plants, but would be interested in using this option for outlying stores.
- Store C - Recycle at the moment, but if this proved to be a 'green route' it would be considered.

Contact with Friends of the Earth

In the past, FOE has shown strong opposition towards mass burn incinerators. If they were to take a similar stance with this technology then planning permission and public opinion could be detrimentally affected. The person contacted at FOE commented that they would be looking into the environmental aspects of these plants but that at present felt that this was a 'promising' technology.

Response to the Faxback Survey

Response Rates

The survey was sent out to 199 contact addresses. Within the allowed three-week time period, 67 responses had been faxed back. Three additional returns were received one week later. The overall survey response rate was therefore 70 out of 199, or 35%. Due to the nature of the project this figure was deemed sufficiently high for assumptions to be made and conclusions drawn.

The responses were divided into the three areas covered by the survey, namely - Waste Disposal Authorities; Unitary Authorities and landfill site operators. Fig. 22.1 below provides a breakdown of responses from each area covered.

Non-response

Reasons for non-response included: pressure of work; staff shortages and instructions not to do any 'form filling'; tie-ins to long contracts; the current environmental policy not fitting with gasification/pyrolysis; or complete disinterest.

The following sections of this chapter looks more closely at the responses of each of the 3 target areas, and follows the format of the questions on the faxback sheets. The Chapter concludes with some of the areas being drawn together to provide a more universal picture.

Overview of Respondents

The Value of a Prior Demonstration Project

Question 1 on the faxback sheet asked respondents to comment on the value of having a prior waste gasification/pyrolysis demonstration project in the UK. There was a fairly unanimous response to this question. The WDAs were in no doubt of its value, with 70% of the opinion that it would be very valuable. The landfill site operators were less enthusiastic, although there was a strong divide between the companies who operated landfill sites alone, and those who operated more diverse operations, particularly recycling. These results suggest, that at present, there is little point in pursuing companies that solely operate landfill sites. The overall response to this question is shown pictorially in Fig. 22.2.

Commissioning Facilities

Questions 2a dealt with interest for commissioning a facility within the next 5 or 10 years and the perceived barriers in doing so. Respondents from all target areas were more interested in doing this within the next 5 years than within 10 years, although there was considerable interest for within 10 years. This is a clear indication that there is an immediate and future market for this type of technology, were the demonstration project to prove successful.

Available Feedstock

The landfill site operators handled a fairly diverse range of potential feedstocks but no figures are available for the amounts of waste managed by them. The WDAs managed large quantities of waste but the feedstock types they had available for processing within a facility were limited. The UAs handled much less waste but had a greater variety of feedstocks available.

If the plant can, as the technology suppliers have suggested, perform economically on a MSW feedstock alone, then the interest of the WDAs, not wishing to form partnerships, will be maintained.

Further work needs to be carried out on levels of feedstock segregation and preparation, in terms of economic viability and emissions.

Barriers

Each of the target areas pinpointed similar barriers to commissioning this type of facility. Comments came from both respondents who had expressed an interest and those who had not. These included:
- Contractual constraints
- Competitiveness
- Planning permission
- Public perception
- Emissions and concentrated pollutants in ash
- Doubts about the technology e.g. long-term performance and its ability to deal with large amounts of MSW
- Complications of having to segregate wastes
- Finding sustainable markets for residues
- Should only be viewed as part of an integrated waste management strategy
- The facility would need to be modular.

Apart from contractual constraints, these barriers would all be addressed were the demonstration project to go ahead. In the mean time more detailed information, on the technology and its capabilities, should be made available to those respondents expressing an interest. This could be achieved via workshops and information put on the Internet.

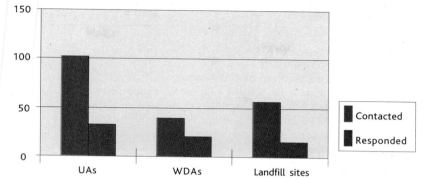

Fig. 22.1. Number of sites contacted and responses.

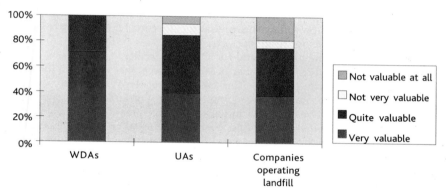

Fig. 22.2. Value placed on having a prior demonstration project in the UK.

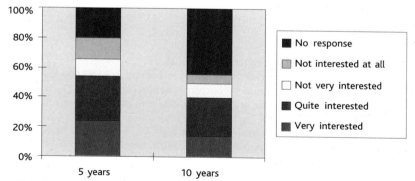

Fig. 22.3. Interest shown for commissioning a facility within the next 5 or ten years.

Typical buyer profiles

WDAs

A typical WDA buyer would:
- be located in the south or midlands areas.
- have an average waste arisings figure of approximately 300,000 tpa. This waste would all be currently sent to landfill or a very small proportion (up to 500 tpa) might be incinerated, either WtE or with metal extraction.
- have access to some source segregated waste and have a system for removing glass and metals.
- have a sufficient amount and variety of high CV feedstocks or be willing to form a partnership to ensure this e.g. with WCAs; private companies collecting high CV commercial/industrial wastes.
- be coming to the end of 'tie-ins' with other long-term contracts and be in the process of writing a new waste management strategy.
- have an environmental policy that accepts gasification/pyrolysis as a form of waste reprocessing and would advertise this to local store chains for PRNs.
- would be capable of putting a proposal together for NFFO contracts and be able to find markets for residues.
- have an appropriate piece of land for siting the facility.

As a means of estimating the number of WDA projects that would go forward a very crude calculation can be made. This takes 50% of the WDAs who said they would be very interested in commissioning a facility and 25% of those who said they would be quite interested. This initial calculation provides a figure of 42%. Nineteen WDAs stated that they would be very interested or quite interested in commissioning a facility, 42% of 19 is equal to 8. As such it can be estimated that, under ideal circumstances, 8 WDA projects could go forward.

In terms of waste figures another crude calculation can be made. This takes the average waste arising figure for all WDAs expressing an interest in commissioning a facility, multiplying this by 8 (the number of projects) and then multiplying that figure by 0.25 (It is presumed that only a quarter of the waste available will be suitable for processing). This calculation produces a figure of 599040 tpa, or 85577 tpa diverted waste per project. This is a rather ambitious figure but it does suggest that the proposed throughput of 30-50,000 tpa is achievable by the WDAs.

UAs

A typical UA buyer would:
- be located anywhere in England and Wales.
- have an average waste arisings figure of approximately 87,000 tpa. All or the majority of this waste would be currently sent to landfill.
- have access to some source segregated waste and have a system for removing glass and metals.
- have a sufficient amount and variety of high CV feedstocks or be willing to form a partnership to ensure this e.g. with neighbouring UAs or WDAs; Commercial/Industrial sectors.
- be coming to the end of 'tie-ins' with other long-term contracts and be in the process of writing a new waste management strategy.

- have an environmental policy that accepts gasification/pyrolysis as a form of waste reprocessing and would advertise this to local store chains for PRNs.
- would be capable of putting a proposal together for NFFO contracts and be able to find markets for residues.
- have an appropriate piece of land for siting the facility.

An estimate of the number of UA projects that would go ahead can be made employing the same method used to calculate this figure for the WDAs. The initial calculation provides a figure of 35%. Twenty-two UAs stated that they would be very interested or quite interested in commissioning a facility, 35% of 22 is equal to 8. As such it can be estimated that, under ideal circumstances, 8 UA projects could go forward.

In terms of waste diverted a figure of 174204 tpa is estimated, or 21776 tpa per project. This suggests that UAs may need to form partnerships in order to achieve a throughput of 30-50,000 tpa or perhaps it would be more appropriate for UAs to use smaller facilities, with a throughput of 15-35,000 tpa. Sizing down of the plant would, perhaps, be the direction to move in as a large majority of the UA respondents did not wish to form partnerships and had available a wide range of high CV feedstocks.

Landfill site operators

A typical landfill site operator buyer would:
- be located in the north (64% of respondents interested in commissioning a facility were based in the north.).
- be involved with other activities as well as landfill e.g. recycling or WtE incineration.
- have access to some source segregated waste and have a system for removing glass and metals.
- have a sufficient amount and variety of high CV feedstocks or be willing to form a partnership to ensure this.
- have an environmental policy that accepts gasification/pyrolysis as a form of waste reprocessing and would advertise this to local store chains for PRNs.
- would be capable of putting a proposal together for NFFO contracts and be able to find markets for residues.
- have an appropriate piece of land for siting the facility.

An estimate of the number of projects that would go ahead can be made employing the same method used to calculate this figure for the WDAs and UAs. The initial calculation provides a figure of 31%. Eleven landfill site operators stated that they would be very interested or quite interested in commissioning a facility, 31% of 11 is equal to 3. As such it can be estimated that, under ideal circumstances, three projects could go forward.

As no figures were available for the amounts of waste handled by these companies, it is not possible to make an estimate of the waste diverted by these three projects.

Conclusions

The aims of this scoping study were to:
- *establish the approximate size and profile of the UK market for small waste to energy projects based on gasification/pyrolysis;*

- *determine what, if any, barriers there were to the take up of this concept.*

Analysis of the data has shown that potentially 19 waste gasification/pyrolysis projects could go forward. Sixteen of these projects would divert a total of 773244 tpa of waste arisings going directly to landfill. Figures are unavailable for the remaining 3 projects. Further work needs to be carried out on the amounts of waste and types of feedstock these facilities can economically process. This will determine whether partnerships are needed e.g. where there are few feedstock types or lower waste arising figures. It was beyond the scope of this study to comment on the capabilities of the technology above and beyond what has been stated by the technology suppliers themselves.

The perceived barriers to commissioning gasification/pyrolysis facilities would all be addressed were access made available to a demonstration plant. That is, all except current contractual obligation, these will need to run the course of time.

BIBLIOGRAPHY

1. IEA Bioenergy/CADDET Renewable Energy *'Advanced thermal conversion technologies for energy from solid waste'*, August 1998.

2. TPS *'IEA Biomass Agreement Task X. Biomass Utilisation Biomass thermal Gasification and Gas Turbines Activity Sub-task 6 - Gasification of Waste'*, Report TPS 96/19.

3. Niessen W.R et al *'Evaluation of Gasification and Novel Thermal Processes for the Treatment of Municipal Solid Waste'*, August 1996 NREL/TP-430-21612.

4. Primenergy *'Comparative Explanation of Gasification and Incineration'*, WEB Paper.

5. Energy Foundation *'Gasification and electricity generation from biomass'*, WEB Paper.

6. EC THERMIE *'Energy from Co-firing of Coal and Waste-derived Fuel - Slough Trading Estate Power Plant'* Draft DG for Energy XVII.

7. Compact Power *'Solutions to Waste'*, Specifications and promotional material, 1998.

8. VENTEC Waste To Energy -Specifications and promotional material including video. 1998.

9. EA/SEPA *'Producer Responsibility Obligations 1997'* First Edition, July 1997.

10. ETSU *'Proposers' Guide For The New & Renewable Energy Programme'*, July 1996, ETSU-N-122 (ISSUE 2).

Index

Abandoned mine 114, 115
   area 114
Accumulation of species 86
Acid 69, 71
   front 111, 112, 125, 126
   rain 28
Adsorbed species 116
Adsorption 118, 120
   capacity 119
Aggregate 19
Agrarian 21
Air emissions 15
Air pollution control technology 16
Alcohol Program 9, 183, 186
Alkaline front 112
Anaerobic 26
   bioreactor systems 75
   digestion 76, 163
   gas plants 26
Angiosperms 55
Anhydrous ethanol 23
Anion 133
   exchangers 128
Anion-exchange membrane 133, 135
Anode purging solution 115, 116, 121
Anthocyanidins 51, 61, 64
Applicable to immobilized cell, three-phase
   fluidi 80
Aqua regia 120
   extraction 122, 123
Ash 15
   recycling 4
Availability of accurate transport
   coefficients 80
Averaging process 85
Averaging theorem 86
Averaging volume 86
Axial dispersion coefficient (Dz) 81
Azaarenes 292

β -O-4 55, 58, 59, 69
β -O-4 linkages 56, 58, 59
Backdiffusion 135
Bailing 6, 10, 223

Baled MSW 223
Bark 51, 67, 68
Bellman optimality principle of dynamic
   programmin 87
Bio- and thermophysical properties of
   heterogeneou 80
Bio-electricity 98
Bio-reactors 6
Bioavailable 125
Biocatalyst 78
Biocatalyst particles 79
Bioethanol 43
Biofilm reactors 78
Biofilm-scale 84
Biofuels 42
Biogas 44, 77, 177, 178, 179
   plants 5
   Production 187
Biokinetics intrinsic coefficient 87
Biological 95
   decomposition 224
   remediation 107
Biomass 6, 7, 22, 26, 33, 95
Biomethanation 8, 155
Biophysical and thermophysical
   properties 75
Bioreactors 75
Biparticle fluidized bed 75
   reactor (BFBR) 79
Bottom ashes 290
Brazil 8, 183
Briquettes 5, 307, 312
Briquetting 312
Bromo-chloro-phenols 285
Bulk flow 84
Bundesgesetzblatt 203
Bunsen "ice calorimeter" 82

Cadmium 286
Caffeate 55
Canada 8
Carbon dioxide 215, 254, 290
Carbon monoxide 290
Carbon sequestration 263

Carbonates   123, 125
Cathode effluent solutions   127
Cation exchange resins   128
Cation exchangers   128
Cation-exchange membrane   133
Cell design   130
Cement kiln   10, 215, 283
CH4   254
Chalcone   61, 62, 64
Chalcone synthase   61
Characteristics of food waste   126
Chemical speciations   120
Chile   8, 177, 180
Chilean   177
Chlorinated Benzenes   285
Chlorogenic acid   61
Cinnamic alcohols   52, 55, 57, 59
Cinnamyl alcohols   59
Clay particles   111
Clean production   297
Clean technologies   8
Cleaner fuel   312
Climate   199
Climate Change   10, 249, 261
Climate-relevant emissions   9
Close substance cycles   255
Co-firing   10, 215
Co-incineration   198
Colloids   111
Comparison and evaluation of results of
    bioprocess   90
Complexing agents   113
Composition   188
Composting   311
Compression wood   57, 58
Coumarate   55
Concentration deviation   86
Concentration of the limiting substrate S and
    the   87
Condensed tannins   61, 65, 66, 67
Coniferyl alcohol   52, 53, 55
Coniferyl aldehyde   55
Continuous monitoring   283
Continuum axioms for the mass and
    momentum of mult   76
Convective effects   86
Cu   293
Current densities   129
Current density   131, 133, 135
Current efficiency   129, 130, 131, 133, 134,
    135

De novo synthesis   280

Decision (control) variables   88
Degradation   226
Dehydrogenative polymer, DHP   55
Dehydrogenative polymerisation   52, 55
Desalination   135
Desorption   112, 116, 118, 120, 125
Determination of which bioreactor
    parameters are m   90
Development and application of sound
    biophysical t   90
Development of analytical models   90
Development of effective system's
    optimization me   90
Diffuse double layer   109
Diffuse layer   110
Diffusion   81, 112
Diffusion processes   84
Diffusion rates   78
Diffusional force   108
Diffusive effects   86
Dihydroflavones   61, 63
Dioxins   279
Direct measurement of volume V of a certain
    mass m   83
Dirigent   58
Dirigent-proteins   58, 59
Discrete hierarchical system   85
Dispersive and diffusive flux   86
Dissolution   112, 116, 125
Distinct polyphasic heterogeneity and
    anisotropici   82
Divergence theorem   86
Domestic wastes   22
Dynamic model   79
Dynamic Programming Optimization of the
    Bioprocess   87
Dynamic programming optimization
    technique   87

E-p-coumaryl   52
EC Air Emission Limit Values   279
EC regulatory limit   283
Ecological fuels   25
Economics   295
Ecosystems   28
Efficiency   217
Effluent solutions   108, 129
Effluent treatment   129
Effluent treatment processes   113
Effluent Treatment Processing   127
EIX   128
Electo-osmotic velocity   117
Electric field (V/m)   108, 117

Electric field intensity   116, 118
Electrical potential   110, 111, 128
Electrical power   173
Electricity   95
Electro-deposition   131
Electro-osmosis   109, 110, 111, 112, 114, 116, 122
Electro-osmotic advection   112
Electro-osmotic flow   112, 114, 118, 122
Electro-osmotic velocity   117, 118, 123
Electroactive species   130
Electrochemical   127
Electrochemical decontamination   107
Electrochemical ion exchange   128
Electrochemical ion exchange process   129
Electrochemical processes   127
Electrochemical technologies   108, 128
Electrode compartments   114
Electrode material   130
Electrode materials and structure   130
Electrode potential   129
Electrodeposition   134, 136
Electrodialysis   108, 132, 134, 135, 137
    cell   132, 133
    process   133, 135
    stack   133
Electrokinetic   7, 109
    phenomena   108, 109, 110
    remediation   107, 120
    remediation technology   108
    soil   111
    soil processing   107, 108, 111, 113, 115, 120
    zeta potential   112
Electrolysis   5, 108, 112, 129, 131, 134
Electrolysis cell   129, 130
Electrolysis medium   130
Electrolysis of water   111, 112, 115, 116, 118, 123, 125, 126
Electrolyte solution reservoirs   114
Electrolyte solutions   114, 115, 116, 122, 123
Electrolytic cell   129
Electrolytic process.   129
Electrolytic recovery   131
Electromigration   109, 110, 111, 112, 118, 125
Electromigrative transport   135
Electroneutrality   108
Electroosmosis   118, 121, 125
Electrophoresis   110, 111
Electroreclamation   107
Electrostatic force   108
Electrostatic repulsion   108

Electrowinning   108, 135, 136
Electrowinning/electrostripping technique   136
Electrowinning/electrostripping   137
Ellagic acid   65
Ellagitannins   65
Emission reduction   215
Emissions   224
Emissions to Air   278
Emissions to Ashes   290
Emissions to Water   290
Energetic Crops   37
Energy   168, 184
Energy consumption   120
Energy needs assessment   310
Energy-intensive branches   262
Enhanced electrokinetic soil processing   112
Enhancement schemes   113
Enhancement techniques   112
Enhancement technologies   113
Enhancing agents   112
Entrapment of microbial cells in polymer gels   78
Environment   61, 63
Environmental conditions   27, 51
Enzyme-catalyzed processes   77
Epidemiological   275
Ethanol   22
EU   95
EU directive   296
Europe   105
European Union   250
Excavation   107
Exchangeable   123, 125
Exchangeable fraction   125
Excretory system   51
Exposure   273
Extraction   123, 125
Extraction methods   120
Extractive polyphenols   51
Extractives   69

Faradaic processes   133
Fe/Mn oxides   123, 125
Fermentation alcohol   23
Ferulate   57, 60
Ferulic   52
Ferulic acid   57
Ferulic acid esters   59, 60
Fibre   56, 57
Fine-grained soil   109
Flavanone   61, 62, 63, 64
Flavones   61, 63

Flavonoids   61, 63, 64, 68
Flavonols   61
Fluidized-bed reactors (FBR)   78
Flushing   107
Fly ashes   290
Foliage   49
Food waste   126, 127
For three-phase fluidized beds   81
Forced convection in bioreactor models   81
Forest residues   49, 50, 51, 69, 70, 71, 72
Forest waste   6
Fossil   21, 28
Fouling   133
Friedensreich Hundertwasser   202
Fuels Production   75
Function of the biomass cells concentration   87
Functional hierarchy   85
Furans   277
Further development of viable mathematical descrip   90

G-S (guaiacyl- syringyl) lignins   55
Gallic acid   65
Gallotannins   65
Galvanic sewage   135
Gas and liquid superficial velocities   80
Gas, liquid and solid holdup   81
Gasification   12, 317
General transport theorem is   86
Germany   9, 195, 197, 198
Gravitational settling   110
Greenhouse emissions   262
Greenhouse gas emissions   3
Greenhouse gases   169, 261
Growth of immobilized microorganisms can   87
Growth stages of immobilised cells   83
Guaiacyl (G)   56
Guaiacyl lignins   55, 57
Guaiacyl type (G)   56
Guaiacyl-syringyl (G-S)   56
Guaiacyl syringyl lignins   55, 56, 57

Hazardous   22
Hazardous waste   197
Heat conductivity   82
Heavy Metals   286
Helmholtz-Smoluchowski equation   117
Hierarchical volume averaging method   76, 85
Hierarchy of transport processes   85
Hierarchy of waste management   295

Higher efficiency   312
Highly mobile   125
Homogeneous reaction rate   87
Household and commercial waste   317
Household waste   211
Human health   11, 273, 280
Humic   68, 71
Humic acids   69, 71, 72
Humic substances   49, 50, 67, 68, 69, 72
Humification   70
Humus   49, 50, 51, 67, 68, 69, 70
Hyacinth   25
Hydraulic conductivity   107, 109
Hydrodynamics of biparticle fluidized bed reactors   81
Hydrogen bromide (HBr)   289
Hydrogen fluoride (HF)   289
Hydrolysable tannins   63
Hydroxy -cinnamic acids   52, 53

Identify the organisms   91
Immobility   119, 120
Immobilization of microorganisms   78
Immobilized biomass   76
Immobilized cells   75
Immobilized-cell biparticle fluidized bed reactor   83
Immobilized-Cell Reactor's Technology   77
Implementation of biofuels production   75
Incineration   9, 11, 197, 253
Incinerator   215
India   8
Industrial Transformation   255
Industrial wastewater   131
Industrialisation   249
Industry   157
Inhibition compounds   126
Interrelated decision processes   87
Intrinsic adherence to relevant constraints   80
Intrinsic micro-macro relations   85
Inventories   280
Ion exchange   108, 128, 136
Ion exchange process   128
Ion exchange resins   128
Ion exchange/electrowinning system   136
Ion selective membrane   112
Ionic Migration   109, 110
Ionic mobility   112
Ionic strength   112
IPCC   262
Is zeta potential (V)   117
Isomorphous substitution   108

Isotropic turbulence theory   81

Kaolinite   115, 120
Kaolinite soil   114, 115, 116, 118, 120
Kyoto   268

L-phenylalanine   52, 57
L-phenylalanine ammonia-lyase   52, 61
L-phenylanine   52
Landfill   177, 181, 253
Landfill Biogas   8
Landfill gas   167
Landfill gas recovery   77
Landfill methane emissions   261
Landfilled   277
Landfilling   15
LCC   71
Leachate   179, 226
Leucoanthocyanidins   61
Lignans   58, 59, 68
Lignification   51, 58, 59
Lignin(s)   50, 51, 56, 57, 58, 67, 69, 70, 71, 72
Lignin biodegradation   67, 68
Lignin contents   51
Lignin related substructures   67
Lignin- carbohydrate complexes   59, 60, 71
Lignin- protein theory   50
Lignin-carbohydrate complexes   70
Lignins   51, 55, 56, 58, 68, 69, 70, 72
Lignocellulosic biomass wastes   76
Liquid dispersion   81
Livelihoods   310
Lo Errazuriz Sanitry Landfill   182
Local constitutive theory   85
Local molar rate of production per unit volume of   87
Longitudinal and lateral dispersive fluxes   86
Low-permeability soils   108
Low-value wastes   310

Management law   204
Market   317
Marketing   12
Mass balance   280
Mass burn   16
    facilities   5
Mathematical Modeling   79
Mercury   287
Mesophilic anaerobic fermentation system   88

Metabolic   52
Methanogens   77
Methanol   22
Methides   66
Michaelis-Menton kinetics   81
Micro-scale bioreaction kinetics and diffusion   83, 91
Microbial growth   85
Microbiological and chemical reactions kinetics   85
Microbiological processes   126
Microencapsulation   78
Migration   110
Migration or sedimentation potential   110
Minimise disposal needs   92
Mobility   112, 119
Modular facilities   5
Molar diffusive flux   86
Molar flux of species A   86
Molar rate of production per unit volume of species   86
Molasses   23
Monitoring   273
Mono-incineration   198
Monolignol glycosides   59
Monolignols   52, 53, 55, 56
MPW   219
MSW   42
Multi-stage continuous   134
    dialyzer   134
    electrodialysis   134
    electrodialyzer   134
Multidimensional vectors   88
Municipal wastewater treatment   120, 122

Nairobi   307, 311, 312
Non-darcian flow effects   81
Non-hazardous   22
Non-Newtonian (power-law) fluids   88
NOx, SOx, HCl   277
Ntal aspe   105

Occurrence and correspondence   90
Odour   224
Of transport phenomena on microbial growth and act   83
One-dimensional steady state Fourier heat conduction   82
Optimal decision values   89
Order and types of physical phenomena   80
Organic carbon   49, 50
    pollutants   233
    /Sulfides   123, 125

OSPAR Convention   295
Ostwald de Waele   83
Overall objective function   88

p-coumaric   52
   acid   57
p-coumaryl   55
   alcohol   52, 53, 55, 57, 58
   alcohol, coniferyl   52
p-hydroxy benzoic acid   57
Packaging waste   320
PAH   233
PAHs   285
Parameters discretisation   88
Parametric data determination   92
Particle Analysis   229
Particle-scale   84
Particles which are absorbing bioreactor
   product   85
Particles with immobilized bacteria   85
Particulates   288
Pb   293
PBrDD/F   233
Pcdd/F   234
PCNs   285
Pellitization   5
Permittivity of the medium   117
Permselectivity   134
Persistent pollutants   273
pH buffering capacity   123, 126
pH, volatile acids concentration, alkalinity,
   gas   79
Phase average   85
Phenylalanine   52
   ammonia-lyase (PAL)   52, 61
Phenylpropanoid   52, 60
   biosynthesis   61
   metabolism   53, 61
Phisico-chemical and measuring-technical
   character   82
Photosynthesis   27
Physically sound experimenting and
   predictive mode   92
Physico-chemical decontamination   107
Pickling wastewaters   131
Plantations for fuel   27
Polarization   133
   regime   135
Pollution   151
Poly cyclic aromatic hydrocarbons   283
Polyaromatic Hydrocarbons   233
Polychlorinated biphenyls   283
   dibenzothiophenes   285

Polyphenol theory   51
Polyphenolic   69
   extractives   70
Polyphenols   49, 50, 51, 60, 61, 63, 67, 69,
   70, 72
Pore water   112
Pore- and particle-size   76
Pore-scale   84
Power plant   215
Power supply   114
Precipitation   112
Principle of the electrodialysis   132
Proanthocyanidin   61, 65, 66, 67, 68, 69, 72
Proteins   58
Pseudo-plastic fluids   83
Pump-and-treat   107
Pyrolysis   317

Quinone   65
   methide   59, 66
Quinone methide intermediates   65

Radicals   55
Rameal chipped wood   50, 69
Ramial chipped wood RCW   68
Random coupling   55, 56
RCW   69
Reaction temperature, retention time, overall
   heat   88
Reactor   76
   kinetics and transport phenomena   75
Recalcitrant   51
   soil organic matter   67
Recovering energy from biomass   5
Recycled material   262
Recycling   15, 253, 295
Reduction   253
Refuse-derived   16
   fuel or RDF plants   5
Renewable energy   21, 24
   sources   33
Repeatable measurement   90
Residual   123, 125
Resonance hybrids   55
Respiratory illnesses   287
Reticulated vitreous carbon (RVC)   131
Retrofit   17
Rheological behaviour index   83
   consistency index   83
   model   83
Risk assessment   275
Round bales   224
Rubber   219

Sclerenchyma cells 51
Secondary metabolites 51
Sedimentation Potential 110
Sequential 120, 123, 125
   extraction 121, 125
   extraction method 121, 122
Serial sequential process 88
Sewage Sludge 198
Shear rates on the wall surface. Concentric-
   cylin 83
Shearing stresses 83
Sinapyl 52
   acid 55
   alcohol 52, 55
Sludge 120, 219
Slurry 26
Small branches 49
Small or intermediate technologies 307
Small scale energy production 310
Smoke 229
Social acceptance 307
Socially–Responsive 11, 309
Sodium ion 126, 127
Soil cell 114
Soil flushing 107
Soil humus 50, 72
Soil organic matter 49, 71, 72
Soil permeability 116
Soil washing 107
Soil 49, 50, 51, 67, 70, 71, 72
Solid Waste 27
Solid waste 15
Solidification and stabilization 107
Solubility 112
Sorption 112
Spatial and temporal uniformities of
   substrates 79
Spatial distribution of porosity and
   heterogenity 85
Speciation 112
Speciation of heavy metals 120
Species A concentration 86
Specific density 83
Specific heat capacity 82
Specific mass productivity 79
Spittelau 201, 207, 210, 213
Square bales 224
Sri Lanka 8, 151
Stack gases 278
Stakeholder 310
State variables of 88
State-transformation functions 88

Steady laminar flow 86
Stormwater 226
Streaming Potential 109, 110
Structural polyphenols 51, 52
Subsidiarity 250
Suitable structuring 88
Surface charge 108
Sustainability 295
Synapic acids 52
Synergetic 76
Synergetic microbiological and
   physicochemical pro 76
Synergetics between biodynamics and
   physicochemica 85
Synergism 78
Syringyl 55, 71
Syringyl lignin 53, 55

Tailing-soil 114, 115, 116, 118, 120
Tannins 62, 63, 68
Temperature 224
The bubble coalescing coefficient (C) 81
The condensed tannins 63
The Netherlands 215
The present-worth" method 88
Thermal conversion 15
Thermal treatment 201
Thermodynamic 6
Thermodynamic and Process Parameters
   75
Thermodynamical and rheological
   properties 82
Thermodynamical and rheological
   properties of agri 76
Thermophysical and biophysical proper-
   ties 90
To be functions of axial position within the
   biore 80
Total synergetic dispersivity tensor 86
Toxic heavy metal 16
Toxicity 273
Toxicity Characteristic Leaching Procedure
   (TCLP) 19
Toxicity fingerprints 238
Trans-cinnamic acid 52, 53
Transportation effect 264
Twigs 49
Two or multiphase reactor media—content
   76

Ultrafine particles 288
Uncertainties 275

United Nations Centre for Human
  Settlements   311
Urban poor   309
Urban solid waste management   309
Urbanization   249

Vacuum extraction   107
Vascular   51
Vascular cells   52
Vascular plants   60, 61
Vessels   56, 57
Vienna   9, 201, 204, 213
Virgin material   262
Viscosity of the   117
Volatile fatty acids   226
Volatile organic compounds   285
Volume averaged biomass concentration
  87
Volume averaged rate of growth of immo-
  bilized biom   87
Volume averaged value   87

Volumetric productivity   79

Waste management   249, 317
  law   203
Waste to energy plant   317
Waste-to-energy   15
Wastes   95, 249
  Fermentation   75
  management and disposal, reducing
  waste "le   92
Wastewater treatment   120, 128
Weakly bound fractions   125
Well-defined matrix   78
White Paper   95
White-rot fungi   70, 71
WTE   15

Zero emissions strategy   297
Zero net emissions   263
Zeta potential   112, 117, 118